METHODS IN MOLECULAR BIOLOGY

For further volumes:
http://www.springer.com/series/7651

Synapse Development

Methods and Protocols

Edited by

Alexandros Poulopoulos

Department of Stem Cell and Regenerative Biology, Harvard University, Cambridge, MA, USA

Humana Press

Editor
Alexandros Poulopoulos
Department of Stem Cell and Regenerative Biology
Harvard University
Cambridge, MA, USA

ISSN 1064-3745 ISSN 1940-6029 (electronic)
Methods in Molecular Biology
ISBN 978-1-4939-6686-8 ISBN 978-1-4939-6688-2 (eBook)
DOI 10.1007/978-1-4939-6688-2

Library of Congress Control Number: 2016956874

Cover Illustration: Cover image design by Massimiliano Stagi

Printed on acid-free paper

This Humana Press imprint is published by Springer Nature
The registered company is Springer Science+Business Media LLC
The registered company address is: 233 Spring Street, New York, NY 10013, U.S.A.

Preface

This book is a collection of protocols for experimentation into how neurons connect to produce the extraordinary functionalities of the nervous system. The complexity underlying the development of synapses allows and warrants an assortment of approaches to probe the many layers of structure, mechanism, and emergent function. Appropriately, neuroscience is populated by researchers across disciplines, and with them comes a range of technologies and methodologies that make the neuroscience toolkit one of characteristic wealth.

The experimental diversity converging onto the synapse has made it one of the most intensely studied and intricately described biological systems. Observing in real-time how a single protein flips between conformational states to regulate ion flows, following how thin cellular protrusions traverse the body to form networks of the highest complexity seen in nature, measuring how circuits interact and morph throughout the span of life to document its story, our understanding of synapse development is arguably unparalleled in its pluralism of perspective. Yet it seems the more detailed our descriptions, the more expansive their heterogeneity and the less inclusive the models. Depending on the system in question, synaptogenic behaviors range anywhere from the preordained pairings of Sperrian chemoaffinity to the play-by-ear modes of promiscuous exuberance and subsequent refinement. Plasticity in turn occurs up or down, fleeting or perduring, and is stirred on by molecules, voltages, and timing. While thankfully all these serve to construct a system able to produce both hiccups and quantum theory, they also do away with comfy notions of a generalizable synapse awaiting full description. The challenge is on, and the design of clever new experimentation ever more opportune.

In putting together this volume, we've had the aim of compiling a laboratory guide that covers a broad spectrum of applicable methods (largely excluding electrophysiology, which has merited its own dedicated volumes in this series), with the benefit of a narrower focus of implementation to the study of synapse development. I've asked experts to present the current state of the art of classic protocols, giving their insider tweaks, tips, and tricks for optimal application to the synapse and its particularities. I've also asked pioneers of highly specialized approaches to contribute detailed guides to methods utilizing sophisticated instrumentation that is often institutionally available to labs but for which methodological expertise is difficult to come by. These 23 chapters by 63 investigators are presented in five parts that deal with synapses: (1) in culture, (2) their constituents, (3) their structures, (4) their dynamics, and (5) the assemblies they form. Our hope is that these guides will minimize the barrier to entry for the integration of new approaches with existing expertise, producing syntheses that will foster novel perspectives on the many ways in which synapses form, transform, and transmit.

Cambridge, MA, USA *Alexandros Poulopoulos*

Contents

Contributors

BEKIR ALTAS • *Department of Molecular Neurobiology, Max Planck Institute of Experimental Medicine, Göttingen, Germany*
SINA V. BARYSCH • *Zentrum für Molekulare Biologie Heidelberg (ZMBH), DKFZ-ZMBH Alliance, Heidelberg, Germany*
SREYA BASU • *Department of Cell Biology and Genetics, Erasmus MC, Rotterdam, The Netherlands*
JOSEPH BERGAN • *Department of Psychological and Brain Sciences, University of Massachusetts, Amherst, MA, USA*
IOANNA BETHANI • *Department of Neurobiology, Max Planck Institute for Biophysical Chemistry, Göttingen, Germany*
CHRISTOPH BIESEMANN • *Department of Molecular Neurobiology, Max Planck Institute of Experimental Medicine, Göttingen, Germany*
MIQUEL BOSCH • *Institute for Bioengineering of Catalonia (IBEC), Barcelona, Spain*
HANS RUDOLF BRENNER • *Department of Biomedicine, University of Basel, Basel, Switzerland*
JORGE CASTRO • *Department of Brain and Cognitive Sciences, The Picower Institute for Learning and Memory, Massachusetts Institute of Technology, Cambridge, MA, USA*
REGINA WAI-YAN CHOY • *Department of Psychiatry, School of Medicine, University of California, San Francisco, CA, USA*
LAURA COCAS • *Department of Neurology, University of California, San Francisco, CA, USA*
BENJAMIN H. COOPER • *Department of Molecular Neurobiology, Max Planck Institute of Experimental Medicine, Göttingen, Germany*
KATALIN CZÖNDÖR • *Interdisciplinary Institute for Neuroscience, University of Bordeaux and CNRS, Bordeaux, France; Interdisciplinary Institute for Neuroscience, CNRS, UMR 5297, Bordeaux, France*
MATHIAS DE ROO • *Department of Neuroscience, School of Medicine, University of Geneva, Geneva, Switzerland*
CATHERINE DULAC • *Department of Molecular and Cellular Biology, Howard Hughes Medical Institute, Harvard University, Cambridge, MA, USA*
ZOHREH FARSI • *Department of Neurobiology, Max Planck Institute for Biophysical Chemistry, Göttingen, Germany*
GLORIA FERNANDEZ • *Department of Biological Sciences, California State University, Los Angeles, CA, USA*
BENJAMIN FLOTTMANN • *Institute of Physical and Theoretical Chemistry, Goethe-University Frankfurt, Frankfurt, Germany*
ANDREAS FRICK • *Neurocentre Magendie, INSERM, Bordeaux, France; University of Bordeaux, Bordeaux, France*
MELANIE GINGER • *Neurocentre Magendie, INSERM, Bordeaux, France; University of Bordeaux, Bordeaux, France*

MATTHIAS G. HABERL • *Neurocentre Magendie, INSERM, Bordeaux, France; University of Bordeaux, Bordeaux, France*
YASUNORI HAYASHI • *Brain Science Institute, RIKEN, Wako, Japan*
MIKE HEILEMANN • *Institute of Physical and Theoretical Chemistry, Goethe-University Frankfurt, Frankfurt, Germany*
JAVIER M. HERNANDEZ • *Department of Structural Biochemistry, Max Planck Institute for Molecular Physiology, Dortmund, Germany*
FRANK HERRMANNSDÖRFER • *Department of Functional Neuroanatomy, Heidelberg University, Heidelberg, Germany*
ETIENNE HERZOG • *Interdisciplinary Institute for Neuroscience, University of Bordeaux, Bordeaux, France; Interdisciplinary Institute for NeuroScience, CNRS, Bordeaux, France*
SOYON HONG • *Department of Neurology, F.M. Kirby Neurobiology Center, Boston Children's Hospital, Harvard Medical School, Boston, MA, USA; Program in Neuroscience, Harvard Medical School, Boston, MA, USA*
MRINALINI HOON • *Department of Biological Structure, University of Washington, Seattle, WA, USA*
HEINZ HORSTMANN • *Department of Functional Neuroanatomy, Heidelberg University, Heidelberg, Germany*
CORDELIA IMIG • *Department of Molecular Neurobiology, Max Planck Institute of Experimental Medicine, Göttingen, Germany*
YOH ISOGAI • *Department of Molecular and Cellular Biology, Howard Hughes Medical Institute, Harvard University, Cambridge, MA, USA; Sainsbury Wellcome Centre for Neural Circuits and Behaviour, University College London, London, UK*
OLAF JAHN • *Proteomics Group, Max Planck Institute of Experimental Medicine, Göttingen, Germany*
LEI JIANG • *Higuchi Biosciences Center, University of Kansas, Lawrence, KS, USA*
HIROSHI KAWABE • *Department of Molecular Neurobiology, Max Planck Institute of Experimental Medicine, Göttingen, Germany*
EVANGELOS KISKINIS • *The Ken & Ruth Davee Department of Neurology & Clinical Neurological Sciences, Department of Physiology, Feinberg School of Medicine, Northwestern University, Chicago, IL, USA*
STEPHAN KÖGER • *Department of Physiological Genomics, Ludwig Maximilians University, Munich, Germany*
THOMAS KUNER • *Department of Functional Neuroanatomy, Heidelberg University, Heidelberg, Germany*
DELAINE D. LARSEN • *Department of Biochemistry and Biophysics, School of Medicine, University of California, San Francisco, CA, USA*
CHRISTIAN LOHMANN • *Department of Synapse and Network Development, Netherlands Institute for Neuroscience, Amsterdam, The Netherlands*
ELISA LUQUET • *Interdisciplinary Institute for Neuroscience, University of Bordeaux, Bordeaux, France; Interdisciplinary Institute for NeuroScience, CNRS, Bordeaux, France*
ELIAS K. MICHAELIS • *Higuchi Biosciences Center, University of Kansas, Lawrence, KS, USA*
MARY L. MICHAELIS • *Higuchi Biosciences Center, University of Kansas, Lawrence, KS, USA*
ANNIE MUNIER • *Institut de Biologie Paris Seine, University Pierre et Marie Curie, Paris, France*
SIDDARTH NANGUNERI • *Department of Functional Neuroanatomy, Heidelberg University, Heidelberg, Germany*

Haruhisa Okawa • *Department of Biological Structure, University of Washington, Seattle, WA, USA*

Nirmala Padmanabhan • *Department of Physiology and Pathophysiology, Max Rady College of Medicine, Rady Faculty of Health Sciences, University of Manitoba, Winnipeg, MB, Canada; Neuroscience Research Program, Kleysen Institute for Advanced Medicine, Health Sciences Centre, University of Manitoba, Winnipeg, MB, Canada*

Minjong Park • *Department of Psychiatry, School of Medicine, University of California, San Francisco, CA, USA*

Karen Perez de Arce • *Department of Neuroscience, School of Medicine, Tufts University, Boston, MA, USA*

Adema Ribic • *Department of Neuroscience, School of Medicine, Tufts University, Boston, MA, USA*

Douglas S. Richardson • *Department of Molecular and Cellular Biology, Harvard University, Cambridge, MA, USA; Harvard Center for Biological Imaging, Harvard University, Cambridge, MA, USA*

David P. Santos • *The Ken & Ruth Davee Department of Neurology & Clinical Neurological Sciences, Department of Physiology, Feinberg School of Medicine, Northwestern University, Chicago, IL, USA*

Nadine Schmidt • *Department of Physiology II, Albert-Ludwigs University, Freiburg, Germany*

Tabrez J. Siddiqui • *Department of Physiology and Pathophysiology, Max Rady College of Medicine, Rady Faculty of Health Sciences; University of Manitoba, Winnipeg, MB, Canada; Neuroscience Research Program, Kleysen Institute for Advanced Medicine, Health Sciences Centre, University of Manitoba, Winnipeg, MB*

Raunak Sinha • *Department of Physiology and Biophysics, University of Washington, Seattle, WA, USA*

Massimiliano Stagi • *Department of Cellular and Molecular Physiology, Institute of Translational Medicine, University of Liverpool, Liverpool, UK; University of Liverpool, Crown Street, Liverpool, UK*

Beth Stevens • *Department of Neurology, F.M. Kirby Neurobiology Center, Boston Children's Hospital, Harvard Medical School, Boston, MA, USA; Program in Neuroscience, Harvard Medical School, Boston, MA, USA*

Mriganka Sur • *Department of Brain and Cognitive Sciences, The Picower Institute for Learning and Memory, Massachusetts Institute of Technology, Cambridge, MA, USA*

Olivier Thoumine • *Interdisciplinary Institute for Neuroscience, University of Bordeaux, Bordeaux, France; CNRS, Bordeaux, France; Interdisciplinary Institute for Neuroscience, CNRS, UMR 5297, Bordeaux, France*

Theodoros Tsetsenis • *Department of Neuroscience, Mahoney Institute for Neurosciences, Perelman School for Medicine, University of Pennsylvania, Philadelphia, PA, USA*

Varun Venkataramani • *Department of Functional Neuroanatomy, Heidelberg University, Heidelberg, Germany*

Daniel K. Wilton • *Department of Neurology, F.M. Kirby Neurobiology Center, Boston Children's Hospital, Harvard Medical School, Boston, MA, USA; Program in Neuroscience, Harvard Medical School, Boston, MA, USA*

Johan Winnubst • *Janelia Research Campus, Howard Hughes Medical Institute, Ashburn, VA, USA*

Andrew Woehler • *Berlin Institute for Medical Systems Biology, Max Delbrueck Center for Molecular Medicine in the Helmholtz Association, Berlin-Buch, Germany*

Part I

Forming Synapses in a Dish

Chapter 1

A Cell Culture System to Investigate the Presynaptic Control of Subsynaptic Membrane Differentiation at the Neuromuscular Junction

Nadine Schmidt, Sreya Basu, Stephan Kröger, and Hans Rudolf Brenner

Abstract

For decades the neuromuscular junction (NMJ) has been a favorite preparation to investigate basic mechanisms of synaptic function and development. As its function is to transmit action potentials in a 1:1 ratio from motor neurons to muscle fibers, the NMJ shows little or no functional plasticity, a property that makes it poorly suited to investigate mechanisms of use-dependent adaptations of synaptic function, which are thought to underlie learning and memory formation in the brain. On the other hand, the NMJ is unique in that the differentiation of the subsynaptic membrane is regulated by one major factor secreted from motor neurons, agrin. As a consequence, myotubes grown on a laminin substrate that is focally impregnated with recombinant neural agrin closely resemble the situation in vivo, where agrin secreted from motor neurons binds to the basal lamina of the NMJ's synaptic cleft to induce and maintain the subsynaptic muscle membrane. We provide here a detailed protocol through which acetylcholine receptor clusters are induced in cultured myotubes contacting laminin-attached agrin, enabling molecular, biochemical and cell biological analyses including high resolution microscopy in 4D. This preparation is ideally suited to investigate the mechanisms involved in the assembly of the postsynaptic muscle membrane, providing distinct advantages over inducing AChR clusters using soluble agrin.

Key words Synapse development, Muscle, Neuromuscular junction, High resolution microscopy

1 Introduction

The formation of synapses is an example of extreme subcellular differentiation that is regulated through reciprocal interactions between the synapsing cells. This process involves the development of the presynaptic growth cone into a presynaptic nerve terminal able to secrete neurotransmitter. At the site of contact its follower cell forms a postsynaptic membrane rich in neurotransmitter receptors.

Although numerous molecular factors involved in the reciprocal regulation of synapse formation have been identified, the cell biological mechanisms and biochemical pathways underlying these

Alexandros Poulopoulos (ed.), *Synapse Development: Methods and Protocols*, Methods in Molecular Biology, vol. 1538, DOI 10.1007/978-1-4939-6688-2_1, © Springer Science+Business Media LLC 2017

processes are poorly understood for many types of synapses. The synapse between motor neurons and skeletal muscle fibers, the neuromuscular junction (NMJ), is particularly well suited to investigate the cell biological mechanisms by which the presynaptic neuron—the motor neuron—controls the accumulation of the neurotransmitter receptors—the acetylcholine receptors (AChRs)—in the postsynaptic muscle membrane because they are mediated by one major factor from the motor nerve terminal—agrin [1–4]. Agrin acts through its receptor/effector LRP4/MuSK in the muscle fiber membrane [5, 6]. When applied to muscle fibers in vivo, agrin induces a nerve-free postsynaptic apparatus containing structural protein complexes, which in unknown ways anchor synaptic signaling molecules to the synaptic cytoskeleton; these include the AChRs as well as those that mediate neuromuscular signaling involved in NMJ development and maintenance (e.g., MuSK itself). Furthermore, muscle nuclei accumulate at the postsynaptic apparatus and become transcriptionally specialized by agrin to express postsynaptic components. However, although many of these components have been identified, their functions in postsynaptic assembly and their mechanistic interactions that induce synaptic AChR accumulation remain poorly understood.

One approach to study postsynaptic differentiation is to observe the dynamics of the development of the postsynaptic apparatus under the control of agrin and of the ensuing AChR clusters in cultured muscle cells in 4D by means of high resolution microscopy. Classically, agrin is applied in a soluble form to cultured myotubes; the alterations in size, shape, and receptor density of the AChR clusters and of the associated components of the synaptic apparatus in response to experimental manipulation of the suspected pathways involved are then observed over time. However, mechanistic interpretation of such data is often difficult because it is not known, first, where along a myotube a cluster is formed; second, what determines the site of its formation, and third, application of soluble agrin in the culture medium is not the physiological form of its presentation to myotubes. Rather, agrin in vivo, upon its release from the motor nerve terminal, stably binds to the basal lamina (BL) between the nerve terminal and the muscle membrane, thus stably acting on a small region of muscle membrane; agrin binds to BL is via its N-terminus to laminin [7]. This difference in agrin presentation may explain, at least in part, why in its soluble form, agrin does not induce the full complement of postsynaptic specializations observed in muscle fibers in vivo.

Here, we provide a simple way to mimic presentation of agrin bound to the laminin substrate of cultured myotubes. The preparation to be described below has been instrumental in establishing agrin's role to induce synaptic AChR gene expression [8], in demonstrating a novel role of NRG/ErbB signaling in agrin-induced AChR clustering [9], and in elucidating the organization of a synaptic network of microtubules involved in the focal trafficking of

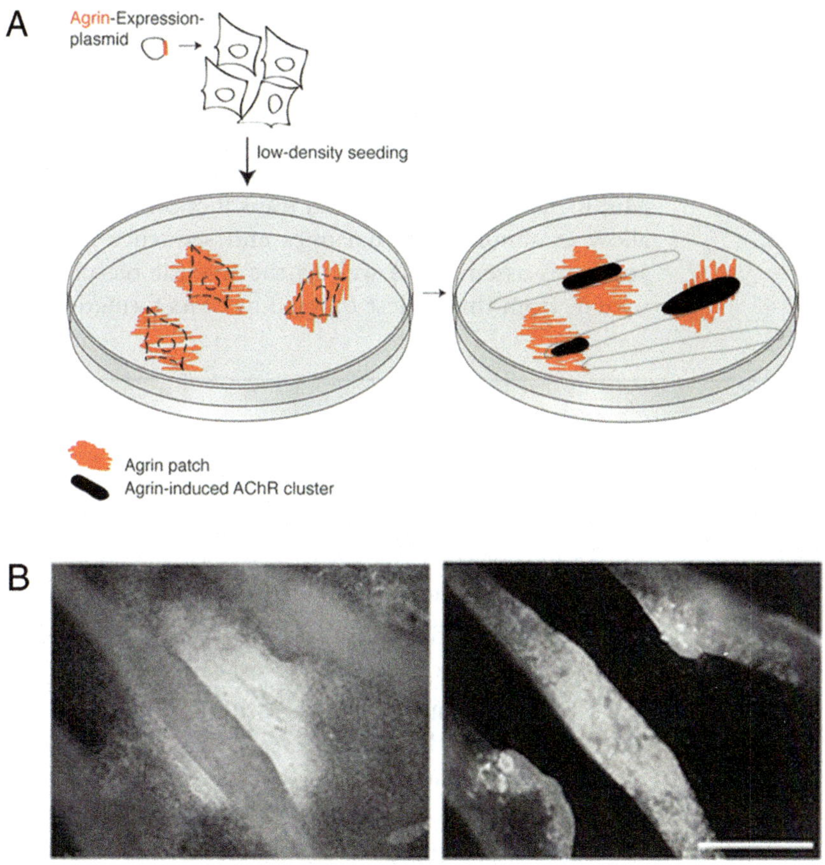

Fig. 1 (**a**) Scheme illustrating focal impregnation of culture substrate with neural agrin fragment. Transfected Cos-7 cells secrete and locally deposit neural agrin on a laminin substrate, and the cells are then lysed. Myotubes are subsequently cultured on this substrate. Stable AChR clusters form where myotubes contact agrin patches. (Scheme reproduced from ref. [10]). (**b**) Example of myotubes cultured on top of a large agrin patch produced as described in panel A. Agrin patch stained with anti-agrin antibody (*left*), and resulting AChR clusters on myotubes stained with rhodamine-α-bungarotoxin. Bar: 100 μm. (Reprinted with permission from Jones G., Herczeg A, Ruegg M.A, Lichtsteiner M, Kröger S, Brenner HR (1996) Substrate-bound agrin induces expression of acetylcholine receptor epsilon-subunit gene in cultured mammalian muscle cells. Proc Natl Acad Sci U S A 93(12), 5985–5990. Copyright (1996) National Academy of Sciences, U.S.A.)

AChRs to the synaptic muscle membrane [10]. All these processes proved relevant for full NMJ development and maturation in vivo. In contrast, application of soluble agrin could not produce these processes in cultured muscle cells, emphasizing the significance of substrate-attached agrin presentation.

The basis of our approach is to generate patches of recombinant agrin deposits on laminin culture substrates; muscle cell precursors are then seeded on this substrate and allowed to fuse to myotubes; where myotubes happen to contact agrin patches, large (20–60 μm diameter) AChR clusters are formed that are amenable to fluorescence microscopic examination in 4D (Fig. 1).

One limitation of this system is that myotubes form *after* agrin is deposited, i.e., on preexisting agrin patches, which prevents microscopic observation of AChR clusters from the earliest stages of their formation. In principle, this problem can be circumvented by using muscle cell lines lacking the agrin effector MuSK. After *Musk–/–* myotubes have formed, they can be induced to express MuSK by adeno- or lentiviral transduction. This extra step should allow examining the dynamics and sequence of the subsynaptic apparatus assembly as well as the nuclear recruitment associated with it, from the earliest time point of its formation.

2 Materials

2.1 Preparation of Culture Dishes

1. 6-well plastic culture plates for transfection of Cos cells.
2. 100 mm plastic dishes for expansion of myoblasts.
3. 35 mm plastic dishes with glass bottom (poly-l-lysine-coated) for microscopic observation of myotube cultures.
4. 35 mm plastic culture dishes for biochemical analysis of myotubes.
5. Matrigel (5 ml): Thaw 24-48 h (4 °C) prior to stock preparation. Then add 155 ml of sterile Ca^{2+}-/Mg^{2+}- free PBS, aliquot 10 ml stocks, and store at −20 °C.
6. Collagen from Calf Skin (100 mg): Add 100 ml of sterile ddH_2O and 200 μl of glacial acetic acid. Mix at RT until dissolved (o/n). Store in 10 ml aliquots at −20 °C.
7. Cell-line medium: DMEM (high Glucose without Ca^{2+}/Mg^{2+}), 10% FCS, 1% Pen/Strep, 1 mM Hepes.
8. Laminin solution (mouse EHS-sarcoma-derived): dilute fresh for every experiment in PBS to 10 μg/ml final concentration.
9. Sterile filters (0.45 μm).
10. Cell line for expression of neural agrin: we use Cos-1 or Cos-7 cells.
11. Transfection reagent of choice: we use Fugene HD (Promega).
12. Agrin plasmid: Full-length neural chicken agrin cagrin7A4B8 [7]. 2 μg used per well of a 6-well plate [8]. For preparation of Agrin, *see* Subheading 3 below.

2.2 Isolation and Purification of Primary Myoblasts from Rodent Hindlimb Muscle

1. Five mouse pups (postnatal day) P1-P3.
2. Collagenase: 10 mg/ml *Clostridium histolyticum* Type 4 collagenase (i.e., Sigma) diluted in sterile ddH_2O. Aliquot and store at −20 °C.
3. Dispase solution: 2,4 units/ml Dispase II (approx. 1 g in 229 ml, depending on lot) in sterile Ca^{2+}-/ Mg^{2+}- free PBS. Store in 9 ml aliquots (2, 4 units/ml) at −20 °C.

4. Collagenase/Dispase mixture: Dilute 1 M $MgSO_4$ (sterile) in 21 ml Hyclone medium (final conc 1 mM), add 2,4 ml sterile Ca^{2+}-/ Mg^{2+}- free PBS, 7 ml Dispase solution and 600 μl Collagenase solution. Sterile filter with a 0.45 μm filter. Mixture should be frozen at −20 °C, but avoid multiple thawings. In total 7 ml are used per pup.
5. Ham F-10 Medium (500 ml): Nutrient Mixture Ham (1×).
6. Human recombinant bFGF (10 μg): Prepare a stock of 5 μg/ml in sterile PBS, aliquot and store at −20 °C. Use at 1:1000.
7. Growth medium 1 (GM-1): Add 20% FCS (fetal calf serum), 1% Pen/Strep, and 4 mM Glutamine (all final concentrations) to Ham-F-10 medium. Supplement with bFGF (1:1000).
8. Trypsin solution (0.05% Trypsin in 0.53 mM EDTA).
9. Cold and sterile filtered Dulbeccos's PBS without Ca^{2+}/ Mg^{2+} + 0.1% BSA (Bovine Serum Albumin):
10. 2 ml Eppendorf tubes, sterile, autoclaved.
11. Anti-mouse-α7-Integrin antibody recognizing extracellular epitope; optional for myoblast purification [11].
12. Dynabeads®Sheep Anti-Rat IgG (Life Technologies).
13. Matrigel-coated 100 mm culture dishes: preparation described in Methods below.
14. Human recombinant bFGF (10 μg): prepare a stock of 5 μg/ml in sterile PBS, aliquot (500 μl), and store at −20 °C. Use at 1:1000.
15. Growth medium 3 (GM-3) *without* Ca^{2+}: Hyclone medium with 20% FCS, 1% P/S, 4 mM Glutamine and 1 mM $MgSO_4$. Supplement with bFGF (1:1000).

2.3 Culturing and Expansion of Myoblasts; Their Fusion to Myotubes

1. Trypsin solution: 0.05% Trypsin in 0.53 mM EDTA.
2. GM-3 medium *with* Ca^{2+}: DMEM supplemented with 20% FCS, 1% Pen/Strep, 4 mM Glutamine and bFGF (1:1000).
3. Horse Serum: heat inactivated for 30 min at 56 °C and stored as 10 ml aliquots at −20 °C.
4. Differentiation medium: DMEM with 5% Horse Serum, 1% Pen/Strep and 4 mM Glutamine. For live-cell imaging, DMEM without phenol red is preferred.
5. Tetrodotoxin (TTX): 1 mM stock.

2.4 Instrumentation

1. Incubator (37 °C/ 5% CO_2).
2. Dissecting set: small scissors, small pointed forceps (sterilized), and sterile surgical blades.
3. Centrifuge with precooling function.
4. Hemocytometer.

5. Dynal-MPC-L magnetic particle concentrator.
6. Spinning wheel at 4 °C (for several mixing steps below).

3 Methods

3.1 Preparation of Culture Dishes

Briefly, for cell biological (microscopy) experiments, laminin-coated culture dishes are focally impregnated with agrin. To this end, Cos-7 cells are transfected with a plasmid expressing an isoform of neural agrin comprising the N-terminal laminin binding domain [7]. The transfected cells are then seeded onto laminin-coated dishes, allowed to secrete agrin for 48 h and lysed. Myoblasts are grown and fused on those dishes. For biochemical assays, laminin-coated dishes are painted on their entire surface with a solution of recombinant neural agrin.

For microscopy: Glass bottom dishes focally impregnated with Agrin (Agrin-Cos Dishes)

1. Seed Cos-7 cells on a 6-well plate (300,000–500,000 cells per well).
2. 24 h later transfect cells with an expression plasmid encoding full length chicken Agrin [7, 8] using the transfection agent of choice. Culture for 24 h.
3. To laminin-coat glass bottom dishes (35 mm), fill dishes with 0.75–1 ml laminin and incubate for 1–2 h at 37 °C. Rinse once with PBS. Add 2 ml PBS per dish. Dishes are best if used immediately. Don't store longer than 2–3 days at 4 °C.
4. Re-seed the transfected Cos cells onto the 35 mm laminin-coated dishes: add 0,5 ml Trypsin per well and incubate until cells detach. Add 2 ml Cell-line medium per well and count resuspended cells in hemocytometer. Seed 300,000–500,000 cells per 35 mm laminin-coated glass bottom dish. Culture for 48 h.
5. After 48 h, cells are lysed for 90 min in 2% Triton X-100 in PBS, followed by extensive washing (6–8× in PBS for 1 h each). Dishes can be stored sterile for 2–3 weeks at 4 °C (*see* **Note 1**).

For biochemical assays: Dishes uniformly coated with purified recombinant Agrin

1. Coat 6-well dishes with laminin (1–2 h at 37 °C) as described above.
2. Aspirate laminin and incubate dishes with 0.5 μg/ml agrin solution for 2 h at 37 °C. Wash once with PBS. Add 2 ml PBS per well and store at 4 °C as described above.

For expansion of primary myoblasts: Matrigel-coated 100 mm-dishes

1. In 500 ml Hyclone medium with 1 mM $MgSO_4$, add 20 ml Matrigel and 20 ml collagen from calf skin. Sterilize by filtration.
2. Add 10 ml per 100 mm dish and incubate at 37 °C for 1–2 h. Best if used fresh.

3.2 Isolation and Purification of Primary Myoblasts from Hindlimb Muscle (According to ref. [12])

1. Isolate hindlimbs from mouse pups in a laminar flow or UV-sterilized chamber, using a standard dissecting microscope, small pointed forceps, and sterile surgical blade. Discard the feet. Peel off the skin, dissect hindlimb muscles, and transfer to a 35 mm dish containing sterile PBS. Use at least four pups (*see* **Note 2**).
2. Mince muscles finely in a petridish, using two sterile surgical blades to avoid squeezing the tissue. Carefully transfer the minced tissue to a sterile glass bottle containing prewarmed Collagenase/Dispase mixture at 4 ml per pup and shake/stir for 25 min at 37 °C for dissociation of cells. Transfer the supernatant cell suspension to a fresh 50 ml Falcon tube containing 1/10th the volume of FCS to inactivate digestion mix. Repeat the digestion of the remaining tissue, this time using 3 ml digestion mixture per pup. Collect the supernatant and inactivate with 1/10th volume of FCS. Pool the supernatants in a 50 ml Falcon and centrifuge at 450 ×*g* at 4 °C for 15 min.
3. Discard supernatant, resuspend the pellet in GM-1 medium (supplemented with bFGF), and seed on 100 mm matrigel-coated dishes prepared as described above. Use five dishes for each pup.
4. Let cells proliferate for 2 days at 37 °C/ 5 % CO_2; cells should reach 70–80 % confluency. Medium change is not necessary.
5. Cell harvesting (per 100 mm dish): Carefully suck off the medium and add 1 ml fresh trypsin solution. Incubate at 37 °C/ 5 % CO_2 until the cells begin to detach. Immediately add 0.5 ml pure FCS. Important: do not trypsinize longer than 30s–1min! Do not harvest more than five 100 mm dishes at a time to prevent cell damage by over-trypsinization.
6. Collect cells by washing each 100 mm dish with 5 ml of cold PBS/ 0,1 % BSA. Store cell suspension in a 50 ml Falcon tube on ice. Use one 50 ml Falcon tube for five 100 mm dishes. To collect the remaining cells on the dish, wash the five 100 mm dishes consecutively with the same 10 ml cold PBS/ 0,1 % BSA. Add the 10 ml to the cell suspension stored on ice.
7. Repeat **steps 5** and **6** for all dishes. Meanwhile, keep the harvested cells on ice.
8. Centrifuge at 4 °C for 15 min at 450 ×*g*.
9. Discard the supernatant and resuspend the cell pellet in 15–20 ml cold PBS/ 0.1 % BSA. Count the cells. In our hands,

roughly half of the total cell count are myoblasts, thus for 20×10^6 counted cells, calculate 10×10^6 myoblasts.

10. Centrifuge the cell suspension once more at 4 °C for 15 min at $450 \times g$. After spinning, remove the supernatant thoroughly. Resuspend the pellet in such a way that 1 ml cold PBS/0.1 % BSA does not contain more than 5×10^6 myoblasts. Transfer this suspension to sterile 2 ml Eppendorf tubes to get no more than 10×10^6 myoblasts per 2 ml in each tube.
11. Add anti-α7-Integrin antibody. Use 5 μl per 10×10^6 myoblasts. Mix by rotating on spinning wheel for 25 min at 4 °C (*see* **Note 3**).
12. Spin down cells in a precooled centrifuge at 4 °C for 3 min at $450 \times g$.
13. Discard the supernatant, resuspend the pellet in 2 ml PBS/ 0.1 % BSA, and spin again as in **step 12**. Repeat this washing **step 3–4** times to remove unbound antibody.
14. Beads preparation (sometime during steps 10 and 12): 0.7 beads per myoblast are used (e.g., for 10×10^6 myoblasts, calculate 7×10^6 beads). Add the total amount of beads required for all cells in 2 ml cold PBS/0.1 % BSA. Wash beads using a Dynal-MPC-L magnetic particle concentrator [11] using 2 ml of fresh PBS/ 0.1 % BSA. Repeat washing three to four times. For each tube of cells, resuspend the total amount of beads in 50 μl PBS/ 0,1 % BSA (e.g., three tubes of cells require 150 μl for resuspension). Store beads on ice.
15. Add 50 μl of beads to each tube of washed cells.
16. Rotate for 25 min at 4 °C to allow the coupling of the Dynabeads®Sheep Anti-Rat IgG to the anti-α7 integrin antibody bound to the myoblasts.
17. As in **step 14** wash the cells with the help of a magnetic particle concentrator three to four times in 2 ml cold PBS/0.1 % BSA.
18. Count cells using the hemocytometer.
19. Seeding of myoblasts: Plate 500,000 cells per 100 mm dish in GM-3 medium without Ca^{2+} supplemented with bFGF (1:1000). Incubate cells at 37 °C/ 5 % CO_2 for 2 days (*see* **Note 4**).

3.3 Further Proliferation of Myoblasts and Fusion to Myotubes

1. After 2 days of culturing, carefully trypsinize the enriched myoblasts (as in Subheading 3.2, **step 5**) and re-seed in GM-3 medium with Ca^{2+} on agrin-impregnated dishes. For a 35 mm dish, seed between 300,000 and 500,000 cells.
2. When cells reach confluency, add differentiation medium and monitor myotube formation daily. Let differentiate as required. Generally, 2–3 days is sufficient for myotube formation. Medium should be changed every alternate day. (*see* **Note 5**).

3.4 Preparation of Recombinant Agrin

We purified Agrin through a nickel-affinity column from supernatants of stably transfected EBNA-HEK293 cell expressing a His-tagged agrin-construct (NtAc95AgrinA4B8) containing 95KD of the C-terminus and the 25kD laminin-binding domain of the N-terminus (NtA) of chick agrin. The N-terminal insert is essential for agrin binding to laminin-coated dishes [7].

4 Notes

1. Risk of contamination! Wash the dishes very carefully under sterile conditions.
2. Younger pups give better yield.
3. If anti-α7-Integrin antibody is not available, myoblasts can be partially enriched by mincing, trypsinization, and trituration of muscle tissue, followed by removing debris by filtering through sterile lens paper. Non-muscle cells are then partially removed by preplating the cell suspension on plastic dish (*see*, e.g., [11]).
4. Check cells under the microscope daily. If cells detach, replace the medium.
5. Myotubes are prone to detach due to their spontaneous contractile activity, a sign of their viability. Addition of 1 μM tetrodotoxin (TTX) prevents contractions and prolongs the duration for which cultures can be maintained.

References

1. McMahan UJ, Horton SE, Werle MJ et al (1992) Agrin isoforms and their role in synaptogenesis. Curr Opin Cell Biol 4(5):869–874
2. Cohen I, Rimer M, Lømo T et al (1997) Agrin-induced postsynaptic-like apparatus in skeletal muscle fibers in vivo. Mol Cell Neurosci 9(4):237–253
3. Jones G, Meier T, Lichtsteiner M et al (1997) Induction by agrin of ectopic and functional postsynaptic-like membrane in innervated muscle. Proc Natl Acad Sci U S A 94(6):2654–2659
4. Wu H, Xiong WC, Mei L (2010) To build a synapse: signaling pathways in neuromuscular junction assembly. Development 137(7):1017–1033
5. Kim N, Stiegler AL, Cameron TO et al (2008) Lrp4 is a receptor for agrin and forms a complex with MuSK. Cell 135(2):334–342
6. Zhang B, Luo S, Wang Q et al (2008) LRP4 serves as a coreceptor of agrin. Neuron 60(2):285–297
7. Denzer AJ, Gesemann M, Schuhmacher B et al (1995) An amino-terminal extension is required for the secretion of chick agrin and its binding to extracellular matrix. J Cell Biol 131:1547–1560
8. Jones G, Herczeg A, Ruegg MA et al (1996) Substrate-bound agrin induces expression of acetylcholine receptor epsilon-subunit gene in cultured mammalian muscle cells. Proc Natl Acad Sci U S A 93(12):5985–5990
9. Schmidt N, Akaaboune M, Gajendran N et al (2011) Neuregulin/ErbB regulate neuromuscular junction development by phosphorylation of α-dystrobrevin. J Cell Biol 195(7):1171–1184
10. Schmidt N, Basu S, Sladecek S et al (2012) Agrin regulates CLASP2-mediated capture of microtubules at the neuromuscular junction synaptic membrane. J Cell Biol 198(3):421–437
11. Blanco-Bose WE, Yao CC, Kramer RH et al (2001) Purification of mouse primary myoblasts based on alpha 7 integrin expression. Exp Cell Res 265(2):212–220
12. Reutenauer J, Dorchies OM, Patthey-Vuadens O et al (2008) Investigation of Debio 025, a cyclophilin inhibitor, in the dystophic *mdx* mouse, a model for Duchenne muscular dystrophy. Br J Pharmacol 155(4):574–584

Chapter 2

Co-culture Synaptogenic Assay: A New Look at Fluorescence Reporters and Technological Devices

Karen Perez de Arce* and Massimiliano Stagi*

Abstract

The mechanism underlying the differentiation of pre- and postsynaptic specifications involves the sequential and dynamic recruitment of specific molecules coordinated by bidirectional signaling across the synaptic cleft. In this chapter, we describe the co-culture assay, a useful method to evaluate cell-surface molecules through its ability to promote the recruitment of proteins required for synapse structure and function. The versatility of this simple and reliable method is illustrated by the wide variety of applications ranging from analysis of synaptogenic activity to evaluation of soluble compounds with therapeutic potential. In addition, we provide a framework to enable the co-culture assay as a tool for high-throughput studies, thereby improving the efficiency and sensitivity of this classic method in neuroscience.

Key words Co-culture assay, Adhesion molecules, Synaptogenesis, Confocal imaging, High throughput

Abbreviations

HTPS	High-throughput studies
CAMs	Cell adhesion molecules
NLG1	Neuroligin 1
EphBs	Ephrin B
LRRTM1	Leucine rich repeat transmembrane neuronal 1
SynCAMs/Necls	Synaptic adhesion molecule
HEK 293	Human embryonic kidney
COS	Cercopithecus aethiops kidney
MFD	Microfluidic devices
TIRF	Total internal reflection fluorescence
CFP	Cyan fluorescent protein
P0	Postnatal day 0
Fiji	ImageJ image processing package (fiji.sc/Fiji)

*Authors contributed equally, correspondence may be addressed to either.

Alexandros Poulopoulos (ed.), *Synapse Development: Methods and Protocols*, Methods in Molecular Biology, vol. 1538, DOI 10.1007/978-1-4939-6688-2_2, © Springer Science+Business Media LLC 2017

Amaxa	Electroporation system for gene delivering (bio.lonza.com)
MATLAB	Matrix laboratory technical computing language (mathworks.com)
GNU Octave	Matrix laboratory technical computing language (gnu.org/software/octave/)
Imaris	Interactive microscopy image analysis software (bitplane.com)

1 Introduction

Development and maintenance of synaptic architecture is considered one of the most fundamental processes underpinning neuronal health and functionality. One technique that has become fundamental for the study of synapse-organizing factors is the co-culture assay, a classic method in neuroscience with a pivotal role in the identification and characterization of transmembrane adhesion complexes as major players in synapse development [1–3]. The main outcome of this experimental approach is to provide a quantitative measurement of synaptogenic activity, based on the extent to which neuronal proteins are recruited to putative hemi-synapses as a result of the cell-surface protein expression on non-neuronal cells [4, 5]. Therefore, this reliable and convenient technique provides a comparative method not only for the analysis of several trans-synaptic molecules in parallel, but also for exploring cooperative actions between them.

The substantial contribution of the co-culture assay revealing the synaptogenic activity of diverse families of neuronal cell adhesion molecules (CAMs), such as Neuroligins and Neurexins [3, 6], EphBs and ephrinBs [7, 8], LRRTMs [9], SynCAMs/Necls [2], and netrin G ligands (NGLs) [10], has been crucial to further our understanding of mechanisms that regulate synapse formation and maintenance. Thus, studies in vivo using knockout mouse models for EphBs/ephrinBs [7, 8, 11, 12] and neurexins/neuroligins [13, 14] have established these trans-membrane proteins are more relevant for maintaining the integrity of synaptic junctions in vivo than they are essential for initiating the synapse formation process. Therefore, complementary studies on whole animal systems offer the opportunity to investigate these proteins in a physiological context enlightening more precise descriptions of their functions during synapse assembly, maturation, and stability rather than focusing only on synaptogenic activity [15].

The co-culture assay's versatility is outlined along this chapter by the detailed description of useful modifications together with the discussion of recent technical tools that improve its precision and sensitivity and also provide enhanced reliability in its adaptation for high-throughput studies (HTPS). Thanks to a new set of technological devices this assay can be upgraded to conduct HTPS with the aim of performing large screens for novel soluble factors, molecules, and drugs with biological and/or biochemical activities that regulate synapse assembly [16–19]. In this context, screens

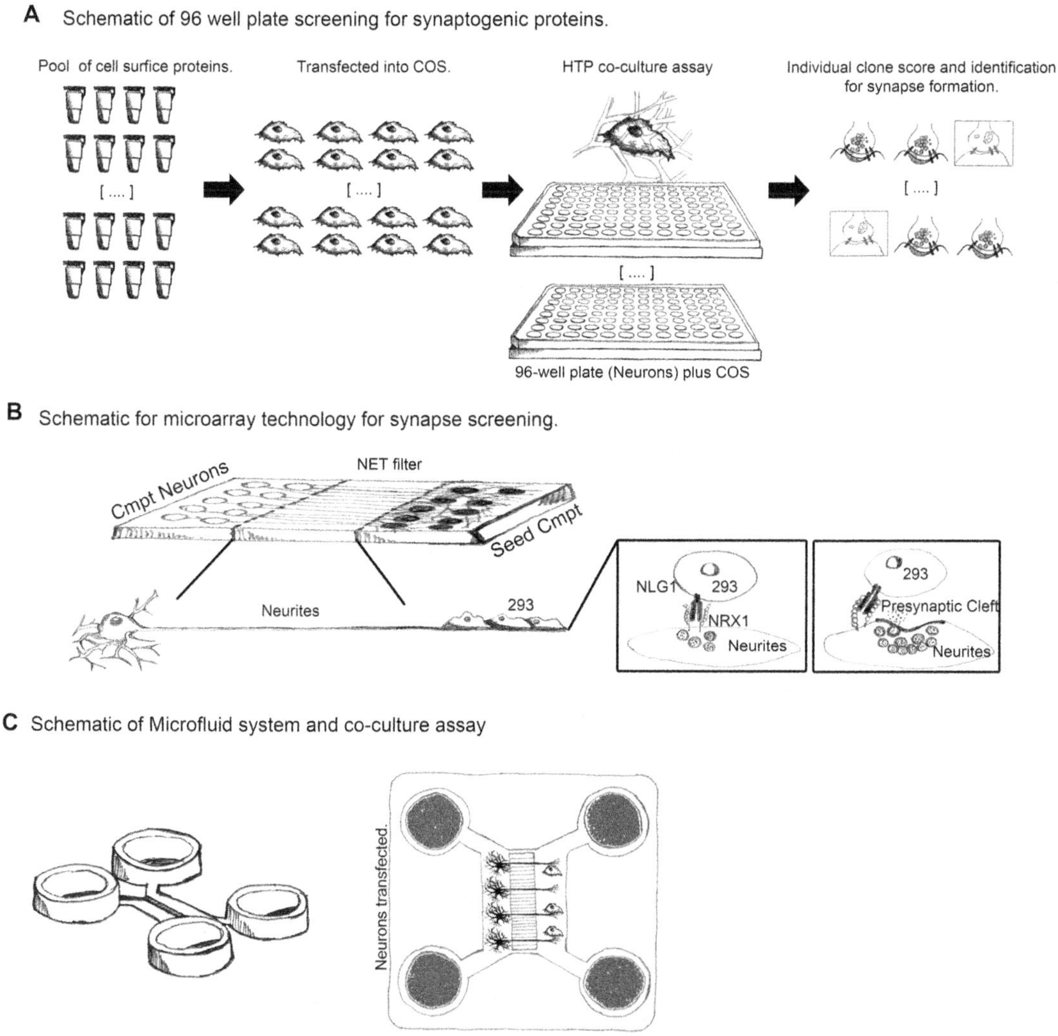

Fig. 1 Devices for large-scale co-culture assays. (**a**) Synapse Microarray Technology. The schematic representation shows the chambers where neurons are loaded (neuron cmpt) and the chamber where the co-culture is placed (seed cmpt). Heterologous expression of Neuroligin 1 (NLG1) in HEK293 will induce presynaptic differentiation in neurons by recruitment of presynaptic proteins in the seed compartment. (**b**) Large screening of synaptogenesis-inducing molecules using a 96-well format and imagining with an HTP screening microscope. Several cell-surface proteins are transfected in HEK293 and seeded as multiple replicates over neurons previously plated in each of the wells. (**c**) Microfluidic culture devices that isolate the neuronal processes from the cell bodies via a series of interconnected axon-guiding microchannels

can be performed using an HTPS microscope in 96-well plate formats to allow large scale imaging and analysis of synaptogenic activity (Fig. 1a). Alternatively, synapse microarray technology enables neurites to extend through microchannels and form dense networks in compartments distinct from those where their cell bodies reside [17]. Since this platform has a controlled array of microwells (Fig. 1b), the induction of presynaptic structures will be selectively established in predetermined positions and therefore will restrict imaging space, decreasing search time for synaptogenic events, and increase the reproducibility of co-culture assays. Other interesting platforms are microfluidic devices (MFD), originally described for the study of axon regeneration [19], and subsequently used as co-culture chambers [16, 18]. A typical MFD has two compartments, the soma and the axon compartment, connected through arrays of axon-guiding microchannels (Fig. 1c). This platform also enables multiple parallel treatments and/or co-cultures, with fluidic isolation as key feature, since axons and non-neuronal cells are placed in different compartments from the neuronal somata.

In parallel to the technological devices described above, there are several new molecular biological tools that are particularly useful as well. New fluorescent probes with improved sensitivity and temporal resolution for studding synaptic activity are available and used intensively in the field. Thus, co-culture assays can be used to assess the induction of new synapses, not only through endogenous synaptic protein localization (Table 1), but also activation of

Table 1
Antibodies for synaptic markers used in co-culture assays

Description	Target antigen	Provider
Axons	Tau GAP-43	Pierce [30] Pierce [32]
Dendrites	MAP2	Chemicon [33]
GABAergic presynaptic sites	GAD65	Chemicon [15]
GABAergic postsynaptic sites	Gephyrin GABAA receptor subunit γ 2	Synaptic system [34] Alomone [33]
Glutamatergic postsynaptic sites	PSD-95 NMDA receptor subunit NR1 Glutamate receptor subunit GluR1	Neuromab [35] Synaptic systems [15] Upstate [26, 33]
Presynaptic actives zones	Bassoon	ENZO [36]
Synaptic vesicles	Synapsin 1 Syntaxin 1 Synaptophysin 1 Synaptotagmin 1 SV2	Chemicon [23] Synaptic systems [37] Synaptic systems [23, 37] Synaptic systems [38] Developmental studies Hybridoma bank [39]
GABAergic synaptic vesicles	vGAT	Synaptic systems [40]
Glutamatergic synaptic vesicles	vGlut1	Neuromab [35]

new synaptic terminals. This has already been done with synaptic proteins tagged with pHluorin and GCaMP2, a pH-sensitive GFP and a genetically encoded calcium sensor, respectively [20–24]. In addition, many new biosensors show great potential for this assay: GCaMP6 [25] an updated calcium sensor for imaging neural activity, iGluSnFR [26] to visualize glutamate dependent synaptic transmission, and ArcLight-S249 to detect single action excitatory potentials in individual neurons and dendrites [27].

The following protocol will describe the complete panorama of this method with detailed descriptions of the classic co-culture assay from the basics of neuronal culture preparation to elements of analysis using the most common software available, including the new technological devices that may be used for HTPS.

2 Materials

2.1 Reagents

1. Heterologous Cells: HEK 293 or similar (*see* **Note 1**).
2. DMEM: Dulbecco's Modified Eagle's Medium.
3. CMF-HBSS Solution: Hanks' Balanced Salt Solution (HBSS), 1 mM Sodium Pyruvate, 50 U/mL Penicillin/Streptomicyn, 10 mM HEPES, filtered through 0.2 μm under sterile conditions.
4. Papain Solution: 25 U/mL Papain, 20 U/mL DNAse, 1.5 mM CaCl2, 0.75 mM EDTA, 0.75 mM NaOH, l-Cysteine 5.5 mM in HBSS, filtered through 0.2 μm under sterile conditions.
5. Trypsin/EDTA: 0.25% Trypsin, 0.5 mM EDTA in HBSS pH = 7.0–7.6.
6. Plating Medium: Minimum Essential Medium (MEM), 10% Fetal Bovine Serum (FBS), 1× Glutamax (e.g., Life technologies), 5 mM HEPES, 50 U/mL Penicillin/Streptomycin, 0.6% glucose, filtered through 0.2 μm under sterile conditions.
7. Neurobasal Medium: Neurobasal (Life Technologies), 1× B27 supplement (Life Technologies), 1× Glutamax, 50 U/mL Penicillin/Streptomicyn, filtered through 0.2 μm under sterile conditions.
8. HEK293 Medium: DMEM, 10% FBS, 1.2 mM Sodium Pyruvate, 50 U/mL Penicillin /Streptomycin, filtered through 0.2 μm under sterile conditions.
9. Neurobasal-ARAC Medium: 2 μM Cytosine-β-d-arabinofuranoside in Neurobasal Medium.
10. Matrigel Solution: CMF-HBSS Solution and Matrigel® (as per lot dilution instructions, BD).
11. Transfection reagent: e.g., FuGENE 6 Transfection Reagent (Promega) or Lipofectamine LTX (Life Technologies).

12. PFA Solution: 4% Paraformaldehyde, 4%Sucrose in PBS, pH to 7.4 and filter, store at 4 °C for no more than 2 weeks
13. Permeabilization Solution: 0.1% Triton X-100 in PBS.
14. Blocking Solution: 3% FBS in PBS.
15. Mounting medium with anti-fade: e.g., Aqua-Mount (Thermo Scientific).

2.2 Instrumentation

1. Dissecting tools (sterilized): fine-tipped forceps (for example, Dumont no. 5), microdissecting scissors (Vannas-style spring scissors).
2. Tissue culture incubator at 35.5 °C with humidified, 5% CO_2 atmosphere.
3. Sterile plastic ware: 100×20 mm tissue culture dishes, 75-cm^2 tissue culture flasks, 24- and 6-well tissue culture plate.
4. Sterilization filter units: syninge-driven, and cups of 250 and 500 mL.

3 Methods

3.1 Hippocampal Cultures

1. Prepare coverslips in 24-well plates by adding 150 μl Matrigel Solution per coverslip and leave for at least 2 h in the incubator. Aspirate the Matrigel Solution leaving a thin layer coating the glass. Dry the coverslips for a few minutes. Wash the coverslips two times with plating medium and finally add 1 mL plating medium. Store the 24-well plate in the incubator until hippocampal neurons are ready to be seeded.
2. Dissect out the brain from one P0 rat pup using an approved method of euthanasia (*see* Fig. 2a); immediately store the brain in CMF-HBSS Solution on ice until you have collected all needed samples. One brain will provide enough cells for 12 coverslips of neuron-glia culture.
3. In the laminar flow hood, collect all brains and place them in a dish with CMF-HBSS Solution on ice. The brains should be completely submerged until the hippocampi are removed from all the samples.
4. Carefully remove the meninges and dissect out the hippocampi with fine forceps under the dissection microscope. Collect the hippocampi in a 15 mL tube with CMF-HBSS Solution on ice (*see* Fig. 2a).
5. In a tissue culture hood, wash the hippocampi with cold CMF-HBSS Solution two times.
6. Aspirate the liquid and add Papain Solution to the hippocampi. Incubate them for 30 min at 37 °C.

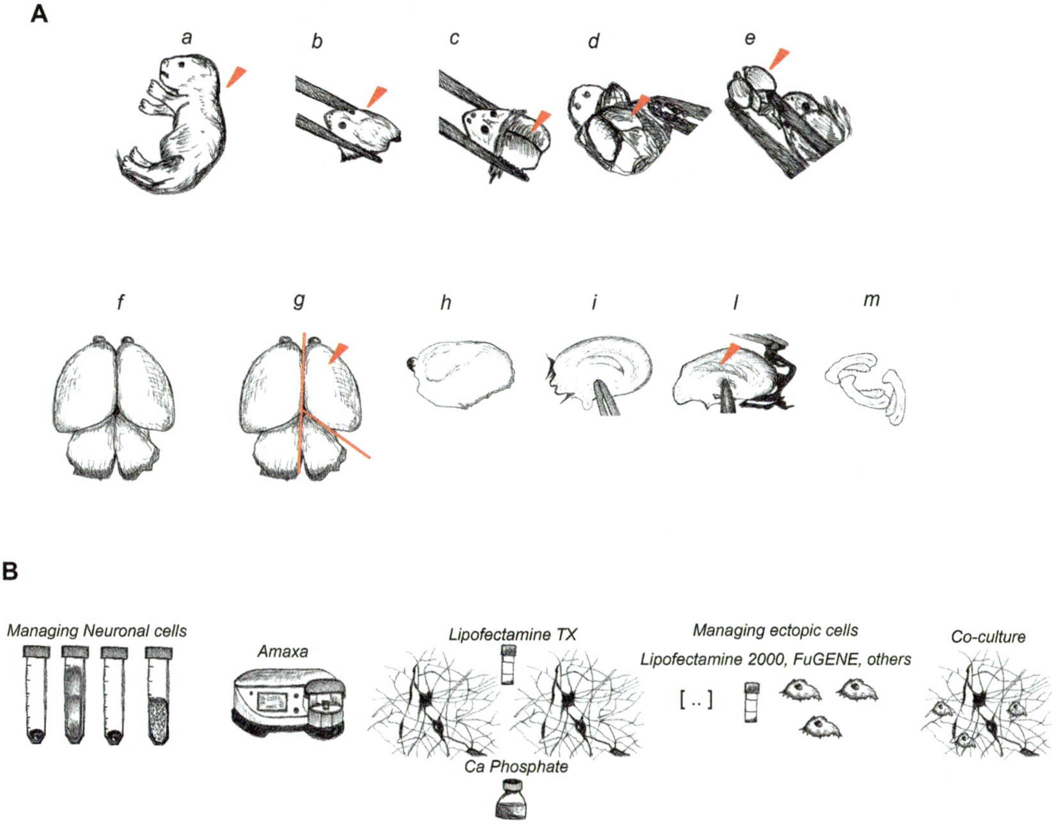

Fig. 2 Schematic representation of hippocampus dissection and cell transfection. (**a**) Flow chart showing dissection of hippocampus from a P0 rat brain. Take out the brain (*steps a, b, c, d,* and *e*) and once the meninges have been removed (*f*), separate the cortical rind from the thalamo-cortical fiber tract (step *g*). Finally, the hippocampi can be removed using forceps (*steps h, i, l,* and *m*). (**b**) Overexpression of synaptic proteins in neurons can be performed using different protocols. Neurons can be transfected according to the Amaxa protocol [31] before plating, or neurons can be transfected using Lipofectamine LTX or Calcium Phosphate at 5–7 DIV. HEC293 cells can be transfected with the commercial regent FuGENE as is described in Subheading 3.2

7. Aspirate the papain solution and wash the hippocampi twice with Plating Medium leaving them finally in 2 mL of medium.
8. Triturate the hippocampi gently by passing the suspension 10–12 times through a glass Pasteur pipette previously fire-polished (hold glass pipette for 20–30 s close to a flame until the edges become smooth (*see* **Note 1**)).
9. Allow un-dissociated pieces of hippocampus to settle by gravity for 2–3 min.
10. Determine cell density using a hemocytometer. The yield should be around 500,000 cells per hippocampus. Plate the dissociated neurons at 30,000–40,000 cells per well (150 μL approx.) on the plate that was prepared in **step 1**. Place the neurons in the cell culture incubator for 20–24 h.

11. Aspirate the Plating Medium and add 1 mL of Neurobasal Medium per well. No washes are required in between the medium change.
12. Neurons at 6–8 Day In Vitro (DIV) are recommended for co-culture. Alternatively, neurons can be transfected for expression of synaptic markers tagged with fluorescent probes (*see* **Note 2** and Fig. 2b).

3.2 HEK293 Cell Culture

1. HEK293 cells are plated with a seeding confluence of 15,000 cell/cm^2. The working volume for one well with a 9.4 cm^2 surface area is 1 mL (*see* **Note 3**). This must be done when neuronal cultures reach 4–6 DIV.
2. 24 h post-seeding, HEK293 cells are transfected using the vector of interest, tagged with a fluorescent protein, or co-expressing soluble fluorescent protein as a marker of transfected cells. Successful transfections require plasmid DNA with a 260/280 nm absorption ratio close to 1.8 in addition to previous optimization of Transfection-reagent:DNA ratio (*see* **Note 4**).
3. For transfection, we will describe the procedure for standard lipofection using the FuGENE transfection reagent. To prepare transfection solutions, first mix FuGENE with pre-warmed DMEM (serum-free, no antibiotics) to a final volume of 50 μL per well of cells that will be transfected (*see* **Note 5**). Vortex and incubate for 5 min.
4. Prepare an additional tube with a mix of plasmid DNA and DMEM to a final volume of 50 μL per each well to be transfected.
5. Add the 50 μL of DNA solution to the 50 μL of diluted FuGENE and incubate at least for 15 min at room temperature (*see* **Note 6**).
6. Add 100 μL of the Fugene:DNA complex solution over each well of cells in a drop-wise manner. To ensure a homogenous distribution along the surface, mix solution with the medium by gentle rotational movements.
7. Return the cells to the incubator for 24–48 h. Check the transfection efficiency on an epifluorescence microscope to visualize fluorescence-expressing cells. Usually, 48 h post-transfection HEK293 cells show high levels of expression.

3.3 Neuron-HEK293 Cell Co-culture

1. Harvest HEK293 cells 24–48 h post-transfection, by adding 500 μL Trypsin/EDTA per well.
2. After a few minutes place the suspension in 4.5 mL of HEK293 Medium and pellet the cells by centrifugation at 500 ×*g* for 5 min.
3. Resuspend the cells in 2 mL Neurobasal Medium.

4. Determine the cell density using a hemocytometer.
5. Dilute the cells in order to have 20,000cells/100 μl and add ARAC (22 μM) to avoid over growth of HEK293 cells after co-culture with neurons (*see* **Note 7**).
6. Seed 20,000 HEK293 cells (100 μl) per well of neurons at 6–8 DIV. Maintain the co-culture in the incubator for 24–48 h (*see* **Note 8**).

3.4 Co-culture Immunostaining

1. Wash the co-culture with PBS at RT, then fix the coverslips by bathing in PFA Solution for 15 min at RT. This step is critical since longer fixations can affect immunofluorescence quality (*see* **Note 9**).
2. Wash the cells carefully three times using PBS.
3. Incubate with Permeabilization Solution for 15 min at. Wash cells three times with PBS.
4. Add Blocking Solution for 1 h at RT.
5. Incubate with primary antibody for synaptic markers diluted appropriately in Blocking Solution for 1 h at RT (*see* **Note 10**). Then wash three times with PBS. Representative images of presynaptic proteins vGluT1 and Bassoon are shown in Fig. 3.
6. Add the secondary antibody diluted appropriately in Blocking Solution and incubate for 1 h at RT. Wash three times after remove the secondary antibodies.
7. Mount the coverslips using a drop of mounting medium with optional anti-fade (i.e., Aqua-Poly/Mount solidifies after 24–48 h/20–50% humidity). The coverslip can be stored in the dark at 4 °C.

3.5 Image Acquisition

The settings for image acquisition depend strongly on the fluorophores selected for the assay and the experimental conditions. Spinning disk or TIRF microscopy is suggested when the hemisynapse must be visualized using live imaging of fluorescent reporters such as pH-sensitive targeted synaptic proteins. The opportunity to combine different biosensors, specific synaptic proteins, and microfluidic device makes the possibilities of the co-culture assay compatible with a large spectrum of applications in neuroscience.

The "classic" co-culture assay experiment using the protocol described in this chapter can be easily analyzed using a confocal microscope with the same image acquisition settings for each sample of an experiment (*see* Fig. 3). In this case, quantification of presynaptic protein recruitment is performed using Z-section images that were converted to maximal projection images (*see* Fig. 3a). Transfected HEK293 cells, with control or Neuroligin 1 overexpression constructs, in close proximity to neurons were selected based on CFP labeling with the aim of measuring the fluo-

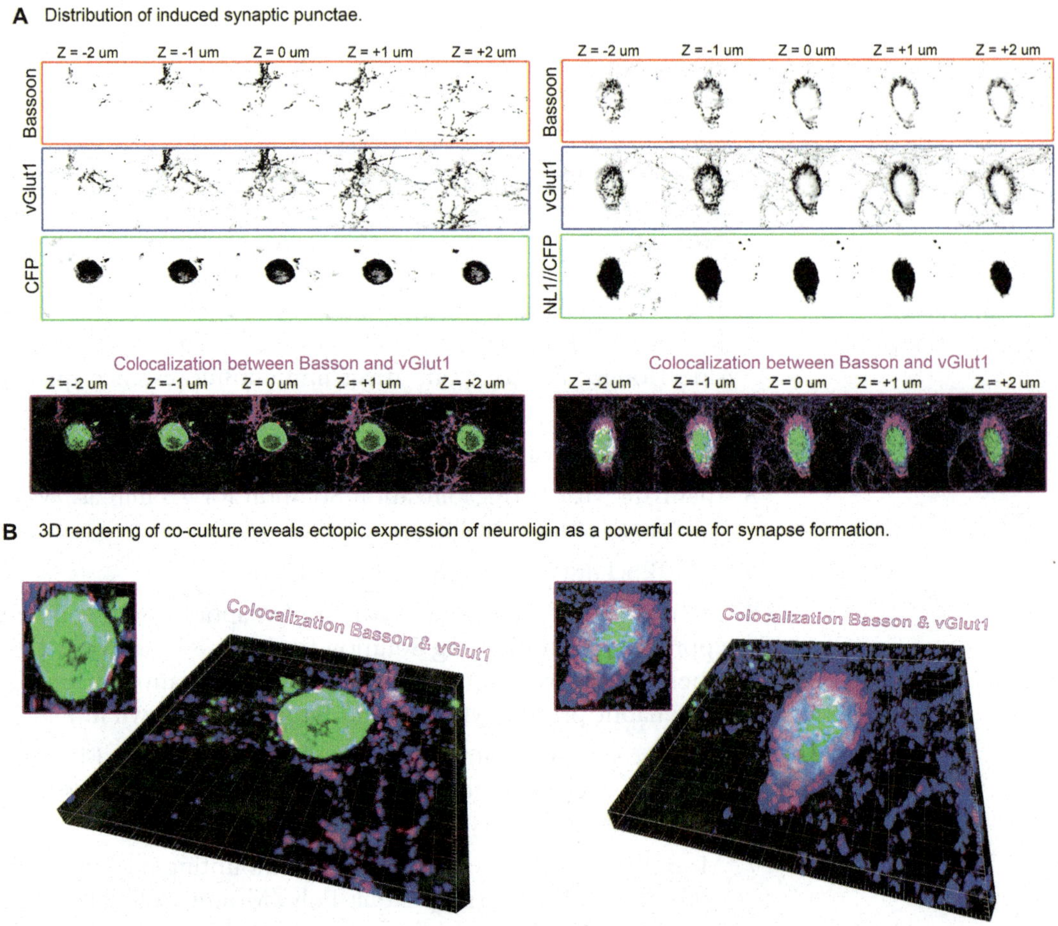

Fig. 3 Co-culture assay image acquisition. (**a**) Visualization of unmerged serial optical sections for presynaptic proteins Bassoon (*red*) and vGlut1 (*blue*) after 48 h of co-culture with HEK293 cells expressing Neuroligin 1 with CFP (CFP/NLG1) and CFP alone as negative control (**a**). *Bottom panel* shows the colocalization of vGlut1 and Bassoon with CFP (*green*) that is color-coded in *violet*; only the expression of NLG1 induces the recruitment of synaptic markers. (**b**) 3D rendering of neuron-HEK293 co-culture showing synaptic proteins that co-localize with heterologous expression of NLG1 (**b**). Presynaptic vGlut1 and Bassoon clusters are distributed along HEK cell surfaces that express NLG1 (*violet*)

rescence intensity for both red (Bassoon) and far-red channels (vGlut1) in the region of interest.

3.6 Analysis

In this chapter, we give a practical example of analysis of co-culture data obtained from a point-scan confocal microscope (*see* Fig. 3). Immunostaining of mixed co-cultures for the presynaptic markers vGlut1 and Bassoon was performed at 10 DIV, and images were acquired by confocal microscopy with the modalities described above. As control, HEK293 expressing the fluorophore probe alone (i.e., CFP) should be analyzed in parallel with HEK293 expressing the candidate protein tagged with a fluorescent probe, i.e., the bicistronic vector encoding both CFP and the postsynaptic adhesion molecule Neuroligin 1 (*see* Fig. 3).

A variety of standard software has been intensively used for co-culture quantifications allowing the automatization of the process, among the most popular are Matlab script (*see* **Note 11**), GNU octave and ImageJ software [28] (*see* **Note 12**). The quantification of the HEK293 cell surface area fraction immune-positive for synaptic markers requires the initial establishment of the proper threshold for the channel with the fluorescence signal from a given HEK293 cell (in this case CFP). This will help to define the perimeter of the cell, which subsequently will be used as a mask for the second and third channels that correspond to the synaptic markers. The readout most commonly used, as we mentioned above, is the fraction area positive for synaptic marker over the total surface area (perimeter of the HEK293 cell). To obtain this value the user should divide the surface area of either channel two or three by the surface area of channel one (CFP) which was previously obtained by applying the threshold and mask steps using the imaging software. Critical for accurate quantitative analysis is using identical acquisition settings for every image in addition to performing the image analysis with constant thresholds. Representative images of the synaptic marker distribution on the ectopic Neuroligin-expressing cell surface after the z-stack montages were loaded (*see* Fig. 3a) and converted to maximal-projection images using imaging software (*see* Fig. 3b, upper left panels).

Optionally, representative images of 3D rendering can be achieved using Imaris software (*see* **Note 13**). One of its main advantages is the relatively easier way to produce 3D rendering compared with other software. In addition, Imaris version 7.4 and newer can access real-time data from Fiji or ImageJ (*see* **Note 12**), making the 3D rendering operation more efficient and less time-consuming (*see* Fig. 3b). The images must be in tiff format to be processed by the software and initially the image sequence should be loaded in Fiji using the RGB matrix to see the different channels. Open the Imaris software and select Fiji task from the main menu, select the image to analyze from Fiji using the menu window. Once the software automatically loads the 3-channel z-stack image from Fiji, the 3D rendering will be immediately build. Although Imaris software is able to import other image formats, the import process itself can lose spatial information. One way to overcome this issue is doing the redefinition of spatial parameters with very simple and intuitive operations.

3.7 High-Throughput Devices

1. HTP screening microscope.

The assay is an extension of the co-culture methodology and analysis described above. The great advantage is that it is possible to use 96- or 384-well plates to screen a large pool of candidates of synaptic inductors simultaneously. The possibility of using different cell types for co-culturing and alternative libraries of chemical compounds, and the possibility of multiple repetitions make the use of an HTP device incredibly useful for the co-culture assay with high statistical power. Expression library pools are now available from many vendors and the cost has become relatively affordable.

The challenges of this approach are related to the large scale transfection cost, antibody and related reagents, but mostly to the hardware, since a microscope with HTP capability may be a significant expensive. There are two main options for microscopes with HTP capabilities: Epifluorescent system (e.g., ImageXpress, Molecular Device) or spinning disk system (e.g., Opera /Operetta, Perkin Elmer). Already some laboratories have used screening approaches successfully, identifying LRRTM1 as a synaptogenic factor [9]. These commercial microscopes are accompanied by strong customer service, with in-house training and the possibility of integrating the screening software to automatize your co-culture assay to your experimental needs.

A useful alternative for a robust and high-content method is culturing neurons on micro-patterned substrates comprising arrays of individual micron-scale dots coated with proteins that trigger neuronal development [29]. The main advantage of this alternative method is that it can be used with traditional microscopy techniques in addition to be useful for the screening of pharmaceutical compounds modulating synaptogenesis.

2. Microfluidic devices.

These devices can be used to enrich the imaging area in pre- or postsynaptic terminals, providing a standard and dense region to assay synapse formation in the co-culture assay. Briefly described, etched soft substrates are attached to the imaging device (slide or chamber), with small wells that will be used to plate neurons/heterologous cells. In addition, fine-etched channels (approx. 250–100 μm) that allow the extension of neuronal processes, but not neuronal cell bodies, mediate the communication between wells. Therefore, we can plate neurons in one or several wells and subsequently these neurons will assemble synapses on heterologous cells located in adjacent wells. Neurons can be prepared as described in Subheading 3.1. The density of the neuronal culture must be around 50,0000 cells/mL, but this device has been used mainly with neuronal cultures from E18 embryos [30]. The platform to implement this assay can be designed manually and fabricated in PDMS using soft lithography and replica molding [18, 19], but is also commercially available (e.g., Millipore, AX45005 and XONA Microfluidics). Neurons and HEK293 cells are seeded in different compartments as is shown in Fig. 1c. For the synaptic markers, immunostaining follows the protocol described in Subheading 3.4, though an additional step may be necessary to block nonspecific binding using 5–10% goat serum or BSA.

4 Notes

1. Dissociation of hippocampi is a critical step; to avoid low viability be very gentle during the pipetting step and do not use pipettes with a very narrow diameter.
2. To analyze synaptic activity-inducing factors, neurons can be transfected with synaptic markers tagged with fluorescent probes as pHlurorin or GCamP (*see* Fig. 2b). For this assay, neurons are transfected via nucleofection (Amaxa) before plating (*see* Subheading 3.1, **step 9**). Alternatively, neurons can be transfected at 7–9 DIV accordingly with published methods with Lipofectamine LTX [15] or Calcium Phosphate [31].
3. Other heterologous cells also can be utilized successfully in this assay. PC12 cells endogenously express several neuronal proteins, which is advantageous if neuron-specific posttranslational modification is required. However, the neuronal proteins expressed by PC12 cells can interfere with the interpretation of results. COS7 cells are also a viable alternative to HEK293 cells as they have greater areas with flatter surfaces.
4. Previous optimization of FuGENE:DNA ratio can be made using the provider guidelines. In particular for HEK93 cells, 3 μL of FuGENE and 1 μl DNA (3:1 ratio) give high transfection efficiency.
5. To avoid adverse effects on transfection efficiency do not allow FuGENE to come into contact with the walls of the tube containing DMEM. It is recommended to make a master mix of FuGENE solution depending on the total number of wells to be transfected. Each well of transfected HEK293 cells in this protocol will provide enough cells for more than four co-culture coverslips.
6. Incubation can continue up to 45 min without affecting the transfection efficiency, although there is some variation depending on the cell line used for the co-culture assay. For HEK293 cells, incubations up to 20 min are enough to get high transfection efficiency at 24–48 h post-transfection.
7. Since high ARAC concentrations can be toxics for neurons, ARAC should be diluted at 2 μM final concentration with Neurobasal Medium.
8. Volumes up to 100 μL of HEK293 cells per 1000 μL well of neurons are optimal for co-culture conditions.
9. Preparation of paraformaldehyde (PFA) solution is a critical step. Always adjust the pH close to 7.4 and filter the solution. Preferably use fresh aliquots of PFA for each experiment, as it is only stable for 2 weeks at 4 °C. Alternatively aliquots can be kept at −20 °C and discarded after use to avoid freeze- thaw cycles.
10. The spatial features and intensity of synaptic puncta revealed by each synaptic marker will directly impact the quality of the

quantification process. Although the staining conditions should be established empirically for each primary antibody, improvements can be made by overnight incubation at 4 °C. A detailed description of primary antibodies useful for co-culture quantifications is shown in Table 1.

11. MathWorks MatLab® (http://www.mathworks.com) or Gnu Octave (http://www.gnu.org/software/octave/) is high-level interpreted language, primarily intended for numerical computations with extensive graphics capabilities for data visualization and manipulation. The Octave language is quite similar to Matlab so that most programs are easily portable. As Matlab is a commercial package, it is richer in functions and imaging analysis capabilities. For both packages an intermediate knowledge of computer programming is required.
12. ImageJ (http://imagej.nih.gov/ij/download.html) is straightforward software for image analysis. Several versions have been developed, and are distributed via different institutions: Fiji, MCMB, MIC, etc. Currently, ImageJ FIJI is certainly the most versatile and open software. If the user has a good knowledge of Java, they can easily build and/or modify the plugins according to their own needs.
13. Bitplane Imaris® (http://www.bitplane.com/imaris) is professional software for bio-imaging analysis with modules for multidimensional rendering, co-visualization, and quantification of particle tracking and morphology. The software can be integrated with Matlab, so the user can build their own specific micro-scripts using MatLab extensions.

Acknowledgments

Thank you to Dr. Laura Swan for reading and suggestions to the manuscript.

References

1. Fu Z et al (2003) Functional excitatory synapses in HEK293 cells expressing neuroligin and glutamate receptors. J Neurophysiol 90(6):3950–3957
2. Biederer T et al (2002) SynCAM, a synaptic adhesion molecule that drives synapse assembly. Science 297(5586):1525–1531
3. Scheiffele P et al (2000) Neuroligin expressed in nonneuronal cells triggers presynaptic development in contacting axons. Cell 101(6):657–669
4. Biederer T, Scheiffele P (2007) Mixed-culture assays for analyzing neuronal synapse formation. Nat Protoc 2(3):670–676
5. Craig AM, Graf ER, Linhoff MW (2006) How to build a central synapse: clues from cell culture. Trends Neurosci 29(1):8–20
6. Graf ER et al (2004) Neurexins induce differentiation of GABA and glutamate postsynaptic specializations via neuroligins. Cell 119(7):1013–1026
7. Aoto J et al (2007) Postsynaptic ephrinB3 promotes shaft glutamatergic synapse formation. J Neurosci 27(28):7508–7519
8. Kayser MS et al (2006) Intracellular and trans-synaptic regulation of glutamatergic synaptogenesis by EphB receptors. J Neurosci 26(47):12152–12164

9. Linhoff MW et al (2009) An unbiased expression screen for synaptogenic proteins identifies the LRRTM protein family as synaptic organizers. Neuron 61(5):734–749
10. Kim JH et al (2005) Retrovirally transduced NCAM140 facilitates neuronal fate choice of hippocampal progenitor cells. J Neurochem 94(2):417–424
11. Henkemeyer M et al (2003) Multiple EphB receptor tyrosine kinases shape dendritic spines in the hippocampus. J Cell Biol 163(6):1313–1326
12. Henderson JT et al (2001) The receptor tyrosine kinase EphB2 regulates NMDA-dependent synaptic function. Neuron 32(6):1041–1056
13. Levinson JN et al (2005) Neuroligins mediate excitatory and inhibitory synapse formation: involvement of PSD-95 and neurexin-1beta in neuroligin-induced synaptic specificity. J Biol Chem 280(17):17312–17319
14. Chih B, Engelman H, Scheiffele P (2005) Control of excitatory and inhibitory synapse formation by neuroligins. Science 307(5713):1324–1328
15. Fogel AI et al (2007) SynCAMs organize synapses through heterophilic adhesion. J Neurosci 27(46):12516–12530
16. Park J et al (2012) Multi-compartment neuron-glia co-culture platform for localized CNS axon-glia interaction study. Lab Chip 12(18):3296–3304
17. Shi P et al (2011) Synapse microarray identification of small molecules that enhance synaptogenesis. Nat Commun 2:510
18. Park J et al (2009) Microfluidic compartmentalized co-culture platform for CNS axon myelination research. Biomed Microdevices 11(6):1145–1153
19. Taylor AM et al (2005) A microfluidic culture platform for CNS axonal injury, regeneration and transport. Nat Methods 2(8):599–605
20. Zhao C, Dreosti E, Lagnado L (2011) Homeostatic synaptic plasticity through changes in presynaptic calcium influx. J Neurosci 31(20):7492–7496
21. Dreosti E et al (2009) A genetically encoded reporter of synaptic activity in vivo. Nat Methods 6(12):883–889
22. Royle SJ et al (2008) Imaging phluorin-based probes at hippocampal synapses. Methods Mol Biol 457:293–303
23. Mao T et al (2008) Characterization and subcellular targeting of GCaMP-type genetically-encoded calcium indicators. PLoS One 3(3):e1796
24. Bozza T et al (2004) In vivo imaging of neuronal activity by targeted expression of a genetically encoded probe in the mouse. Neuron 42(1):9–21
25. Chen TW et al (2013) Ultrasensitive fluorescent proteins for imaging neuronal activity. Nature 499(7458):295–300
26. Marvin JS et al (2013) An optimized fluorescent probe for visualizing glutamate neurotransmission. Nat Methods 10(2):162–170
27. Jin L et al (2012) Single action potentials and subthreshold electrical events imaged in neurons with a fluorescent protein voltage probe. Neuron 75(5):779–785
28. Schneider CA, Rasband WS, Eliceiri KW (2012) NIH Image to ImageJ: 25 years of image analysis. Nat Methods 9(7):671–675
29. Czondor K et al (2013) Micropatterned substrates coated with neuronal adhesion molecules for high-content study of synapse formation. Nat Commun 4:2252
30. Hamilton LK et al (2010) Widespread deficits in adult neurogenesis precede plaque and tangle formation in the 3xTg mouse model of Alzheimer's disease. Eur J Neurosci 32(6):905–920
31. Zeitelhofer M et al (2007) High-efficiency transfection of mammalian neurons via nucleofection. Nat Protoc 2(7):1692–1704
32. Mercken M et al (1992) Immunocytochemical detection of the growth-associated protein B-50 by newly characterized monoclonal antibodies in human brain and muscle. J Neurobiol 23(3):309–321
33. Granseth B et al (2006) Clathrin-mediated endocytosis is the dominant mechanism of vesicle retrieval at hippocampal synapses. Neuron 51(6):773–786
34. Winkelmann A et al (2015) Identification of a new genomic hot spot of evolutionary diversification of protein function. PLoS One 10(5):e0125413
35. Perez de Arce K et al (2010) Synaptic clustering of PSD-95 is regulated by c-Abl through tyrosine phosphorylation. J Neurosci 30(10):3728–3738
36. Hirbec H et al (2003) Rapid and differential regulation of AMPA and kainate receptors at hippocampal mossy fibre synapses by PICK1 and GRIP. Neuron 37(4):625–638
37. Voglmaier SM et al (2006) Distinct endocytic pathways control the rate and extent of synaptic vesicle protein recycling. Neuron 51(1):71–84
38. Willig KI et al (2006) STED microscopy reveals that synaptotagmin remains clustered after synaptic vesicle exocytosis. Nature 440(7086):935–939
39. Umemori H et al (2004) FGF22 and its close relatives are presynaptic organizing molecules in the mammalian brain. Cell 118(2):257–270
40. Prange O et al (2004) A balance between excitatory and inhibitory synapses is controlled by PSD-95 and neuroligin. Proc Natl Acad Sci U S A 101(38):13915–13920

Chapter 3

Synaptogenic Assays Using Neurons Cultured on Micropatterned Substrates

Katalin Czöndör and Olivier Thoumine

Abstract

One of the difficulties for studying the mechanisms of synaptogenesis stems from the spatial unpredictability of contact formation between neurons, and the involvement of many parallel adhesive pathways mediating axon/dendrite recognition. To circumvent these limitations, we describe here a method allowing the investigation of synaptic contacts at controlled locations with high precision and statistics. Specifically, primary neurons are cultured on micropatterned substrates comprising arrays of micron-scale dots coated with purified synaptogenic adhesion molecules. Coating the substrates with the homophilic adhesion molecule SynCAM triggers the formation of functional presynaptic structures in axons, while neurexin elicits postsynapses in dendrites from neurons expressing the counter receptor neuroligin. This assay can be combined with various imaging techniques including immunocytochemistry to screen the accumulation of synaptic components, long-term live cell recordings to probe the kinetics of neurite growth and synapse differentiation, as well as high resolution single molecule tracking.

Key words Synapse formation, Micropatterned substrates, Adhesion molecules, Neurexin, Neuroligin, SynCAM, N-cadherin

1 Introduction

Assembly of the myriads of synapses between neurons during development of the central nervous system (CNS) represents a pivotal process that determines the generation, maintenance, and function of neural circuitries. At the level of individual axon/dendrite contacts, synaptogenesis is a complex multistep process initiated by specific adhesion proteins [1]. Once the contact is stabilized, scaffolding molecules and functional receptors that are essential for the transmission of information from one neuron to the other are recruited to contact sites [2, 3]. Importantly, the type of the presynaptic neurotransmitter (glutamate versus GABA) machinery, and the corresponding postsynaptic receptors and scaffolding molecules assembled at the synaptic contact, determines whether the synapse is excitatory or inhibitory. Once formed, synapses remain

Alexandros Poulopoulos (ed.), *Synapse Development: Methods and Protocols*, Methods in Molecular Biology, vol. 1538,
DOI 10.1007/978-1-4939-6688-2_3, © Springer Science+Business Media LLC 2017

very dynamic, as they undergo both structural and functional alterations in response to activity changes in the neuronal network [4]. Given such molecular complexity, heterogeneity, and dynamics, unraveling the mechanisms underlying synaptogenesis and synapse function is highly challenging. Thus, the development of simplified experimental systems that aim at triggering synaptogenesis in a controlled way, and where the role of specific molecules can be isolated, is highly needed.

Adhesion molecules play an important role in the initial stages of synapse formation, by strengthening the contact between the pre- and the postsynaptic membranes. However, certain adhesion molecules, like the heterophilic neurexin/neuroligin complex or the homophilic adhesion molecule SynCAM, were also found to play a key role in inducing synapse assembly by recruiting essential synaptic molecules to contact sites [5]. In the last decade, several biomimetic assays were developed including co-cultures [6–10], synapse microarrays [11], microspheres [7, 12, 13], and protein clusters [14–16], demonstrating the ability of certain adhesion molecules to induce synapse formation. These existing methods share the principle of using synaptogenic adhesion molecules to induce the formation of hemi-synapses, thus reducing the complexity of native synaptic contacts to a great extent. However, the spatial and temporal resolutions provided by these assays often do not allow high statistical production at the level of individual synaptic structures.

Here, we describe in technical detail a robust method recently developed to spatially control synaptic differentiation, by culturing neurons on micropatterned substrates comprising arrays of individual micron-scale dots coated with adhesion proteins, and surrounded by a cytophobic agent [17]. The dots are specifically coated with SynCAM1 or neurexin1β (Nrx1β), which triggers the selective differentiation of functional pre- or postsynapses, respectively. Coating of recombinant proteins is made via fluorescently labeled secondary antibodies, allowing both proper orientation of the adhesive ectodomain and fluorescent visualization of the micropatterns. After this coating step, dissociated neurons are plated and cultured on the substrates as regular neuronal cultures. Neurite architecture is highly determined by the specific adhesion to the micron-scale dots, and the development of hemi-synapses is thus defined by the topology of the patterns, leading to the spatial control of synapse formation at a micron-scale resolution. Importantly, micropatterned substrates can be manipulated just as regular glass coverslips, and can therefore be combined with a wide range of microscopy techniques including immunocytochemistry to screen the accumulation of synaptic components, long-term live cell recordings to probe the kinetics of synapse differentiation, local photo-activation to probe functional synaptic properties, and high resolution single molecule tracking of membrane receptors.

Considering the simple coating protocol, this assay can theoretically be adapted to a wide range of other newly reported synaptogenic adhesion proteins, including leucine rich repeat transmembrane proteins that also bind neurexin [18–20], opening the way to the targeted and selective study of the mechanisms underlying synapse formation with unprecedented spatial resolution and statistical power.

2 Materials

2.1 Culturing Neurons on Micro patterned Substrates Coated with Adhesion Proteins

1. Coating buffer: 0.2 M boric acid, pH 8.5.
2. Goat Anti-Human IgG, Fcγ fragment-specific, minimal cross reactivity with other species, unlabeled and conjugated to Alexa647 (Jackson ImmunoResearch).
3. Recombinant proteins: Fc-tagged Neurexin-1β for inducing postsynapses, and SynCAM1-Fc for inducing presynapses (R&D System).
4. Plating medium: Minimum Essential Medium (MEM), 2 mM L-glutamine, 4.5 g/l glucose, 10% horse serum.
5. Culture medium: Neurobasal medium (without phenol red), Neuromix (PAA; 50×), 2 mM L-glutamine.
6. Effectene transfection reagent (Qiagen), if overexpression of the receptor for the coated ligand is necessary.

2.2 Experiments on Fixed Samples

1. Phosphate Buffer Solution (PBS).
2. Fixation buffer: Paraformaldehyde 4% (PFA), 4% sucrose in PBS.
3. Quenching buffer: 50 mM NH_4Cl in PBS.
4. Triton X-100.
5. Bovine Serum Albumin (BSA) (Sigma).
6. Primary and secondary antibodies of interest: rat anti-HA (Roche), mouse anti-PSD-95 (Thermo Scientific), mouse anti-synaptotagmin1 lumenal domain conjugated to Oyster 550 (Synaptic Systems), mouse anti-synapsin (Synaptic Systems), rabbit anti-SynCAM1 (Abcam), anti-rat-FITC (Jackson Immunoresearch), anti-mouse-Rhodamine (Jackson Immunoresearch).
7. Mounting medium: Mowiol (Calbiochem) or equivalent.
8. Fluorescence microscope (Nikon Eclipse TiE or equivalent) equipped with a 60×/1.40 NA oil immersion objective, appropriate filter sets from Semrock/Chroma, and a digital CMOS camera (ORCA-Flash 4.0, Hamamatsu).

2.3 Live Cell Imaging

1. Ludin open sample holder (Life Imaging Services), manufactured to fit the 20×20 mm square micropatterned coverslips.

2. Penicillin/Streptomycin (Gibco).
3. For long-term imaging experiments: BioStation CT (Nikon).
4. For short-term imaging experiments (up to 5 h): an inverted microscope (Nikon TiE or equivalent) equipped with filter sets appropriate for the fluorophores used, 40× or 60× oil immersion objectives, a thermostatic box providing 37 °C and 5% CO_2 (Life Imaging Systems) and CCD camera (CoolSnap, Roper Scientific, Evry, France).

2.4 Universal Point Accumulation in Nanoscopic Topography (UPAINT):

1. Tyrode solution: 108 mM NaCl, 5 mM KCl, 2 mM $MgCl_2$, 2 mM $CaCl_2$, 10 mM HEPES, 15 mM Glucose, pH 7.4.
2. Monoclonal anti-GluA2 antibodies conjugated to ATTO 647N (Attotec) according to the manufacturer protocol.
3. Microscope equipped with a 50 mW, 633 nm laser line and TIRF illumination arm (Nikon Ti-E Eclipse), 100× /1.49 NA oil immersion objective, dichroic mirror (Di01-R635, SemRock), 676/29 nm emission filter (SemRock) placed on a filter wheel (Suter) and an EM-CCD Camera (Evolve, Roper Scientific, Evry, France).

3 Methods

3.1 Coating Micropatterned Substrates with Synaptogenic Proteins

Micropatterned substrates used for this assay are specifically designed with the aim of locally inducing synaptogenesis at the micron-scale, comparable with the size of native synapses. Specifically, the substrates made of 20×20 mm square coverslips contain a hexagonal array of chemically activated dots of 1.5 μm diameter, separated by 5 μm and surrounded by a nonadhesive background (Fig. 1). The substrates are commercially available (CYTOO SA), but can be also produced in the laboratory (*see* **Note 1**). The dots of the substrates are coated with 50 μg/ml poly-L-lysine by the manufacturer, dried, and shipped at room temperature. They can be stored for up to 2 months at 4 °C after production. The specific coating of the substrates is done 1 day before plating the neurons, as described in the following steps (*see* **Note 2**).

1. Substrates are first incubated with goat anti-human Fc antibody at 0.023 mg/ml final concentration diluted in coating buffer. This secondary antibody layer ensures the proper orientation of the adhesive ectodomain of the protein. Importantly, a combination of unlabeled and Alexa-647 conjugated antibodies in a 9:1 ratio is used to allow the fluorescent visualization of the micropatterns. Before incubation, the diluted antibodies are centrifuged for 10 min at 20,000×*g* and 4 °C. To minimize antibody consumption, substrates are flipped onto a 160 μl drop of the antibody mixture on Parafilm

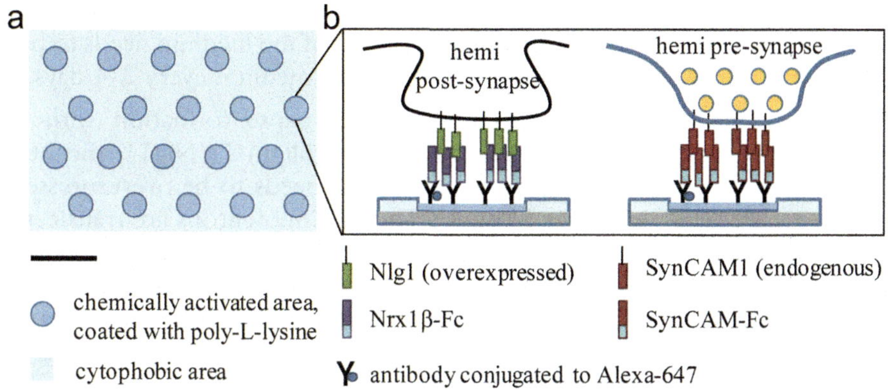

Fig. 1 Schematic diagram of the molecular interactions at contacts between neurons and micropatterned substrates. **(a)** Schematic view of the micropattern geometry. Micropatterned substrates contain a hexagonal array of adhesive dots (1.5 μm), coated with PLL (*dark blue*) and surrounded by a cytophobic agent (*light blue*). Scale bar: 5 μm. **(b)** Schematic view of the molecular interactions between neurons and one dot. Micropatterns are first coated with anti-human-Fc antibodies conjugated to Alexa-647, than specifically with either Nrx1β-Fc (left) or SynCAM1-Fc (*right*). Nlg1 located at the membrane of Nlg1-expressing neurons interact with immobilized Nrx1β-Fc at the dot, inducing postsynapse development. On SynCAM1-Fc coated micropatterns, SynCAM1-Fc interact with endogenous SynCAM1 of untransfected neurons and induce presynapse formation

(previously sterilized with UV for 1 h). The substrates are than incubated with the antibody mix at RT for 2–5 h (*see* **Note 3**).

2. The substrates are flipped into 6-well plates and carefully washed with coating buffer (*see* **Note 4**).
3. After washing, the substrates are incubated with purified Nrx1β-Fc for inducing postsynaptic differentiation or with SynCAM1-Fc for presynaptic differentiation (0.04 mg/ml in coating buffer; centrifuged for 10 min at 20,000×*g*, 4 °C) overnight at 4 °C (*see* **Notes 5** and **6**).
4. Second day, the substrates are placed into a 6-well plate, washed three times with coating buffer and once with plating medium, and left for 1–2 h in the incubator (37 °C, 5% CO_2) in plating medium. We found that the adsorption of serum proteins was important for neuronal development and could not be omitted.

3.2 Plating and Culturing Neurons on Micropatterned Substrates

Primary hippocampal neurons are prepared from embryonic day 18 (E18) Sprague–Dawley rats, as described previously [21].

1. Dissociated neurons are plated at density of 100,000 cells/substrate (250 cells/mm^2 of coverslip) and are allowed to attach to the substrate for 2–4 h in plating medium.
2. 2–4 h after plating, the substrates are carefully washed with conditioned Neurobasal medium (culture medium), which was previously incubated for 4 days in dishes containing glial

cell layer (*see* **Note 7**). The cells can be kept in this medium up to 10 days; however, a part of the medium needs to be replaced by new conditioned culture medium every 3–4 days.

3. In order to induce postsynapse formation with Nrx1β-Fc coated micropatterns, neuroligin-1 (Nlg1), the heterophilic binding partner of Nrx1β, needs to be overexpressed in neurons (*see* **Note 8**). To this aim, neurons are transfected 3 days after plating using Effectene, according to the manufacturer's instructions. 1 μg DNA per substrate is used for a single transfection, and 2 μg total DNA are used for co-transfections. Conditioned culture medium is used for the transfection procedure (*see* **Note 9**).
4. In the case of SynCAM1-Fc coated substrates, the cells do not need to be transfected, since endogenous SynCAM levels in neurons seem to be sufficient to induce presynaptic development.

3.3 Live Cell and Fixed Cell Image Acquisition

Micropatterned substrates are suitable for several types of microscopy observations, just as regular 170 μm-thick glass coverslips. In the following section, we will focus on some of the possible applications aiming at studying the dynamics and the structure of synapse formation and function.

3.3.1 Phase Contrast Imaging of Neurite Development on Micropatterned Substrates

In order to monitor the growth of neurons on micropatterned substrates or to investigate the effect of the adhesion molecule specifically bound to the patterns on neurite development, the substrates are kept in a compact 37 °C/CO_2 cell incubator and monitoring system (BioStation) after plating. This system provides consistent environmental control of temperature, humidity, and gas concentration in combination with live cell imaging (Fig. 2).

1. After neurons have fully attached to the substrate (i.e., 4–6 h after plating), the substrates are mounted in a chamber suitable for imaging.
2. The cultures are kept in the plating medium containing antibiotics (Penicillin-Streptomycin), since the sterility of the culture in an open chamber cannot be ensured.
3. Images are taken every 10 min to monitor short-scale growing of neurites, while images taken every 1 h allow observation of overall neurite elaboration over several days.

Studying the isolated role of specific interactions, like for example Nrx/Nlg adhesion, is also possible by electroporating the neurons with HA-Nlg1 before plating on substrates coated with Nrx1β-Fc (Fig. 2b). By co-expressing a fluorescent protein, such as eGFP, the growth of neurons overexpressing HA-Nlg1 can be selectively monitored by fluorescent imaging and the effect on neuronal development can be directly compared to non-expressing neurons from the same cultures.

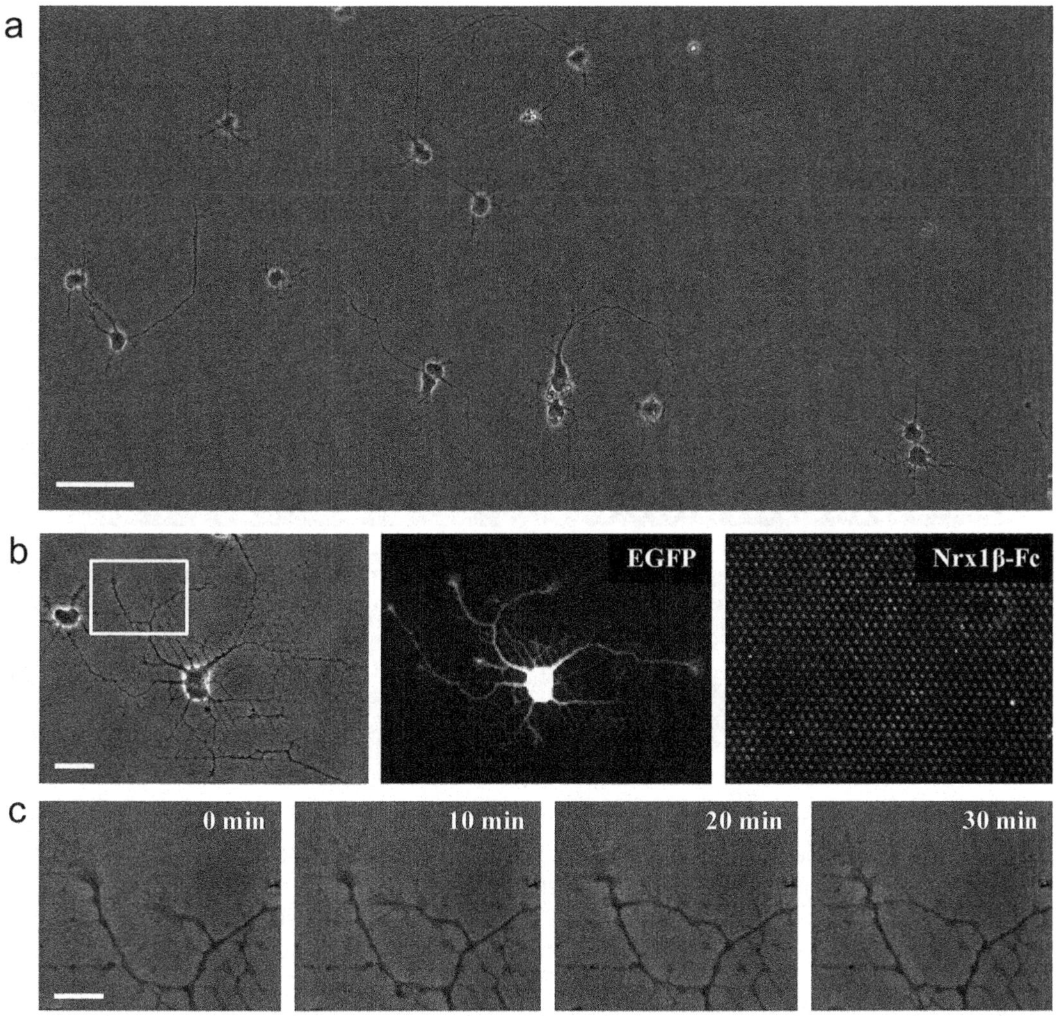

Fig. 2 Neurite growth of neurons cultured on Nrx1β coated micropatterned substrates. **(a)** Phase contrast mosaic image of a neuronal culture growing on Nrx1β-Fc-coated micropatterned substrates for 3 days in vitro. Scale bar: 50 μm. **(b)** Phase contrast and fluorescence image of a neuron expressing GFP and HA-Nlg1, and the corresponding image showing the Nrx1β-Fc coated micropattern. Scale bar: 20 μm. **(c)** Time lapse images of neurite growth, taking the area highlighted in (**b**). Note the motion of leading dendritic growth cones along the dots. Scale bar: 10 μm

3.3.2 Studying Synapse Assembly by Immunostaining of Synaptic Proteins on Fixed Samples

In the following part, we describe an example protocol for investigating the development of excitatory postsynapses in neurons cultured on Nrx1β-Fc-coated micropatterned substrates. Three days after plating neurons are transfected with HA-Nlg1, then fixed and immunostained 4 days later (Fig. 3).

1. Cells are fixed in Fixation buffer at RT for 10 min, and carefully washed three times with PBS (*see* **Note 4**). To saturate the remaining active sites of the PFA, the culture is incubated for 15 min in Quenching buffer, followed by careful washing with PBS.

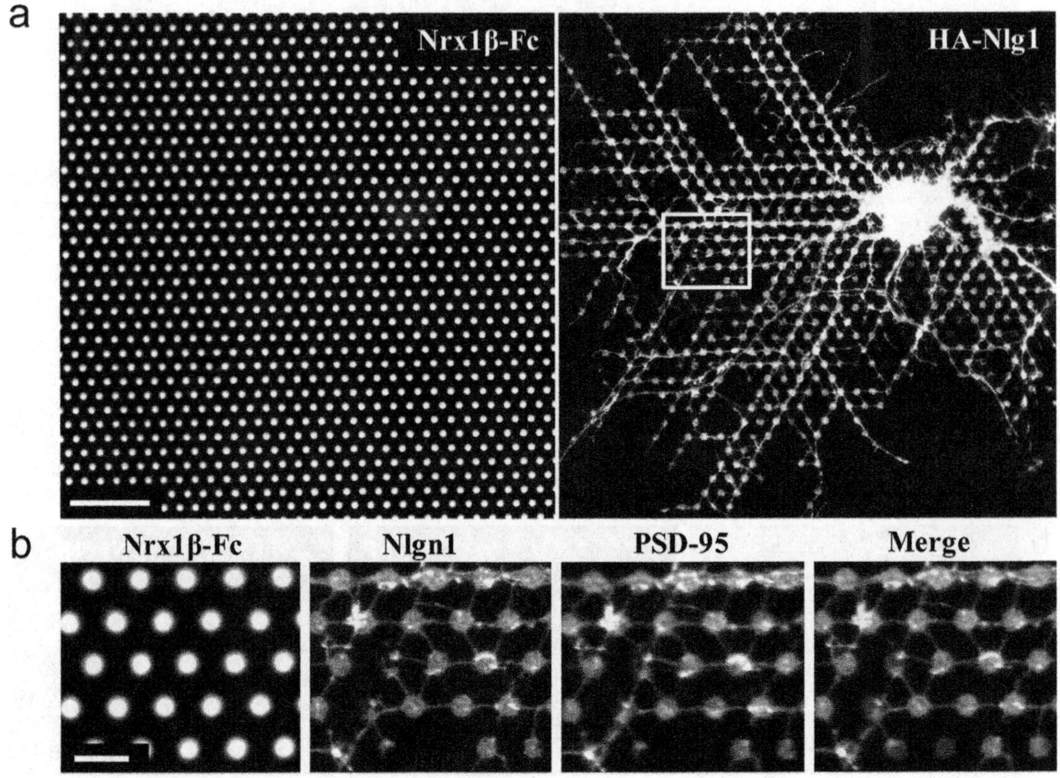

Fig. 3 Dendritic arborization and postsynaptic differentiation in neurons cultured on Nrx1β-Fc-coated micropatterned substrates. **(a)** Neurons were plated on micropatterned substrates coated with Nrx1β-Fc, transfected with HA-Nlg1, and immunostained for anti-HA and endogenous PSD-95 at DIV10. Note the development of a highly patterned dendritic arborization. Scale bar: 20 μm. **(b)** *Insets* show the selective accumulation of Nlg1 (*green*) and PSD-95 (*red*) at Nrx1β-Fc-coated dots (*blue*). Scale bar: 5 μm

2. To label intracellular proteins, cultures are permeabilized with 0.1 % Triton X-100 in PBS for 5 min, followed by careful washing with PBS.
3. Nonspecific binding is blocked by incubation with PBS containing 1 % BSA, for 1 h.
4. Neurons are first incubated for 1 h with primary antibodies, including rat anti-HA to stain HA-Nlg1 and mouse anti-PSD-95 to stain the major scaffolding protein of excitatory synapses. Primary antibodies are diluted 1:400 in PBS containing 1 % BSA (*see* **Note 10**).
5. After washing carefully with PBS containing 1 % BSA, neurons are incubated for 1 h with secondary antibodies: anti-rat-FITC and anti-mouse-Rhodamine at a 1:800 dilution (*see* **Note 11**).
6. Neurons are washed carefully once with PBS containing 1 % BSA, then at least three times with PBS.

7. Substrates are mounted on glass slides in approximately 20 μl Mowiol.
8. 16-bit images of the three fluorescence channels are acquired on an inverted epifluorescence microscope, equipped with the appropriate filter sets. A confocal laser scanning microscope can also be used here.

This protocol can be easily adapted to different kinds of synapse types or to study presynaptic differentiation, like for example staining SynCAM1 and synapsin (presynaptic marker) in neurons cultured on SynCAM-Fc coated substrates (Fig. 4).

3.3.3 Studying Presynapse Function with Live Antibody-Feeding Assay

Immunostaining reveals the presence of molecules in the assembled hemi-synapse, but does not give information about the functionality. One way to address whether for example the presynapses developed on SynCAM-Fc-coated micropatterns are functional is to label the active synapses by antibody-feeding assay (Fig. 4c).

1. Cells are incubated for 10 min at 37 °C with anti-synaptotagmin1 lumenal domain conjugated to Oyster 550 fluorophore, diluted 1:200 in culture medium.
2. Cells are carefully washed with warm conditioned medium to remove the excess antibodies that were not taken up by the active presynapses.
3. Labeling of active synapses can be either visualized immediately under the microscope, by mounting the substrate in Tyrode into the open chamber, or it can be fixed and observed later, with the possibility to immunostain another protein in parallel.

3.3.4 Single Molecule Tracking of AMPA Receptors

Dynamic aspects of synapse organization, such as receptor diffusion, can also be studied by carrying out single molecule tracking experiments in neurons cultured on micropatterned substrates. As an example, we describe the protocol for tracking the lateral movement of endogenous AMPA receptors (AMPARs) in relation to Nrx/Nlg adhesions (Fig. 5).

In general, we incubate neurons with antibodies against the N-terminal domain of the GluA2 subunit of AMPA receptors, which are conjugated to photostable fluorophores. We keep the antibodies in solution to allow constant labeling, while imaging neurons in oblique illumination to focus on antibodies bound to the cell surface, as described previously [22, 23]. In this experiment, primary hippocampal neurons growing on Nrx1β-Fc-coated micropatterned substrate are transfected with HA-Nlg1 and Homer1c-GFP at Days in vitro (DIV) 3–4, and used for single molecule tracking experiments at DIV7-8.

1. The substrate is mounted in Tyrode solution in an open observation chamber and placed on an inverted microscope equipped

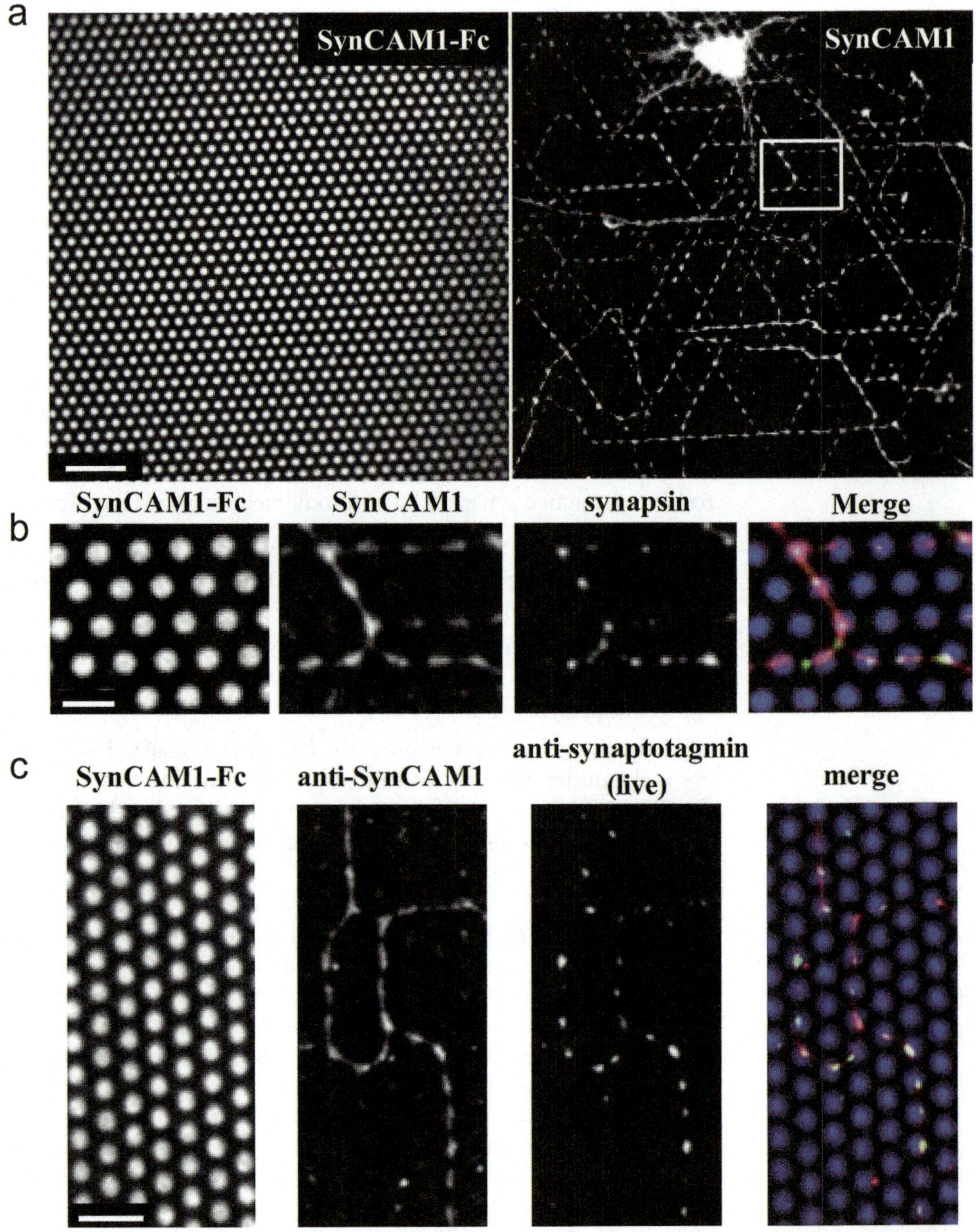

Fig. 4 Axonal arborization and presynaptic differentiation in neurons cultured on SynCAM1-coated micropatterns. **(a)** Neurons were plated on micropatterned substrates coated with SynCAM1-Fc, and immunostained for endogenous SynCAM1 and synapsin at DIV10. Note the development of a highly patterned axonal network. Scale bar: 20 μm. **(b)** *Insets* show the selective accumulation of SynCAM1 (*red*) and synapsin (*green*) at SynCAM1-Fc coated dots (*blue*). Scale bar: 5 μm. **(c)** Live labeling with anti-synaptotagmin–Oyster550 indicates the location of active synapses. Cells were immunostained for endogenous SynCAM1 after fixation. Note the strong accumulation of endogenous SynCAM1 (*red*) and synaptotagmin (*green*) at the dots (*blue*). Scale bar: 10 μm

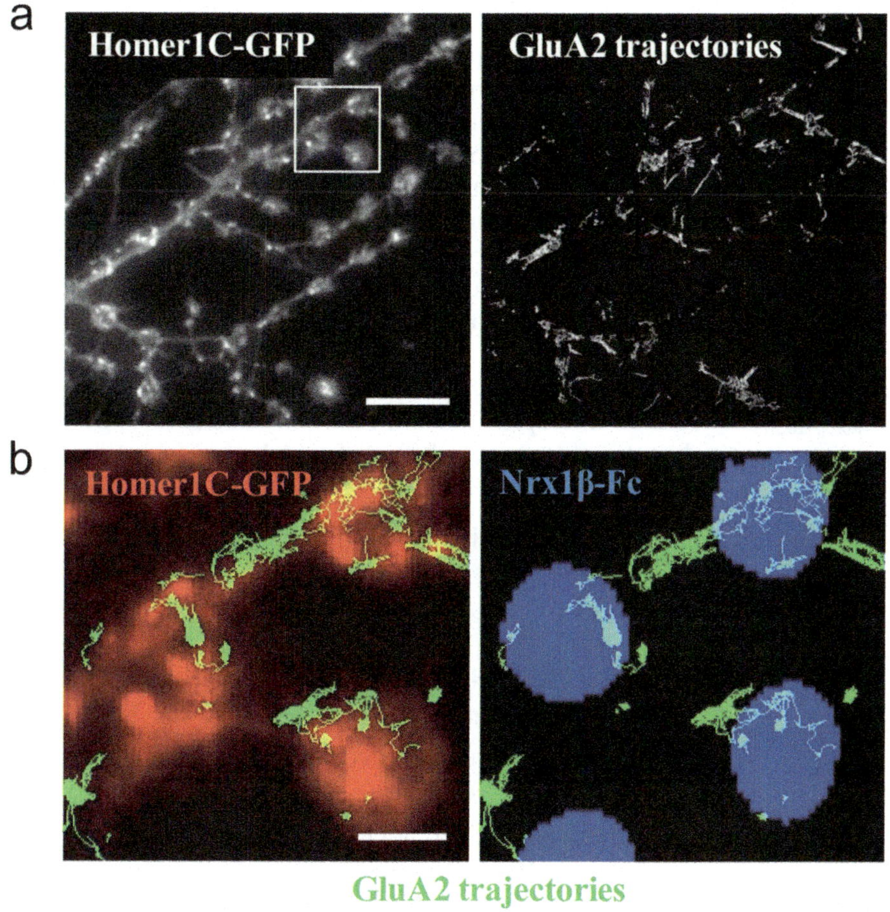

Fig. 5 Trapping of surface-diffusing AMPA receptors at Nrx1β coated micropatterns. **(a)** Dendritic segment of a neuron co-transfected with HA-Nlg1 and Homer1C-GFP, cultured for 7 DIV on micropatterned substrates coated with Nrx1β-Fc (*left*). The diffusion of surface AMPARs was followed by stochastic labeling of GluA2 subunits, and trajectories of individual AMPARs were reconstructed based on recordings of 4000 consecutive images (*right*). Scale bar: 10 μm. **(b)** Enlarged region taken from (**a**) (*white box*) showing reconstructed trajectories of GluA2 diffusion. Note the immobilization of AMPARs when entering the region of dendrites (*left*) overlapping the Nrx1β-Fc coated dots (*right*), suggesting the trapping of AMPARs at postsynapses, similarly as was shown in native synapses. Scale bar: 1 μm

with an EMCCD camera, a thermostatic box providing air at 37 °C, and an APO TIRF 100×/1.49 NA oil objective.

2. Antibodies against GluA2 subunit conjugated to ATTO 647N are added directly in the chamber to the Tyrode solution at a 1:6000 dilution.
3. Samples are imaged using oblique illumination performed with the 633-nm line of a HeNe laser to allow the selective imaging of surface diffusing receptors.
4. Five stacks of 4000 consecutive frames are obtained per cell, with an integration time of 50 ms.

3.4 Image Analysis

3.4.1 Quantification of Synaptic Protein Accumulation at Micropatterns in Fixed Samples

We developed two independent criteria to measure the recruitment of synaptic proteins at micropatterns [15]. The first procedure relies on quantifying an enrichment index measuring the ratio between fluorescence intensities of a given protein at micropatterns, compared to nearby regions on the same piece of neurite. It was written as a program in Metamorph, but could be written in any other image analysis software. The second procedure quantifies the degree of colocalization of synaptic clusters with micropatterns, and uses a custom program developed as a plugin for Image J (NIH). Both programs are available upon request.

Enrichment of Synaptic Proteins at Micropatterns

The first technique is described for neurons growing on Nrx1β-Fc-coated micropatterns and immunostained for Nlg1 and PSD-95, but this protocol can be equally applied to any other synaptic protein. Here are the basic steps:

1. The potential background in the Nlg1, PSD-95, and micropattern images is subtracted.
2. A dendritic region is selected on the fluorescent image of Nlg1 staining and the corresponding area is also selected on the images highlighting the PSD-95 staining and the micropatterns.
3. The neurite outline is determined using the threshold function on the Nlg1 image.
4. A threshold is also applied to the image of the micropatterns, yielding the outlines as individual regions, which are then transferred to the Nlg1 and PSD-95 images.
5. Only micropatterns that overlap with the neurite area above threshold are used for further quantification, while patterns outside the neuron are eliminated.
6. The fluorescence intensity within each pattern for the threshold area is determined on both Nlg1 and PSD-95 stained images, as well as the average fluorescence of the neuronal regions located outside the patterns.
7. An enrichment index is calculated by dividing the fluorescent intensities measured in neurite parts on the patterns by the average intensity measured from dendritic parts located outside patterns.
8. This procedure is repeated for three randomly chosen dendritic areas of each neuron, and values are averaged for each neuron. 10–20 neurons are quantified per condition. Enrichment factors above 1.2 are usually considered significant, and can go up to 1.8 or 2 in the case of very strong recruitments.

Apposition of Synaptic Clusters with Micropatterns

The second procedure is explained for the localization of synapsin or PSD-95 clusters at SynCAM-Fc-coated micropatterns.

1. Synapsin or PSD-95 signals of axonal or dendritic segments, respectively, and Alexa-647 signals corresponding to micropatterns are separately segmented using a wavelet-analysis program [24].
2. The segmented images of synapsin or PSD-95 staining are overlayed with the segmented image of the micropattern.
3. The overlapping area of synaptic and micropattern signals is then calculated on the overlaid image, and the percentage of synapsin or PSD-95 signal overlapping more than 50 % with a micropattern is determined and averaged on a per cell basis.
4. The value for a random distribution of puncta is calculated as the fractional area of micropatterns occupying the field of view, i.e., around 35 %. Values well above this threshold are considered significant.

3.4.2 Computation of AMPA Receptor Diffusion Coefficients from Single Molecule Trajectories

Detection of single AMPARs in the 4000-image stacks is performed using an algorithm based on image wavelet segmentation and single particle centroid determination, written as a program running in the Metamorph® software (Molecular Devices), and described earlier [24]. This program allows the identification of each molecule as an individual object, and connects successive positions within a certain diffusion distance. The instantaneous diffusion coefficient, D, is calculated for each trajectory, from linear fits of the first four points of the mean square displacement (MSD) function versus time [25]. The program also allows reconstruction of super-resolved density maps of AMPARs, where the number of AMPARs visiting each 50 nm pixel of the image is integrated over time for the whole image stack [23].

4 Notes

1. The production of micropatterns used in this assay can be implemented in the laboratory, according to the protocol described previously [26]. Other "soft lithographic" methods can be used to produce micropatterned substrates compatible with high resolution live cell imaging, and with similar geometries as ours. These include printing of adhesion proteins on silanized glass coverslips using a microfabricated unalterable master pattern such as an elastomeric stamp [27], or UV cross-linking [28]. Simpler and inexpensive techniques can be found, such as ink-jet printing methods, which consists in depositing microscopic adhesive droplets on a nonadhesive substrate such as one coated with polyethylene glycol, but the size resolution is usually poor (100–200 μm), although recent improvements reach a few microns precision [29]. An alternative approach called micro-photo-patterning

used computer-controlled bi-photon laser photo-ablation of an anti-adhesive poly-vinyl-alcohol film, giving access to many different types of geometries [30].

2. In the case when fabrication of the substrates is not conducted in sterile conditions, the substrates need to be sterilized before coating, for example by exposing them to a UV lamp (365 nm) for 40 min under a laminar flow hood.
3. Attempts to directly conjugate a Cy5 fluorophore to accessible lysine groups on the Nrx1β-Fc protein resulted in a loss of function of Nrx1β, losing the ability to bind neuroligin-1 on the cell surface.
4. It is very important to keep the substrates always wet, i.e., the liquid should never be completely removed from the substrates while washing. Because of the hydrophobic background, the dewetting of the surface generates strong forces that can lead to the disruption of the antibody layer on the substrate.
5. The recombinant proteins used for coating can alternatively be produced and purified in the laboratory. However, one has to keep in mind that this procedure can take several months, because the production has to be made in heterologous cells to provide the posttranslational changes needed for proper functioning of the proteins. A detailed protocol was described previously [31].
6. Alternatively, the substrates can be incubated with the antibody overnight at 4 °C, and the next day with the adhesion protein for 2–4 h at RT.
7. Since the density of the culture is relatively low, it is very important to keep the neurons in conditioned medium obtained from glial cell cultures. Although the medium contains the Neuromix supplement, additional factors secreted by glial cells are essential for cell survival.
8. In our E18 rat hippocampal Banker cultures observed at DIV7-15, the expression levels of endogenous Nlg1 seem fairly weak, as judged by the low level of surface staining obtained with Nrx1β-Fc, used in place of an antibody against Nlg1 suitable for live cell labeling [15]. These expression levels seem insufficient to mediate a strong recognition of Nrx1β-Fc immobilized on micropatterns, and thereby postsynaptic differentiation. Another possibility is that the amount of Nrx1β-Fc that is adsorbed to the secondary antibody layer is not high enough to bind enough endogenous Nlg1 and mediate synaptic differentiation. Thus, there seems to be a limiting amount of either exogenous Nrx1β or endogenous Nlg1 that prevents postsynaptic differentiation, whereas endogenous SynCAM seems to be in sufficient amounts to mediate presynaptic differentiation.

9. Alternatively, neurons can be electroporated prior to plating, which will typically result in a lower expression level and a higher ratio of cells overexpressing the given protein.
10. In order to minimize the volume of antibodies used, substrates can be placed on Parafilm, so that a drop of 200 μl is sufficient to cover their surface.
11. When choosing secondary antibodies, it is important to avoid using the same fluorescent label that was used for visualizing the micropattern. In the given example, one also has to be careful to use secondary antibodies for rat and mouse primary antibodies which are specifically developed for minimal cross-reactivity between host species. Such antibodies are available at Jackson Immunoresearch Labs.

Acknowledgments

We thank P. Scheiffele and S. Okabe for the generous gift of plasmids; B. Tessier, F. Neca, and Z. Karatas for molecular biology; the Cell culture facility of the Institute; A. Azioune for help on coating protocols, J.B. Sibarita and F. Levet for image analysis; and S. Lafosse for the temporary loan of the Nikon BioStation.
This work received funding from the Centre National de la Recherche Scientifique, Agence Nationale de la Recherche (grants Synapse-2Dt and SynAdh), Conseil Régional Aquitaine and Fondation pour la Recherche Médicale.

References

1. Garner CC, Zhai RG, Gundelfinger ED, Ziv NE (2002) Molecular mechanisms of CNS synaptogenesis. Trends Neurosci 25(5):243–251
2. Friedman HV, Bresler T, Garner CC, Ziv NE (2000) Assembly of new individual excitatory synapses: time course and temporal order of synaptic molecule recruitment. Neuron 27(1):57–69
3. Bresler T, Shapira M, Boeckers T, Dresbach T, Futter M, Garner CC et al (2004) Postsynaptic density assembly is fundamentally different from presynaptic active zone assembly. J Neurosci 24(6):1507–1520
4. Ziv NE, Smith SJ (1996) Evidence for a role of dendritic filopodia in synaptogenesis and spine formation. Neuron 17(1):91–102
5. Biederer T, Stagi M (2008) Signaling by synaptogenic molecules. Curr Opin Neurobiol 18(3):261–269
6. Biederer T, Sara Y, Mozhayeva M, Atasoy D, Liu X, Kavalali ET et al (2002) SynCAM, a synaptic adhesion molecule that drives synapse assembly. Science 297(5586):1525–1531
7. Graf ER, Zhang X, Jin SX, Linhoff MW, Craig AM (2004) Neurexins induce differentiation of GABA and glutamate postsynaptic specializations via neuroligins. Cell 119(7):1013–1026
8. Nam CI, Chen L (2005) Postsynaptic assembly induced by neurexin-neuroligin interaction and neurotransmitter. Proc Natl Acad Sci U S A 102(17):6137–6142
9. Scheiffele P, Fan J, Choih J, Fetter R, Serafini T (2000) Neuroligin expressed in nonneuronal cells triggers presynaptic development in contacting axons. Cell 101(6):657–669
10. Ko J, Zhang C, Arac D, Boucard AA, Brunger AT, Sudhof TC (2009) Neuroligin-1 performs neurexin-dependent and neurexin-independent functions in synapse validation. EMBO J 28(20):3244–3255
11. Shi P, Scott MA, Ghosh B, Wan D, Wissner-Gross Z, Mazitschek R et al (2011) Synapse microarray identification of small molecules that enhance synaptogenesis. Nat Commun 2:510
12. Dean C, Scholl FG, Choih J, DeMaria S, Berger J, Isacoff E et al (2003) Neurexin mediates the

assembly of presynaptic terminals. Nat Neurosci 6(7):708–716

13. Heine M, Thoumine O, Mondin M, Tessier B, Giannone G, Choquet D (2008) Activity-independent and subunit-specific recruitment of functional AMPA receptors at neurexin/neuroligin contacts. Proc Natl Acad Sci U S A 105(52):20947–20952
14. Barrow SL, Constable JR, Clark E, El-Sabeawy F, McAllister AK, Washbourne P (2009) Neuroligin1: a cell adhesion molecule that recruits PSD-95 and NMDA receptors by distinct mechanisms during synaptogenesis. Neural Dev 4:17
15. Mondin M, Labrousse V, Hosy E, Heine M, Tessier B, Levet F et al (2011) Neurexin-neuroligin adhesions capture surface-diffusing AMPA receptors through PSD-95 scaffolds. J Neurosci 31(38):13500–13515
16. Poulopoulos A, Aramuni G, Meyer G, Soykan T, Hoon M, Papadopoulos T et al (2009) Neuroligin 2 drives postsynaptic assembly at perisomatic inhibitory synapses through gephyrin and collybistin. Neuron 63(5):628–642
17. Czondor K, Garcia M, Argento A, Constals A, Breillat C, Tessier B et al (2013) Micropatterned substrates coated with neuronal adhesion molecules for high-content study of synapse formation. Nat Commun 4:2252
18. de Wit J, Sylwestrak E, O'Sullivan ML, Otto S, Tiglio K, Savas JN et al (2009) LRRTM2 interacts with neurexin1 and regulates excitatory synapse formation. Neuron 64(6):799–806
19. Ko J, Fuccillo MV, Malenka RC, Sudhof TC (2009) LRRTM2 functions as a neurexin ligand in promoting excitatory synapse formation. Neuron 64(6):791–798
20. Linhoff MW, Lauren J, Cassidy RM, Dobie FA, Takahashi H, Nygaard HB et al (2009) An unbiased expression screen for synaptogenic ssproteins identifies the LRRTM protein family as synaptic organizers. Neuron 61(5):734–749
21. Kaech S, Banker G (2006) Culturing hippocampal neurons. Nat Protoc 1(5):2406–2415
22. Giannone G, Hosy E, Levet F, Constals A, Schulze K, Sobolevsky AI et al (2010) Dynamic superresolution imaging of endogenous proteins on living cells at ultra-high density. Biophys J 99(4):1303–1310
23. Nair D, Hosy E, Petersen JD, Constals A, Giannone G, Choquet D et al (2013) Super-resolution imaging reveals that AMPA receptors inside synapses are dynamically organized in nanodomains regulated by PSD95. J Neurosci 33(32):13204–13224
24. Izeddin I, Boulanger J, Racine V, Specht CG, Kechkar A, Nair D et al (2012) Wavelet analysis for single molecule localization microscopy. Opt Express 20(3):2081–2095
25. Giannone G, Hosy E, Sibarita JB, Choquet D, Cognet L (2013) High-content super-resolution imaging of live cell by uPAINT. Methods Mol Biol 950:95–110
26. Azioune A, Carpi N, Tseng Q, Thery M, Piel M (2010) Protein micropatterns: a direct printing protocol using deep UVs. Methods Cell Biol 97:133–146
27. Rozkiewicz DI, Kraan Y, Werten MW, de Wolf FA, Subramaniam V, Ravoo BJ et al (2006) Covalent microcontact printing of proteins for cell patterning. Chemistry 12(24):6290–6297
28. Fink J, Thery M, Azioune A, Dupont R, Chatelain F, Bornens M et al (2007) Comparative study and improvement of current cell micro-patterning techniques. Lab Chip 7(6):672–680
29. Sanjana NE, Fuller SB (2004) A fast flexible ink-jet printing method for patterning dissociated neurons in culture. J Neurosci Methods 136(2):151–163
30. Doyle AD, Wang FW, Matsumoto K, Yamada KM (2009) One-dimensional topography underlies three-dimensional fibrillar cell migration. J Cell Biol 184(4):481–490
31. Mondin M, Tessier B, Thoumine O (2013) Assembly of synapses: biomimetic assays to control neurexin/neuroligin interactions at the neuronal surface. Curr Protoc Neurosci Chapter 2:Unit 2.19

Chapter 4

Monitoring Synapses Via Trans-Synaptic GFP Complementation

Theodoros Tsetsenis

Abstract

Over the last years, the analysis of synaptic connectivity in the mammalian brain has been accelerated by the use of techniques combining electrophysiology, light microscopy, viral tracing, and genetic manipulations in animal models. Of particular interest are methods that aim to label synapses by tethering complementary split GFP fragments in opposing sites of the synaptic cleft. Here, I describe SynView, a method for monitoring synapse formation based on GFP complementation, and provide a detailed protocol for use in neuronal cultures from mouse hippocampus.

Key words SynView, Neurexin, Neuroligin, Split-GFP, Synapse formation

1 Introduction

A major challenge in neuroscience research has always been the mapping of synaptic connectivity of the complex neuronal networks in the mammalian brain. Toward this goal, many techniques have been developed over the last years with the aim of creating links between neuronal circuitry and behavior. These range from high-throughput electron microscopy methods [1] to viral tracing strategies [2–4]. While the utilization of these approaches has managed to dissect a number of neural circuits, they are not designed to provide functional information regarding structural synaptic plasticity, which is crucial to understanding how the brain controls complex neural processes. To achieve this, we need methods that can enable the direct visualization of synapses in real time.

There is an integral physical interaction that occurs across the synapse, which is mediated by synaptic cell adhesion molecules and is one of the fundamental events during synaptogenesis. Among the class of synaptic cell adhesion molecules, the trans-synaptic interaction between the presynaptic neurexins and the postsynaptic neuroligins is one of the strongest and best characterized [5]. Several recent studies have attempted to take advantage of the

Alexandros Poulopoulos (ed.), *Synapse Development: Methods and Protocols*, Methods in Molecular Biology, vol. 1538, DOI 10.1007/978-1-4939-6688-2_4,

trans-synaptic interaction of neuroligins and neurexins and develop tools for fluorescent tagging of synapses using split-GFP approaches [6–8]. However, the major concern with all these approaches is that they may not actually report a physiological interaction of a trans-synaptic cell-adhesion complex, but instead monitor the close approximation of two membranes at any location, not only in synaptic specializations.

In order to overcome the above-mentioned limitations, we have recently developed SynView [9], a new method for monitoring synapse formation. In this approach, we have generated novel neurexin-1β and neuroligin-1 fusion proteins with complementary "split" GFP moieties positioned in such a fashion that binding of neurexin-1β to neuroligin-1 permits GFP reconstitution without severely changing their binding affinities. To achieve this, we based the design of all fusions on the atomic structure of the neurexin-1β/neuroligin-1 complex that was previously described [10, 11]. According to this model and differently from prior similar attempts, we inserted split-GFP fragments into neurexin-1β and neuroligin-1 specifically at positions that neither impair their normal folding nor impede their interactions. Moreover, based on our design [9], the split-GFP moieties would come into close enough proximity to reconstitute GFP fluorescence only upon neurexin/neuroligin binding.

Using trans-cellular reconstitution of GFP-fluorescence as an assay, we have successfully validated the SynView methodology and demonstrated that neurexin-1β forms a trans-synaptic complex with neuroligin-1 and that this interaction can be used to label synapses in a specific fashion in vivo. Here, I provide a detailed step-by-step protocol for use of the SynView approach to label synapses in cultured mouse hippocampal neurons.

2 Materials

2.1 Equipment

1. Mammalian culture hood.
2. 5% CO_2 incubator.
3. Confocal microscope.

2.2 Lentivirus Production

1. HEK293T cells (ATCC, CRL-3216).
2. Cell Line Medium: 10% Fetal Bovine Serum in Dulbecco's Modified Eagle Medium.
3. pCMV-VSV-G (Plasmid #8454, Addgene).
4. pMDLg/pRRE (Plasmid #12251, Addgene).
5. pRSV-Rev (Plasmid #12253, Addgene).
6. Lipofection reagent: e.g., X-tremeGENE HP DNA Transfection Reagent (Roche).

2.3 Primary Hippocampal Culture

1. Basic Culture Medium: 0.5 % (w/v) glucose, 0.2 mg/ml transferrin in Minimum Essential Medium (MEM), filter and store at 4 °C.
2. Plating Medium: 10 % Fetal Bovine Serum, 2 mM l-Glutamine, 0.025 mg/ml Insulin in Basic Culture Medium. Filter and store at 4 °C. Expires after 1 month.
3. Growth Medium: 5 % Fetal Bovine Serum, 0.5 mM l-Glutamine, 1 % (v/v) B-27 supplement (Gibco) in Basic Culture Medium. Filter and store at 4 °C. Expires after 1 month.
4. 4 ARA-C Medium: 4 μM Arabionside (Ara-C) in Growth Medium.
5. Hank's Balanced Salt Solution (HBSS): 350 mg/L $NaHCO_3$, 1 mM HEPES in standard Hank's Solution without calcium or magnesium (Sigma),pH with NaOH to 7.3–7.4.
6. Matrigel: 1:50 Matrigel® in sterile MEM, store at 4 °C.
7. Papain Solution: 10 U/ml papain, 1 μM $CaCl_2$, 0.5 μM EDTA in HBSS.
8. DNase Solution : 50 mg/ml DNase I in water. Aliquot and store at −20 °C.
9. Calcium Phosphate Precipitation Mammalian Transfection Reagents or kit.
10. Circular glass coverslips No 1.5, 12 mm diameter.

3 Methods

A timeline for a typical SynView experiment is illustrated in Fig. 1.

3.1 Preparation of Lentiviruses Expressing Nrx1β-GFP1-10^S2^

All procedures for preparing and handling lentiviruses should be performed in Biological Safety Level 2 (BSL-2) environments. Make sure that BSL-2 work practices and personal protective equipment are used at all times.

1. The day before transfection, seed HEK 293 T cells in 6-well plates (*see* **Note 1**). Grow cells in Cell Line Medium. One fully confluent T75 flask can be used to make four 6-well plates.
2. The next day start the transfection procedure when HEK293T cells have reached 80 % confluence. Replace the medium with 1.5 ml of MEM for each well.
3. For every 6-well plate prepare the following mix. In a tube containing 1.2 ml of MEM add the following DNA plasmids and mix:

   6 μg pCMV-VSV-G plasmid.

   6 μg pMDLg/pRRE plasmid.

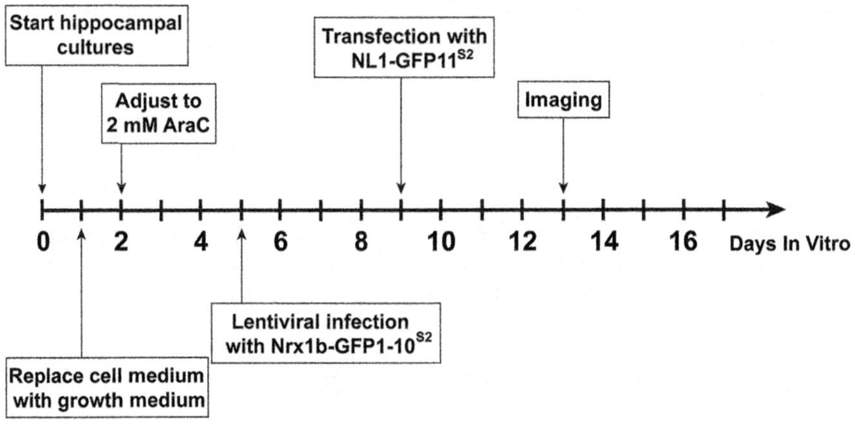

Fig. 1 Experimental timeline for the SynView approach

6 μg pRSV-Rev plasmid.

12 μg FUW-Nrx1β-GFP1-10^{S2}.

4. Add 90 μl of X-tremeGENE HP DNA Transfection Reagent (*see* **Note 2**) and mix thoroughly.
5. Incubate the transfection mix at room temperature for 20 min.
6. Add 200 μl of the transfection mix drop-wise on each well of HEK 293T cells. Return plates in the incubator for 48 h.
7. Collect and pool the supernatant from all wells and centrifuge at 1000 ×*g* for 5 min to precipitate any cells and debris.
8. Use immediately to infect primary hippocampal neurons at Day In Vitro 5 (DIV5) or aliquot and freeze at −80 °C (*see* **Note 3**).

3.2 Primary Culture of Hippocampal Neurons

1. Place autoclaved glass coverslips in each of the middle 12 wells of a 24-well plate. Apply 150 μl of Matrigel on each coverslip. Incubate at 37 °C for at least 30 min.
2. Aliquot 10–14 ml of cold HBSS into a 15 ml conical tube and place on ice (*see* **Note 4**).
3. Prepare the papain solution and keep warm at 37 °C. Prepare 10 ml for every 15 ml conical tube of HBSS.
4. Using sterile techniques, dissect the hippocampi from newborn (P0–P1) mouse pups (*see* **Note 5**). Collect four hippocampi in one 15 ml tube of cold HBSS on ice.
5. Aspirate HBSS from hippocampi and wash once with 14 ml of ice cold HBSS.
6. Aspirate HBSS and filter 10 ml of Papain Solution using a 0.2 μm syringe filter into the tube. Incubate for 20 min at 37 °C water bath.

7. Remove papain and wash hippocampi twice with 10–12 ml Plating Medium.
8. After the second wash, add 1 ml of Plating Medium and triturate using a 1 ml pipette tip, until no tissue clumps are visually present. Add another 12 ml of Plating medium and mix.
9. Count cell density in a hemocytometer and adjust to 75,000 cells/ml.
10. Take plates out of the incubator and aspirate Matrigel from all cover slips in each 24-well plate.
11. Add 1 ml of the neuronal cell suspension in each well. Return plates to the incubator.
12. 12 h after initial plating remove 900 μl of plating medium from each well and replace it with Growth Medium.
13. 48 h after plating (DIV2) remove 500 μl medium and replace with 4 ARA-C Medium (*see* **Note 6**). The cultures are now ready for lentiviral infection at DIV5.
14. When cultures reach DIV5, remove 300 μl of medium and replace them with Lentiviral Medium as prepared in Subheading 3.1.

3.3 Transfection of Primary Neuronal Cultures with NL1-GFP11^{S2} and Imaging

1. At DIV9, remove selected coverslips containing infected cultured neurons, immerse briefly in MEM and transfer in a new 24-well plate containing 500 μl of prewarmed MEM in each well. Return both the original and the new plate to the incubator (5 % CO2).
2. For each coverslip prepare in two separate tubes the following solutions using standard Calcium Phosphate Precipitation Transfection reagents (or commercially available reagents, e.g., Clontech CalPhos Mammalian Transfection Kit):

Solution A	2 μg FUW-NL1-GFP11^{S2} plasmid
	3.1 μl 2 M $CaCl_2$
	Sterile H_2O to adjust total volume to 25 μl
Solution B	25 μl 2× HEPES-Buffered saline

3. Prepare the DNA–Ca_2PO4 precipitate by mixing solution A with solution B. Add about 1/8 volume of solution A at a time into solution B by quickly pipetting several times and gently vortexing at 600 rpm for 2–3 s.
4. Incubate the mixture at room temperature for 15–20 min without any further mixing to allow formation of fine particles of precipitate. Overmixing will cause the formation of large clusters of precipitate and decrease transfection efficiency.
5. Add the DNA–Ca-phosphate suspension solution drop-wise to each coverslip (50 μl per coverslip).

6. Incubate cells with the precipitate for 15–20 min in a 5 % CO_2 incubator at 37 °C (*see* **Note 7**).
7. After incubation, transfer coverslips to new wells containing 1 ml HBSS that was pre-equilibrated in a 5 % CO_2 incubator at 37 °C. Return coverslips to the incubator and incubate from 1 to 5 min until no clusters of precipitate are visible under the microscope (*see* **Note 8**).
8. Transfer the transfected coverslips back to their original wells containing the original neuronal culture medium.
9. The SynView signal appears after 2–3 days as punctate green fluorescence on the transfected neurons (Fig. 2). After this point, neurons can be imaged either live or fixed under a confocal microscope (*see* **Note 9**). The SynView signal is compatible with immunocytochemistry and can be combined with counterstaining using antibodies against synaptic proteins.

4 Notes

1. We have routinely observed that transfection was more efficient when HEK cells were seeded either in 6-well plates or on 10-cm culture dishes. Growth area is comparable in either case and no adjustments are necessary (6 wells = 10-cm dish). Generally, avoid transfection in culture flasks.

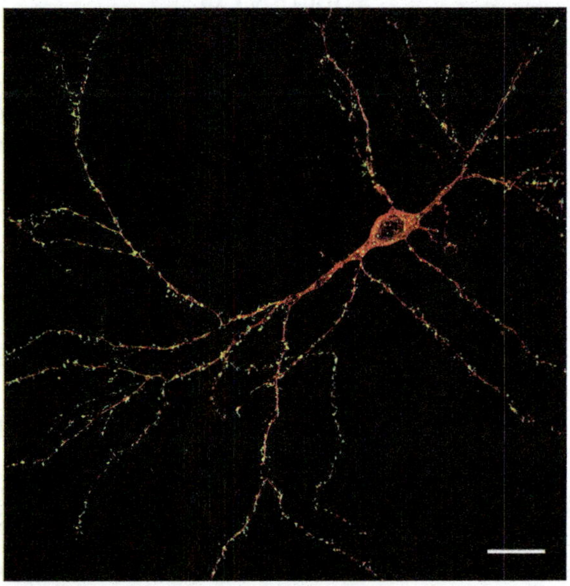

Fig. 2 Confocal image of a hippocampal neuron in culture using the SynView approach. Reconstituted split-GFP molecules (in *green*) identify positions of trans-synaptic neuroligin-1 bound to neurexin-1β that represent synaptic connections. *Scale bar*, 10 μm

2. The current protocol was optimized based on the use of FUGENE6 or X-tremeGENE HP DNA Transfection Reagent as a transfection method for lentiviral production. Use of alternative methods like calcium phosphate will require troubleshooting as they do not end up producing robust final results and are not recommended.
3. When using nonconcentrated lentiviral preparations, best results are obtained when infection is performed using freshly made virus. Aliquoting and storing the lentiviral prep at −80 °C will result in a decrease of the infectious titer. However, if virus production is done in bulk for use in multiple experiments, one freeze-thaw cycle of the virus can be performed. Use of concentrated lentiviral preparations of Nrx1β-GFP1-10^{S2} has not produced robust results in our hands.
4. The volumes in this protocol refer to a hippocampal culture preparation of 12 wells of a 24-well dish. For larger preparations, adjust volumes accordingly.
5. Hippocampi should be collected from newborn mouse pups up to 1 days of age. Preparations from older pups do not usually result in healthy cultures and neuronal viability is compromised. Perform the dissections fast at room temperature and collect and keep the hippocampi on ice until the whole procedure is finished.
6. ARA-C is an inhibitor of glial cell growth and the timing of the addition of 4 ARA-C should be carefully chosen. If added too soon, this will prevent the glial layer that supports and nourishes neurons to be fully formed. However, if added very late, this will allow glial cells to overtake the culture area in expense of neuronal growth. In both cases, neuronal health and viability will be severely compromised. Cultures should be monitored carefully 24 h after addition of Growth Medium to observe the formation of the glial layer on the coverslip surface, which will indicate the exact time point for switching to ARA-C Medium.
7. Incubation times greater than 20 min result in increased neuronal death in the culture. This is due to toxicity caused by the calcium phosphate. Incubation times between 15 and 20 min result in a sufficient number of transfected hippocampal neurons.
8. It is important to ensure that the precipitate is completely dissolved, as it can affect neuronal survival. Repeat the HBSS wash if any clusters remain visible under the microscope.
9. For imaging of fixed cells, perform fixation using electron microscopy grade 4% paraformaldehyde. Avoid using methanol or methanol-containing fixatives as they can cause quenching of the GFP fluorescence and reduce or abolish the SynView signal.

References

1. Helmstaedter M, Briggman KL, Turaga SC, Jain V, Seung HS, Denk W (2013) Connectomic reconstruction of the inner plexiform layer in the mouse retina. Nature 500:168–174
2. Wickersham IR, Lyon DC, Barnard RJ, Mori T, Finke S, Conzelmann KK, Young JA, Callaway EM (2007) Monosynaptic restriction of transsynaptic tracing from single, genetically targeted neurons. Neuron 53:639–647
3. Callaway EM, Luo L (2015) Monosynaptic circuit tracing with glycoprotein-deleted rabies viruses. J Neurosci 35:8979–8985
4. Beier KT, Steinberg EE, DeLoach KE, Xie S, Miyamichi K, Schwarz L, Gao XJ, Kremer EJ, Malenka RC, Luo L (2015) Circuit architecture of VTA dopamine neurons revealed by systematic input–output mapping. Cell 162:622–634
5. Sudhof TC (2008) Neuroligins and neurexins link synaptic function to cognitive disease. Nature 455:903–911
6. Yamagata M, Sanes JR (2012) Transgenic strategy for identifying synaptic connections in mice by fluorescence complementation (GRASP). Front Mol Neurosci 5:18
7. Feinberg EH, Vanhoven MK, Bendesky A, Wang G, Fetter RD, Shen K, Bargmann CI (2008) GFP Reconstitution Across Synaptic Partners (GRASP) defines cell contacts and synapses in living nervous systems. Neuron 57:353–363
8. Kim J, Zhao T, Petralia RS, Yu Y, Peng H, Myers E, Magee JC (2012) mGRASP enables mapping mammalian synaptic connectivity with light microscopy. Nat Methods 9:96–102
9. Tsetsenis T, Boucard AA, Arac D, Brunger AT, Sudhof TC (2014) Direct visualization of trans-synaptic neurexin-neuroligin interactions during synapse formation. J Neurosci 34:15083–15096
10. Araç D, Boucard AA, Ozkan E, Strop P, Newell E, Südhof TC, Brunger AT (2007) Structures of neuroligin-1 and the neuroligin-1/neurexin-1 beta complex reveal specific protein-protein and protein-Ca2+ interactions. Neuron 56:992–1003
11. Ko J, Zhang C, Arac D, Boucard AA, Brunger AT, Südhof TC (2009) Neuroligin-1 performs neurexin-dependent and neurexin-independent functions in synapse validation. EMBO J 28:3244–3255

Chapter 5

Generation of Spinal Motor Neurons from Human Pluripotent Stem Cells

David P. Santos and Evangelos Kiskinis

Abstract

Human embryonic stem cells (ESCs) are characterized by their unique ability to self-renew indefinitely, as well as to differentiate into any cell type of the human body. Induced pluripotent stem cells (iPSCs) share these salient characteristics with ESCs and can easily be generated from any given individual by reprogramming somatic cell types such as fibroblasts or blood cells. The spinal motor neuron (MN) is a specialized neuronal subtype that synapses with muscle to control movement. Here, we present a method to generate functional, postmitotic, spinal motor neurons through the directed differentiation of ESCs and iPSCs by the use of small molecules. These cells can be utilized to study the development and function of human motor neurons in healthy and disease states.

Key words Induced pluripotent stem cells (iPSCs), Neurons, Spinal motor neurons, Differentiation, Stem cells, Dual SMAD inhibition, Neural progenitor cells

1 Introduction

Human embryonic stem cells (ESCs) and induced pluripotent stem cells (iPSCs) are capable—under defined conditions—of propagating indefinitely, as well as of differentiating into all derivatives of the three germ layers: endoderm, mesoderm, and ectoderm [1, 2]. Thus, they can provide an unlimited source for the myriads of different somatic cell types that make up the human body. The technical ease with which iPSCs can be generated from any given individual [3] has created an unprecedented opportunity to study human disease in the context of each person's unique genetic profile [4, 5]. Patient-specific iPSCs can be readily differentiated into the cell types that are relevant for a particular disease (e.g., neurons or cardiomyocytes), in order to gain insights into disease mechanisms and provide a cellular platform for drug screening. This approach provides a human, in vitro cell-based system that can be used to probe disease mechanisms under physiological expression levels of proteins, and in the context of each unique individual.

Alexandros Poulopoulos (ed.), *Synapse Development: Methods and Protocols*, Methods in Molecular Biology, vol. 1538, DOI 10.1007/978-1-4939-6688-2_5,

Motor neurons (MNs) provide a functional connection between the brain and the musculature. This connection mediates our ability to control our movement. The motor neuron circuit includes two broad subclasses of MNs: upper and lower MNs. Upper MNs reside in the cortex and transmit information through their long axons to synapse with lower MNs in the spinal cord. The spinal MNs project their axons outside the CNS to connect with muscle through a specialized synapse known as the neuromuscular junction (NMJ) [6, 7]. MNs are known to become dysfunctional progressively and relatively selectively in a number of neurological diseases including adult-onset amyotrophic lateral sclerosis (ALS), the childhood genetic disease spinal muscular atrophy (SMA), and progressive bulbar palsy [8–10]. The physiological relevance of MNs, as well as their clinical significance for disease, has fuelled research efforts centered on defining the principles of their development and patterning [11, 12]. The knowledge from these invaluable studies has been successfully utilized to recapitulate developmental cues in vitro and effectively differentiate functional MNs from pluripotent stem cells [6].

Here, we describe a technical protocol for the production of spinal MNs from human pluripotent stem cells. This protocol is based on the sequential exposure of cells to small molecules, different versions of which have previously been described in a number of original research studies [13–17]. Figure 1 provides an overview of the major steps in the process. The two major principles of the protocol are an initial production of neural progenitors followed by their appropriate pattering towards the MN lineage. Neuralization is achieved through exposure of undifferentiated stem cells to SB431542 and dorsomorphin, small molecule inhibitors of TGF-β, and BMP signaling, respectively. Neural progenitor cells are appropriately patterned to produce MNs through

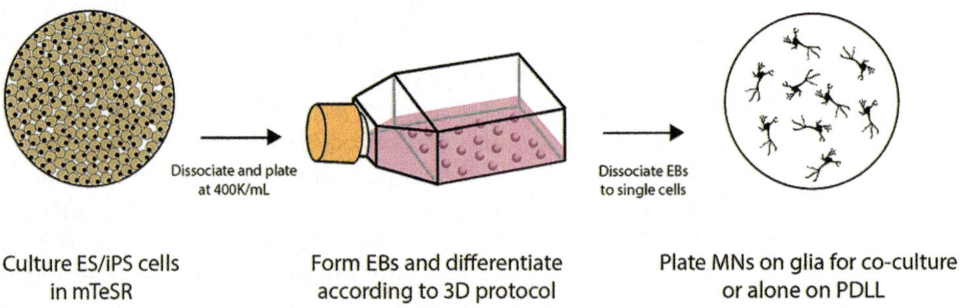

Fig. 1 Overview of MN differentiation protocol. Pluripotent stem cells are expanded until enough cells are ready for the differentiation protocol (400 K cells/mL). Cells are transferred to a low attachment tissue culture flask where they gradually aggregate to form embryoid bodies (EBs). EBs are neuralized and patterned to differentiate into MNs using a combination of small molecules. After 24 days, EBs can be dissociated and MN-enriched cultures are plated as single cells on Poly-d-Lysine/Laminin (PDLL) coated surfaces with or without primary glial cells

exposure to retinoic acid (RA), and smoothened agonist (SAG 1.3), which promote caudalization and ventralization, respectively. This protocol can be applied across any human ESC or iPSC line and typically results in a heterogeneous neuronal population of neural cells (TUJ1 > 98 %) with significant numbers of postmitotic MNs (ISL/HB9 > 15–25 %). We and others have previously described the molecular, physiological, and functional properties of these MNs [13, 15, 17, 18].

2 Materials

All solutions should be prepared in a fume hood under standard sterile technique. Small molecules should be weighed out in an area designated for tissue culture and all microfuge tubes should be autoclaved prior to aliquoting. For all solutions, unless otherwise noted, sterile filter using a 0.22 μm PES filter, store at 4 °C, light protected, and use within 2 weeks. Before feeding cells, all media should be warmed up to 37 °C.

2.1 General Tissue Culture

1. Dissociation Reagent: Accutase® (Innovative Cell Technologies #AT-104), or 1 mM EDTA in TC-grade PBS, sterile filtered (*see* **Note 1**).
2. Matrigel: hESC-qualified Matrigel® (e.g., Corning #354277) in DMEM/F12 at manufacturer-recommended dilution (*see* **Note 2**).
3. Poly-d-Lysine (PDL): 0.1 mg/mL Poly-d-Lysine in H_2O (*see* **Note 3**).
4. Laminin: 20 μg/mL natural Laminin mouse protein in PBS, prepare fresh (*see* **Note 4**).
5. 6-well tissue culture treated plate.
6. 10 cm tissue culture treated plate.
7. 6-well ultra-low attachment multiple well plate (e.g., Corning #3471).
8. Ultra-low attachment culture flask (T25) (e.g., Corning #3815).

2.2 Media Recipes

1. DMEM/F12: 50% Dulbecco's Modified Eagle Medium (DMEM), 50% Nutrient Mixture F-12 (commercially available as a ready mix, e.g., Life Technologies #11320033).
2. Neural Induction Media (NIM): 1% (v/v) Glutamax (e.g., Life Technologies #35050079), 1% (v/v) MEM nonessential amino acids (NEAA) (e.g., Life Technologies 10370088), 8.88 mM D-(+)-glucose (*see* **Note 5**) in DMEM/F12 (e.g., Life Technologies #11320033). After sterile filtering add 1% (v/v) of N2 supplement (e.g., Life Technologies #17502048) and 2 μg/mL heparin sulfate.

3. Knockout Serum Replacement Medium (KOSR): 15 % (v/v) KnockOut™ Serum Replacement (e.g., Life Technologies #10828028) (KSR), 1 % (v/v) Glutamax, 1 % (v/v) NEAA in DMEM/F12 (e.g., Life Technologies #11320033).
4. Neurobasal Medium (NBM): 1 % (v/v) Glutamax, 1 % (v/v) NEAA, 1 % (v/v) N2 supplement, 2 % (v/v) B27 supplement in Neurobasal medium (e.g. Life Technologies #21103049). Add N2 and B27 supplements after sterile filtering.
5. Complete Neurobasal Medium (cNBM): 10 ng/mL BDNF, 10 ng/mL GDNF, 10 ng/mL CNTF, 0.2 μg/mL Ascorbic acid (AA) in NBM. Prepare fresh from stock solutions of BDNF, GDNF, CNTF, and AA (*see* **Note 6**).
6. mTeSR: mTeSR™1 medium (Stem Cell Technologies #05850).
7. mTeSR+RI: 10 μM ROCK inhibitor Y27632 in mTeSR. Prepare fresh from RI stock solution.
8. Glia Medium: 33.3 mM D-(+)-glucose in H_2O, 10 % (v/v) Horse Serum (heat inactivated) in Minimum Essential Media (MEM) (e.g., Life Technologies #10370021).

2.3 Small Molecules

1. ROCK inhibitor (RI) 1000× stock solution: 10 mM Y27632 (e.g., DNSK International #129830-38-2) in water.
2. SB431542 1000× stock solution: 10 mM SB431542 (e.g., DNSK International #301836-41-9) in DMSO.
3. Dorsmorphin 10,000x stock solution: 10 mM Dorsmorphin (e.g., EMD Millipore #171261) in DMSO.
4. BDNF 10,000× stock solution: 100 μg/mL Brain-derived neurotrophic factor (BDNF, e.g., R&D Systems #248-BD), 0.1 % BSA in PBS.
5. GDNF 10,000× stock solution: 100 μg/mL Glial cell-line derived neurotrophic factor (GDNF, e.g., R&D Systems #212-GD), 0.1 % BSA in PBS.
6. CNTF 10,000× stock solution: 100 μg/mL Ciliary neurotrophic factor (CNTF, e.g., R&D Systems #257NT/CF), 0.1 % BSA in PBS.
7. AA 1000× stock solution: 0.2 mg/mL Ascorbic acid (AA, e.g., Sigma Aldrich #A4403) in H_2O.
8. RA 10,000× stock solution: 10 mM Retinoic acid (RA, e.g. Sigma Aldrich #R2625) in DMSO (*see* **Note 7**).
9. Smoothened agonist 10,000× stock solution: 10 mM SAG 1.3 (e.g., DNSK International #364590-63-6) in DMSO.

3 Methods

All procedures should be carried out under standard sterile technique. Ensure all media are warmed up to 37 °C. Any reagent/equipment that is brought into the hood should be thoroughly sprayed with 70% ethanol.

3.1 Culture of Stem Cells

The initial step of starting a stem cell culture will depend on the source of the stem cells. Here, we will assume a vial of frozen cells was obtained from a commercial source.

1. Coat a 6-well tissue culture plate with Matrigel and allow the Matrigel to settle at room temperature for one hour or 37 °C for 30 min (*see* **Note 8**).
2. Once Matrigel has settled, aspirate the residual Matrigel, wash with PBS, and add 2 mL of mTeSR+RI to each well of the 6-well plate.
3. Remove vial of frozen cells from dry ice or liquid nitrogen and thaw out vial in a 37 °C water bath until almost completely thawed (*see* **Note 9**).
4. Transfer contents of the vial with a P1000 pipette into a 15 mL conical tube prefilled with 8 mL DMEM/F12 (*see* **Note 10**).
5. Spin down the tube of cells at 200×*g* for 5 min.
6. Aspirate medium, being careful not to disturb the cell pellet (*see* **Note 11**).
7. Resuspend cell pellet in about 1 mL DMEM/F12.
8. Divide the 1 mL of resuspended cells into two to three wells of a 6-well plate (*see* **Note 12**).
9. The following day look at the cells to ensure there is good attachment and survival. There is no need to feed the cells the day after plating (*see* **Note 13**).
10. Two days after plating, aspirate the medium and replace with mTeSR.
11. Replace with fresh mTeSR on a daily basis.
12. Once the cells have reached 80–90% confluency in a well of a 6-well plate they will need to be split into a new plate. Aspirate medium, wash once with PBS, and add Dissociation Reagent (*see* **Note 14**).
13. Wash the well with an additional 8 mL DMEM/F12 and collect in a 15 mL conical tube.
14. Spin down the tube at 200×*g* for 5 min.
15. Aspirate medium, being careful not to disturb the cell pellet.

16. Resuspend cell pellet in 1 mL of DMEM/F12 and add an appropriate amount of cells to a 10 cm Matrigel-coated plate with 10 mL pre-warmed mTeSR+RI (*see* **Note 15**).
17. As before, 2 days later, aspirate mTeSR+RI and replace with mTeSR.
18. Continue feeding every day with mTeSR and once confluent passage cells as necessary.

3.2 Formation of EBs and Differentiation Steps

Before starting the differentiation of stem cells to MNs, ensure that the stem cell colonies exhibit a uniformly undifferentiated morphology. Any differentiation in the stem cells will hinder differentiation to MNs, decreasing the efficiency and resulting in a more heterogeneous population of cells. A detailed overview of the media and small molecules at each stage of differentiation can be seen in Fig. 2, while Fig. 3 shows cells at all stages of differentiation (hiPSC, EB, MN).

1. *Day 0*: Aspirate mTeSR from the plate of stem cells and wash once with warm PBS. Add 5 mL of warm Accutase to a 10 cm plate and allow the plate to incubate at 37 °C for 5 min.
2. After 5 min triturate briefly using a P1000 to obtain a single cell suspension and transfer the dissociated cells to a 50 mL conical tube.
3. Wash the plate two times with ~15 mL DMEM/F12. Again, transfer these contents to the same 50 mL conical tube.
4. Spin the cells down at 200 ×*g* for 5 min.
5. Aspirate medium, being careful not to disturb the cell pellet and resuspend the cells in ~5–10 mL warm mTeSR+RI (*see* **Note 16**).
6. Take a sample of this cell suspension and count the total amount of cells in the suspension (*see* **Note 17**).

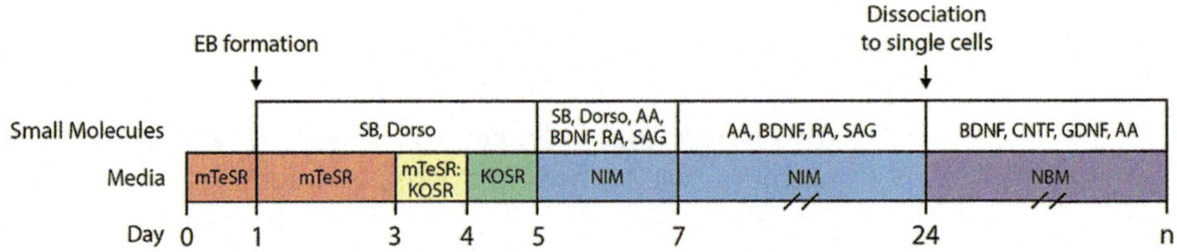

Fig. 2 Schematic with timeline of MN differentiation protocol. Timeline in days is indicated at the *bottom* (0–24), each medium formulation is indicated in the *first row* and each combination of small molecules is indicated in the *top row*. Stem cell cultures are plated in mTeSR for 24 h in low attachment flasks or plates (400 K cells/mL). *SB* SB431542, *Dorso* Dorsomorphin, *AA* ascorbic acid, *BDNF* brain-derived neurotrophic factor, *RA*, retinoic acid, *SAG* smoothened agonist 1.3, *CNTF* ciliary neurotrophic factor, *GDNF* glial cell-derived neurotrophic factor, *KOSR* KOSR medium, *NIM* neural induction medium, *MN* motor neuron medium

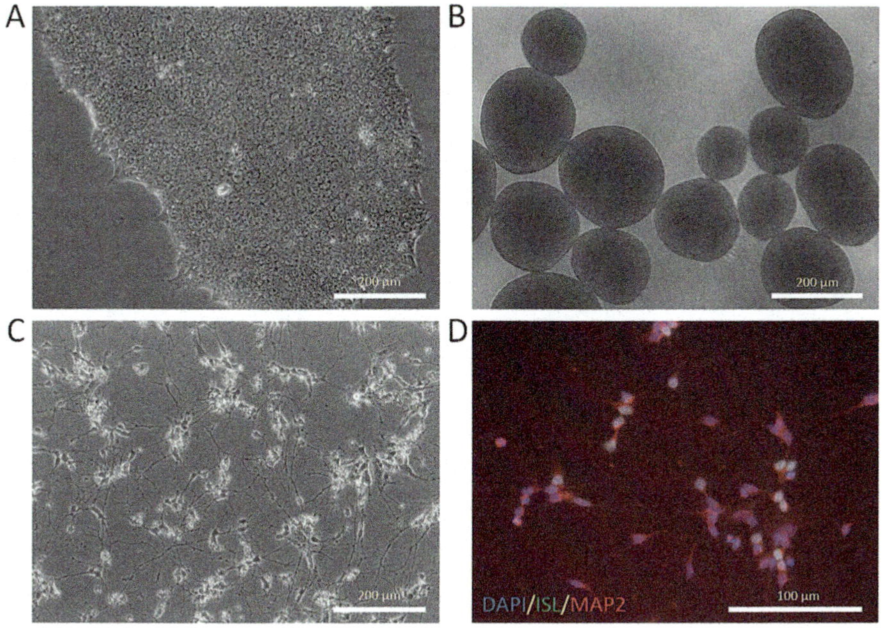

Fig. 3 Example images of cells during the differentiation protocol. (**a**) Human iPSC stem cell colony exhibiting uniformly undifferentiated morphology. Cells resembling this morphology are an ideal starting point for initiating MN differentiation. (**b**) Human iPSC-derived EB MNs on day 24. EBs should exhibit uniform size throughout the differentiation process with defined edges. (**c**) d26 iPSC-derived MNs plated on PDLL. MNs should start to form networks of connections after replating that increase in complexity over time. (**d**) Immunostaining of d26 iPSC-derived MNs. Note that about 15–20 % of neurons ($MAP2^{+}$) are MNs (ISL^{+})

7. Further dilute the cell suspension in mTeSR+RI to a concentration of 400 K cells/mL and transfer 10 mL (4 million cells) to a T25 ultra-low attachment culture flask or 2 mL (800 K cells) to a well of a 6-well ultra-low attachment multiple well plate.
8. *Day 1*: The next day, add 50 % more mTeSR with the following small molecule additives from the stock solutions so that concentrations correspond to the entire medium volume: 10 μM SB431542 and 1 μM Dorsmorphin (*see* **Note 18**).
9. *Day 3*: Two days later, aspirate all the medium, being careful not to disturb the EBs (*see* **Note 19**) and feed with mTeSR:KOSR medium (1:1 ratio) containing 10 μM SB431542 and 1 μM Dorsmorphin.
10. *Day 4*: The next day aspirate medium and replace with 100 % KOSR medium supplemented with 10 μM SB431542 and 1 μM Dorsmorphin
11. *Day 5*: The next day replace medium with NIM supplemented with 10 μM SB431542, 1 μM Dorsmorphin, 10 ng/mL BDNF, 0.2 μg/mL AA, 1 μM RA, and 1 μM SAG 1.3 (*see* **Note 20**).

12. *Day 7*: Two days later replace medium with NIM containing 10 ng/mL BDNF, 0.2 μg/mL AA, 1 μM RA, and 1 μM SAG 1.3.
13. *Day 9–23:* Two days later, and every other day until day 24 refresh medium with the day 7 composition (NIM + BDNF + AA + RA + SAG 1.3) (*see* **Note 21**).
14. *Day 24*: Dissociation and plating of single cells takes place (*see* **Note 22**).

3.3 Preparation of Plates for Plating Stem Cell Derived MNs

Stem cell-derived MNs can be plated on a variety of surfaces. The most commonly used are: poly-d-lysine and laminin (PDLL), poly-l-ornithine and laminin (PLOL), matrigel, or a monolayer of primary glial cells. The choice of surface will depend on the downstream application (*see* **Note 23**). Here, we will describe the protocols for plating MNs on PDLL and a glial monolayer. In addition to preparing the plates/slides as described below, several commercial options are available for purchasing precoated PDLL dishes, chamberslides, etc.

3.3.1 Coating Plates with PDLL

1. 48 h prior to the dissociation of MNs, plate enough poly-d-lysine to cover the surface of your plate/coverslide/etc.
2. Allow the PDL to incubate on the surface at 4 °C for 48 h.
3. Several hours before plating the MNs, aspirtate the PDL and wash three times with PBS. After the final wash add just enough Laminin to cover the surface of the plate and put the plate at 37 °C for about 2 h.
4. When ready to plate MNs, aspirate the laminin and, without washing, add cNBM. The desired amount of neurons can then be added to your plating surface (*see* **Note 24**).

3.3.2 Coating Plates with Glia

1. Mouse glia can be prepared as described previously [19]. Begin procedure of plating glia 2 days prior to plating MNs.
2. Add Glia Medium to the wells of a precoated or freshly prepared PDLL chamberslide/coverslide/etc.
3. Dissociate glia from flask and plate about 90 K glia/cm^2 surface area (*see* **Note 25**).
5. When ready to plate MNs, aspirate the Glia Medium and, without washing, add cNBM.
6. MNs can be plated at a density suitable for the particular downstream experiments to be carried out (*see* **Note 24**).

3.4 Dissociation of EBs and Plating of Single Cells

1. Collect EBs with a 10 mL serological pipette and transfer to a 15 mL conical tube (*see* **Note 26**).
2. Allow the EBs to settle to the bottom of the tube. This will take about 5 min.

3. Aspirate any excess medium and wash two times with warm PBS. Each time allows the EBs to settle before aspirating away the PBS.
4. After washing with PBS, resuspend the EBs in Papain/DNase (*see* **Note 27**).
5. Transfer the Papain/DNase/EB mixture using a P1000 into a well of a 6-well plate.
6. At this point, gently pipette up and down about 10–15 times with a P1000 to dissociate the EBs.
7. Incubate at 37 °C for 30 min.
8. After 30 min, pipette up and down 10–15 times using a P1000 to completely triturate the cells of the EBs (*see* **Note 28**).
9. Add 10 mL more DMEM/F12 to dilute out Papain/DNase and spin at 200 × *g* for 5 min.
10. Aspirate medium and resuspend cell pellet in 1 mL NBM. Pipette with P1000 to break down the MN pellet. Add an additional 1 mL of NBM. Triturate until a homogenous suspension is obtained with no visible cellular aggregates.
11. Filter cell solution through a 40 μm nylon cell strainer to get rid of any large chunks of cells. Wash residual cells in 15 mL conical tube with an additional 1 mL of NBM and pass this through the same filter (*see* **Note 29**).
12. Further dilute the cell suspension in 5–6 mL of NBM.
13. Determine the number of cells and concentration using a hemocytometer or automated cell counter.
14. Plate appropriate amount of cells for desired application (*see* **Note 24**) in cNMB supplemented with RI and 2% FBS (*see* **Note 30**).
15. The day after plating MNs replace half the media with fresh cNBM.
16. Maintain cultures in cNBM and replace medium with 75% fresh cNBM every 2 days.

4 Notes

1. Diluting a 500 mM EDTA stock in H_2O to 1 mM in PBS makes the solution used as a dissociation reagent. EDTA is very difficult to get into solution. In order to effectively make a 500 mM stock solution of EDTA, weigh out 93.05 g EDTA powder. Dissolve in 200 mL of autoclaved sterile H_2O using a sterile magnetic stirrer to mix. Slowly add concentrated NaOH until the EDTA solution reaches a pH of 8.0 (EDTA will not go fully into solution until pH 8.0 is reached—this will take

some time). Adjust the volume to 500 mL with autoclaved sterile H_2O. The EDTA solution can then be sterile-filtered before use.

2. Each Matrigel stock varies in its concentration and as such, the dilution of each lot needs to be calculated. This calculation is done as follows: Take 5000 (5 mL—amount of undiluted Matrigel in this example) and divide by the dilution factor (e.g., 250) to give you the number of aliquots the lot of Matrigel will yield (e.g., 20). Matrigel aliquots are 600 μL. Using the number of aliquots and aliquot volume, calculate the amount of ice-cold DMEM/F12 (e.g., (0.6 mL×20)–5 mL=7 mL) needed to adjust the concentration. Freeze aliquots at −20 °C. Prior to diluting, the Matrigel stock needs to be thawed overnight at 4 °C in an ice bucket. When preparing to coat plates with Matrigel, thaw aliquots on ice and dilute in 24 mL ice-cold DMEM/F12 to make working stock.
3. Poly-d-Lysine may be supplied as a lyophilized powder. In a tissue-culture hood reconstitute the powder in autoclaved sterile H_2O to a concentration of 0.1 mg/mL. This solution can be aliquoted and stored at −20 °C until ready to use. Prior to use, the poly-d-lysine stock solution should be thawed at 4 °C overnight.
4. Laminin may be supplied at a stock concentration of ~1 mg/mL. Immediately prior to use, laminin stock solution should be diluted 50× in PBS to a working concentration of 20 μg/mL. Plates can be coated with just enough laminin to cover the bottom of the well.
5. D-(+)-glucose may be supplied as a 45 % solution. Dilute glucose in autoclaved, sterile filtered H_2O to 20 % before use. Add 4 mL of 20 % w/v glucose solution to 496 mL medium for a final concentration of 8.88 mM.
6. On plating day MN medium is supplemented with 10 μM ROCK inhibitor from stock solution and 2 % FBS to help with survival post-dissociation and attachment.
7. Retinoic acid is extremely unstable. Once thawed, aliquots should be used once and discarded. Frozen stock aliquots of retinoic acid should be kept for no longer than 3 months.
8. The amount of Matrigel used to coat a surface depends on the size of the plate. Generally, just enough is needed to cover the surface of the dish (i.e., a 10 cm dish should be coated with ~8 mL, each well of a 6-well dish should be coated with ~1.5 mL). Matrigel-coated plates can be stored at 4 °C for up to one week before plating cells. If the plates are any older than this, they should be discarded.

9. Swirl the vial of cells around in a 37 °C water bath until there exists only a small sphere of ice. At this point, transfer the vial to the hood and decant into medium-containing dish to minimize the time cells are exposed to the high DMSO concentrations of the cryoprotectant suspension.
10. Be extremely careful when adding the cells to the DMEM/F12, this needs to be done slowly, adding the cells drop-wise. While care should be taken not too add the cells to the DMEM/F12 too quickly, this step needs to be done fast to minimize the time cells are exposed to the high DMSO concentration of the cryoprotectant.
11. The cell pellet should appear milky white.
12. The number of wells that the frozen aliquot of cells should be split into is dependent on the size of the pellet and the characteristics of the cell line and needs to be determined empirically. Without growing the cells and becoming accustomed to their properties, it is difficult to define how this should be done. When working with cells for the first time, plate more rather than less.
13. In most instances cells the culture media of the cells does not need to be replaced the day after plating. However, if the cells are plated very dense, such that the day after plating they are over 75 % confluent, the cells should be fed with fresh mTeSR.
14. There are several common dissociation reagents that can be used for the passaging of stem cells. We typically use EDTA or Accutase. EDTA will produce chunks of cells that cannot be counted, while Accutase will produce single cells that can be counted. When using these dissociation reagents, add just enough to cover the surface of the well (about 5 mL per 10 cm plate or 1 mL per well of a 6-well plate). Depending on the reagent used, protocols are slightly different. If EDTA is the dissociation reagent used, return the plate to the incubator and let it incubate for ~5 min. After 5 min, check that the cells have detached from the surface of the well by gently triturating on the edge of the well (the cells should also have a rounded appearance when viewed under a microscope). If the cells become detached from the surface, gently aspirate the EDTA and wash the cells off the surface of the well with 5 mL DMEM/F12 and collect in a 15 mL conical tube. Alternatively, if Accutase is being used, add Accutase to the well and allow it to incubate at 37 °C for 5 min. Instead of aspirating the dissociation reagent, triturate to obtain a single-well suspension and transfer suspension to a 15 mL conical. An additional 4 mL of DMEM/F12 can be used to wash the plate and dilute out the Accutase.
15. When plating cells it is important to plate at a density that will be dense enough for survival to be high, but not so dense that

the cells will need to be split immediately. Generally, we use a split ratio of 1:10–1:30. Be aware that when going from a well of one size to a well of a different size, you will need to take into account the differences in plating area when calculating the amount of cells to plate.

16. Due to the necessity of cells being plated at a density of 400 K cells/mL for the initiation of the differentiation, it is important that the cells be resuspended to a density that is higher than 400 K cells/mL; otherwise, the cells will need to be spun down again and resuspended in a smaller volume. A confluent 10 cm plate will generally yield about ten million cells or more.
17. Cells can be counted using either a hemocytometer or an automated cell counter.
18. The concentration of SB431542 (10 μM) and Dorsmorphin (1 μM) must be for the whole volume of mTeSR present. In a T25 flask, cells are plated in 10 mL. The next day the 5 mL that will be added should contain enough SB431542 and Dorsomorphin for 15 mL of medium (i.e., 15 μL of SB431542 and 1.5 μL of Dorsmorphin stock solutions).
19. When aspirating media from flasks containing EBs, tilt the flask and rest it sideways at a 30°–45° angle. Once the cells have settled to the bottom of the flask (initially, the cells will take at least 10 min to settle but will settle much more quickly as the differentiation progresses and the EBs become larger in size), slowly aspirate the medium. A small amount of medium will always remain, but attempt to aspirate as much as possible without disrupting any EBs.
20. RA is extremely unstable; aliquots should be used fresh and discarded after one use.
21. During the course of the differentiation, do not let the EBs become too large. If they are clumping use a 10 mL serological pipette to gently break up the EBs to a more uniform small size before allowing them to settle and changing the medium. One potential way to keep the EBs small and uniform in size throughout the course of the differentiation is to keep them gently shaking in the incubator while differentiating.
22. Dissociation takes place on day 24 of this differentiation protocol because it is at this point that the peak generation of post-mitotic $Ki67^-/ISL^+/HB9^+$ MNs has occurred.
23. If MNs are going to be plated and cultured for a long period of time, it is best to plate MNs on a glial monolayer so that clumping of MNs does not occur. If immunostaining is going to be done shortly after plating cells, cells can be plated on a surface of PDLL. This is beneficial because there is no need to worry about cross reactivity of glia with the antibodies used.

However, this is not possible for long-term cultures, as MNs plated on PDLL will tend to aggregate after long periods in culture.

24. The amount of MNs to be plated will depend on the downstream applications. Typically, a density of 20–200 K cells/cm^2 will be sufficient. If cells are going to be used for electrophysiological recordings, a lower density should be used such that isolated, single cells can be interrogated. If, however, cells will be collected for purification of RNA or protein, a higher density should be used such that a sufficient yield of protein/RNA will be acquired.
25. Glial cells should be plated at a density such that on the day of plating MNs (2 days following plating of glia), they have formed a uniform monolayer on the surface of the well.
26. Ensure that the particular brand of serological pipettes being used has a wide enough tip that the EBs are not damaged during the transfer process.
27. Papain may be supplied as a lyophilized powder in a glass vial (e.g., Worthington Biochemical #LK003176). Prior to use, resuspend papain in 5 mL pre-warmed DMEM/F12. DNase may be supplied as a lyophilized powder in a glass vial (e.g., Worthington Biochemical #LK003170). Prior to use, resuspend DNase in 1 mL pre-warmed DMEM/F12. Papain and DNase should be mixed at a ratio of 5 mL Papain:1 mL DNase. Use 1–2 mL of Papain/DNase mix per 10 mL worth of EBs.
28. This dissociation step requires a bit more vigorous pipetting. The EBs need to be broken down to single cells as best as possible. Once the mixture is free of chunks and looks homogenous it is ready to move onto the next step.
29. To get as many cells out of the cell strainer as possible, after allowing the cells to pass through by gravity, lift the strainer up and pipette from the bottom of the strainer. Add these cells to the cell solution that has passed through from gravity filtration.
30. Only on the day of dissociation and plating (d24) is the cNBM supplemented with 10 μM ROCK inhibitor and 2 % FBS.

References

1. Takahashi K et al (2007) Induction of pluripotent stem cells from adult human fibroblasts by defined factors. Cell 131:861–872
2. Thomson JA et al (1998) Embryonic stem cell lines derived from human blastocysts. Science 282:1145–1147
3. Malik N, Rao MS (2013) A review of the methods for human iPSC derivation. Methods MolBiol997:23–33.doi:10.1007/978-1-62703-348-0_3
4. Ichida JK, Kiskinis E (2015) Probing disorders of the nervous system using reprogramming approaches. EMBO J 34:1456–1477. doi:10.15252/embj.201591267
5. Kiskinis E, Eggan K (2010) Progress toward the clinical application of patient-specific pluripotent stem cells. J Clin Invest 120:51–59. doi:10.1172/jci40553
6. Davis-Dusenbery BN, Williams LA, Klim JR, Eggan K (2014) How to make spinal motor neurons.

Development 141:491–501. doi:10.1242/dev.097410

7. Kanning KC, Kaplan A, Henderson CE (2010) Motor neuron diversity in development and disease. Annu Rev Neurosci 33:409–440. doi:10.1146/annurev.neuro.051508.135722
8. Burghes AH, Beattie CE (2009) Spinal muscular atrophy: why do low levels of survival motor neuron protein make motor neurons sick? Nat Rev Neurosci 10:597–609. doi:10.1038/nrn2670
9. Cleveland DW, Rothstein JD (2001) From charcot to lou gehrig: deciphering selective motor neuron death in ALS. Nat Rev Neurosci 2:806–819. doi:10.1038/35097565
10. Cole N, Siddique T (1999) Genetic disorders of motor neurons. Semin Neurol 19:407–418
11. Jessell TM, Surmeli G, Kelly JS (2011) Motor neurons and the sense of place. Neuron 72:419–424. doi:10.1016/j.neuron.2011.10.021
12. Tanabe Y, Jessell TM (1996) Diversity and pattern in the developing spinal cord. Science 274:1115–1123
13. Amoroso MW et al (2013) Accelerated high-yield generation of limb-innervating motor neurons from human stem cells. J Neurosci 33:574–586. doi:10.1523/jneurosci.0906-12.2013
14. Wichterle H, Lieberam I, Porter JA, Jessell TM (2002) Directed differentiation of embryonic stem cells into motor neurons. Cell 110:385–397, doi: S0092867402008358 [pii]
15. Boulting GL et al (2011) A functionally characterized test set of human induced pluripotent stem cells. Nat Biotechnol 29:279–286. doi:10.1038/nbt.1783
16. Chambers SM et al (2009) Highly efficient neural conversion of human ES and iPS cells by dual inhibition of SMAD signaling. Nat Biotechnol 27:275–280. doi:10.1038/nbt.1529
17. Kiskinis E et al (2014) Pathways disrupted in human ALS motor neurons identified through genetic correction of mutant SOD1. Cell Stem Cell 14:781–795. doi:10.1016/j.stem.2014.03.004
18. Wainger BJ et al (2014) Intrinsic membrane hyperexcitability of amyotrophic lateral sclerosis patient-derived motor neurons. Cell Rep 7:1–11. doi:10.1016/j.celrep.2014.03.019
19. Di Giorgio FP, Carrasco MA, Siao MC, Maniatis T, Eggan K (2007) Non-cell autonomous effect of glia on motor neurons in an embryonic stem cell-based ALS model. Nat Neurosci 10:608–614

Part II

Analyzing Synaptic Constituents

Chapter 6

Biochemical Purification of Binding Partners of Synaptic Scaffold Proteins

Bekir Altas*, Olaf Jahn, and Hiroshi Kawabe*

Abstract

The chemical synapse displays specialized intercellular adhesion between pre- and potsynaptic plasma membranes mediated by synaptic cell adhesion proteins. In this asymmetric cell adhesion, pre- and postsynapses have their own unique functions; the presynaptic terminal releases neurotransmitter, which diffuses through the synaptic cleft and is received by receptors accumulated at the postsynapse. Such distinct modes of actions of pre- and postsynapses in synaptic neurotransmission are the rate-limiting factors in signal processing in the brain, and thus protein-protein interactions within the pre- and postsynaptic scaffold are of particular importance for brain function by regulating the pre- and postsynaptic function. In the present paper, we outline a method to screen for binding partners of synaptic scaffold proteins biochemically.

Key words Affinity purification, Synaptic scaffold proteins, Binding partners, Glutathione-S-Transferase (GST) -fused proteins, Mouse brain extract, Mass spectrometry

1 Introduction

Numerous studies on the ultrastructure of chemical synapses have depicted electron-dense architecture apposed to pre- and postsynaptic plasma membranes, referred to as the cytomatrix of the active zone (CAZ) and the postsynaptic density (PSD), respectively, first reported in 1954 [1–4]. The electron-dense structure of the CAZ is an approximately 50–75 nm thick structure [5] where more than ten types of CAZ-specific proteins are highly enriched, binding to each other to form a complex protein meshwork [6]. Precise protein-protein interactions at the CAZ are critical for the temporal and spatial regulation of neurotransmitter release. The PSD, on the other side of the synaptic cleft, is a 30–60 nm thick electron-dense structure [7] that functions as a scaffold to cluster postsynaptic receptors and cell adhesion molecules [8]. Prominent components of the PSD, such as membrane-associated guanylate

*Correspondence may be addressed to either author.

Alexandros Poulopoulos (ed.), *Synapse Development: Methods and Protocols*, Methods in Molecular Biology, vol. 1538, DOI 10.1007/978-1-4939-6688-2_6, © Springer Science+Business Media LLC 2017

kinases (MAGUKs) and Shank family members, are multidomain proteins that interact with each other through individual domains. The morphology of the PSD changes upon repetitive activation of postsynaptic receptors [9] and such changes are caused partially by the dynamic rearrangement of protein-protein interactions [10].

The current protocol describes how to purify binding partners of a CAZ or PSD protein biochemically. First, GST-fused CAZ or PSD proteins are purified and immobilized on glutathione (GSH) sepharose beads, where the detergent extract of mouse brains is loaded. Proteins displaying high affinity for the recombinant CAZ or PSD protein are allowed to bind during incubation. After extensive wash of the beads to remove nonspecifically bound proteins, binding partners of GST-fused proteins are eluted with high salt solution and subsequently with unbound GSH. The eluate is subject to mass spectrometric analysis for the identification of proteins that bound to the recombinant CAZ or PSD protein-of-interest. Alternative approaches to identify candidate CAZ or PSD binding proteins include yeast two-hybrid screening, protein microarrays, and the split ubiquitin method [11–13]. Each technique has its own advantages and disadvantages. However, the biochemical affinity purification approach provides specific benefits over others. Various starting materials could be used as sources of binding partners, the stoichiometry of protein-protein interaction could be quantified, and number of buffer conditions could be applied (e.g., with high or low concentration of salts and with different detergents). Given that binding partners will localize to the synapse, biochemical fractions that specifically enrich for synaptic proteins, such as the mouse brain synaptosome fraction, could be used as starting material for purification in this protocol. Such modification of the current protocol would further enhance signal-to-noise ratios of the readout, and increase the chances of identifying physiologically relevant binding partners.

We provide an example of the protocol to purify binding partners of CAZ and PSD proteins using GST-fused Postsynaptic Density Protein 95 (PSD95, DLG4), Synapse-Associated Protein 97 (SAP97, DLG1), and Regulating Synaptic Membrane Exocytosis 2 (RIM2, RIMS2). We demonstrate that this method is a reliable high-throughput screening to identify binding partners of synaptic proteins by providing numbers of binding partners, some of which have been confirmed and characterized as CAZ and PSD components in the literature.

2 Materials

Ultrapure water (with resistivity of 18.2 MΩ cm at 25 °C) and analytical grade reagents were used when preparing all solutions and buffers. All reagents were stored at 4 °C unless otherwise indicated. Volumetric flasks were used to adjust the final volumes of buffers, unless otherwise indicated.

2.1 Purification of GST-Fused Proteins

1. Electro-competent cells (e.g., BL21 Rosetta, Merck Millipore, 70953).
2. Bacterial expression vector for GST-fused protein of interest: pGEX4T-1-PSD95N (NM_019621; aa 2–76), pGEX4T-1-SAP97 (NM_012788; aa 2–407), pGEX6P-1-RIM2 (NM_001256383; aa 1–155).
3. Electroporator (e.g., *E. coli* Pulser, BioRad, #165-2103) and 0.1 cm Cuvettes (e.g. *E. coli* Pulser Cuvettes BioRad, #165-2089).
4. Lysogeny Broth (LB): 10 g/L Tryptone, 10 g/L NaCl, 5 g/L Yeast Extract in water, autoclave at 121 °C for 25 min.
5. Lysogeny Broth-Ampicillin (LB-Amp): 50 μg/mL Ampicillin in LB, added after cooling down to room temperature.
6. Isopropyl β-d-1-thiogalactopyranoside (IPTG): 0.5 M IPTG in water. Store as 1 mL aliquots at −20 °C.
7. 2× Resuspension Buffer: 100 mM Tris–HCl, 1 M NaCl, 10 mM $MgCl_2$, pH with HCl to 8.0 at 4 °C.
8. Resuspension Buffer: 50 mM Tris–HCl, 500 mM NaCl, 5 mM $MgCl_2$ 10 μg/mL Aprotinin, 1 μg/mL Leupeptin, and 0.1 mM PMSF (or equivalent protease inhibitors) in water, pH with HCl to 8.0 at 4 °C.
9. CHAPS Resuspension Buffer: 1.6% 3-[(3-Cholamidopropyl) dimethylammonio]-1-propanesulfonate hydrate (CHAPS) in Resuspension Buffer.
10. Lysozyme Solution: 100 mg/mL Lysozyme in Resuspension Buffer.
11. DNase I Solution: 5 mg/mL DNase I in Resuspension Buffer.
12. Sonicator and Probe (e.g., Sonopuls HD 2200, and MS73, Bandelin).
13. High-volume centrifuge.

2.2 Immobilization and Yield Estimation of GST-Fused Proteins

1. GSH Sepharose 4B Beads (e.g., GE Healthcare, #17-0756-05).
2. Custom-made columns: Cut off the tip of a 200 μL pipet tip at 0.3 cm distance. Place cotton at the end of the pipet tip (Fig. 1a.1-2-3).
3. Colum holder: Prepare 5 mm-diameter circles with aluminum foil to hold the column onto microcentrifuge tubes (Fig. 1b).
4. Washing Buffer: 50 mM Tris–HCl, 300 mM NaCl, 0.8% CHAPS, pH with HCl to 8.0 at 4 °C.
5. 2× CHAPS Binding Buffer: 2% CHAPS, 40 mM Tris–HCl, 2 mM ethylenediaminetetraacetic acid (EDTA), 300 mM NaCl, 4 mM Dithiothreitol (DTT), pH with HCl to 7.5 at 4 °C.

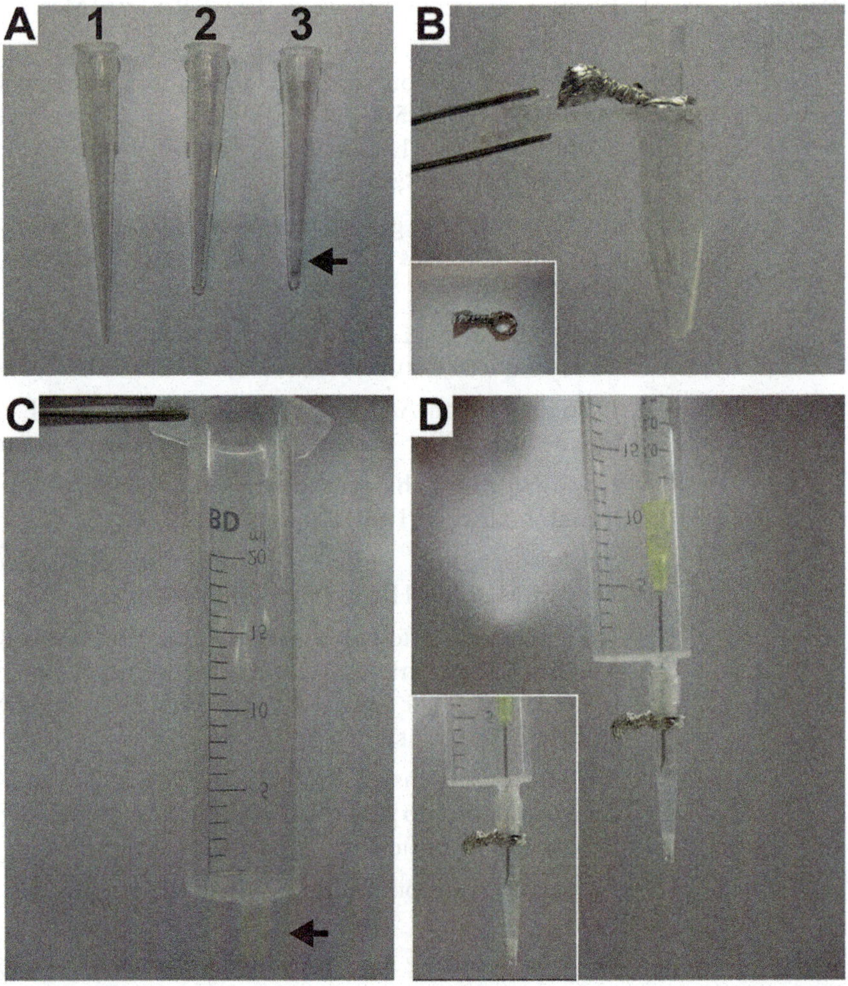

Fig. 1 (**a**) Custom-made columns. 200 μL pipet tip (1) was cut close to the edge (2). Then, a small amount of the cotton was set and placed at the end of the cut tip [*arrow* in (3)]. (**b**) Column holder. Small circular loops were made with aluminum foil (*inset*) to hold the custom-made column onto the microcentrifuge tube. (**c**) A thin layer of parafilm was placed around the tip of a 20 mL syringe to achieve a tight connection between that syringe and the custom-made column [*arrow* in (**c**)]. (**d**) Custom-made column was connected to 20 mL syringe and the mouse brain extract was loaded using a 2 mL syringe and 20G needle

6. CHAPS Binding Buffer: 1% CHAPS, 20 mM Tris–HCl, 150 mM NaCl, 1 mM EDTA, 2 mM DTT, in water, pH with HCl to 7.5 at 4 °C. Prepare this buffer by diluting 2× CHAPS Binding Buffer with water.
7. GSH Elution Buffer: 40 mM GSH in CHAPS Binding Buffer, pH with NaOH to 7.5–8.0 at 4 °C.
8. Infrared Gel Imaging System (Odyssey 9201, LI-COR Biosciences).
9. High Salt Buffer: 50 mM Tris–HCl, 1 M NaCl, 0.8% CHAPS in water, pH with HCl to 8.0 at 4 °C.

2.3 Preparation of Whole Mouse Brain Extract

1. Teflon Homogenizer with Potter homogenizer rotor.
2. Sucrose Solution: 0.32 M sucrose in water and make up to 250 mL. Solution can be stored at 4 °C for one overnight.
3. Total Protein Quantification reagent: e.g., Bradford or BCA.
4. Filter Units, 0.22 μm pore size.
5. Ultracentrifuge with fixed-angle rotor.

2.4 Loading Brain Extracts onto Columns and Elution of Binding Partners

1. NaCl Elution Buffer: 20 mM Tris–HCl, 1 mM EDTA, 1 M NaCl, 1 % CHAPS in water, pH with HCl to 7.5 at 4 °C.
2. 20 mL plastic syringes covered with parafilm at the tip (Fig. 1c).
3. 2 mL plastic syringes.
4. 20G, $1^{1/2''}$ Regular Bevel Needles.

2.5 Sodium Dodecyl Sulfate Polyacrylamide Gel Electrophoresis (SDS-PAGE), Coomassie Brilliant Blue (CBB) Staining and Identification of Specific Binding Partners

1. 3× Laemmli Buffer: 150 mM Tris–HCl pH 6.8 at RT, 30% (v/v) Glycerol, 6 mM EDTA, 10% SDS, 150 mM DTT, 0.2% (w/v) Bromophenol Blue, mix well. If storing aliquots, make 9 mL aliquots without DTT in 15 mL tubes and store at −20 °C. Add 1 mL of 1.5 M DTT into the 9 mL aliquots before use.
2. MOPS-SDS Running Buffer: 50 mM 3-(N-Morpholino)propanesulfonic Acid (MOPS), 50 mM Tris Base, pH 7.7, 0.1% (w/v) SDS, 1 mM EDTA, confirm that pH is around 7.7 at room temperature.
3. 4–12% Bis-Tris Polyacrylamide Protein Gels.
4. Gel Fixation Solution: 40% (v/v) ethanol, 10%(v/v) acetic acid in water.
5. Colloidal Coomassie Working Solution: 0.1% (w/v) Coomassie Brilliant Blue G250, 2% (w/v) ortho-phosphoric acid, 10% (w/v) ammonium sulfate, 20% (v/v) methanol. Do not filter the solution. Make the stock solution without the methanol and keep at room temperature in the dark.
6. Destaining Solution: 1% (v/v) acetic acid, in water, mix solution well.

3 Methods

3.1 Purification of GST-Fused Proteins

1. Adjust the DNA concentration of an expression vector to 20–40 ng/μL with water. Add 1 μL of the DNA to 50 μL of BL21 Rosetta competent *E. coli* or equivalent strain for electroporation. Transfer the bacteria-DNA mixture into an Electrotransformation cuvette. Set the cuvette into electroporator and apply electric pulse at 1.8 kV (*see* **Note 1**).
2. After electroporation, recover *E. coli* with 1 mL of LB (without antibiotic) and transfer into a microcentrifuge tube. Incubate

the bacteria at 37 °C for 30 min for recovery. Spin down *E. coli* using bench-top centrifuge at 13,800 × *g* for 1 min at room temperature. Remove the LB by decantation. Resuspend the bacterial pellet with the remaining LB in the tube and plate cells on LB agar plates supplemented with Ampicillin. Incubate the plate overnight at 37 °C. Next day, pick a single colony and inoculate an overnight liquid bacterial culture in 50 mL of LB-Amp.

3. Transfer 50 mL of cultures into 500 mL of LB-Amp and grow cultures at 20 °C until OD_{600} reaches 1.0 (*see* **Note 2**).
4. Add 0.55 mL IPTG (so the final concentration in the 550 mL culture becomes 0.5 mM) to induce protein expression. Incubate the culture for 20 h at 20 °C (*see* **Note 3**).
5. Spin down the bacteria at 4000 × *g* for 30 min at 4 °C. Remove the supernatant and wash the surface of the pellet with 15 mL ice cold Resuspension Buffer. Do not disturb the pellet at this step.
6. Resuspend the pellet with 50 mL of Resuspension Buffer supplemented with protease inhibitors by pipetting (*see* **Note 4**).
7. Make 5 mL aliquots of the bacterial suspension in 15 mL plastic tubes and snap freeze them in liquid nitrogen. Store the aliquots at −80 °C.
8. Warm up one of the aliquots in a water bath at room temperature. Mix the aliquot occasionally by vortexing. Do not thaw the cell suspension completely at room temperature. Transfer the aliquot on ice to thaw it completely below 4 °C. After thawing, keep the bacterial suspension on ice.
9. Sonicate the bacterial suspension with 30% of maximum power for 20 s in ice cold water using Sonopuls HD 2200 Sonicator equipped with Probe MS73 or equivalent setup. Incubate the suspension for 20 s on ice to cool down the sample. Repeat this process once.
10. After sonication, put 4.8 mL of the bacterial suspension into a new 15 mL plastic tube. Add 100 μL of Lysozyme Solution, 100 μL of DNase I Solution, and 5 mL CHAPS Resuspension Buffer (*see* **Note 5**). Incubate the suspension at 4 °C for 30 min with constant rotation. Avoid forming bubbles at this step, as they may partially denature the recombinant protein.
11. Spin down any insoluble particles at 10,000 × *g* for 10 min at 4 °C.
12. Harvest the supernatant. This is the starting material used to purify the GST-fused protein.

3.2 Yield Estimation of GST-Fused Proteins Purified from E. coli Extract

In this step, we estimate the volume of *E. coli* extract required to purify 60 μg of GST-fused proteins.

1. Pack 50 μL of the GSH sepharose beads in the column (Fig. 1a, b) and equilibrate the column three times with 250 μL of Resuspension Buffer. Prepare three columns for each GST-fused protein (*see* **Note 6**).
2. Load 500 μL, 1 mL, or 2 mL of the bacterial suspension from part 3.1 into the column.
3. Wash the column three times with 250 μL of CHAPS Binding Buffer. Elute the GST-fused proteins with 250 μL of GSH Elution Buffer. Add 25 μL of 3× Laemmli buffer to 50 μL of the eluate. Incubate the samples at 65 °C for 15 min. Load the eluate samples and BSA standards (0.2, 0.4, 0.8, and 1.0 μg) on SDS-PAGE gels. After separating the proteins by SDS-PAGE and staining with colloidal CBB, scan the gel with an Infrared Imaging System able to detect the Coomassie fluorescence in the 700 nm channel. Estimate the protein amount by comparing signals from protein bands of the samples to the BSA standards (Fig. 2). This information will help to calculate the exact amount of the bacterial suspension needed from part 3.1 to purify 60 μg of GST-fused proteins.

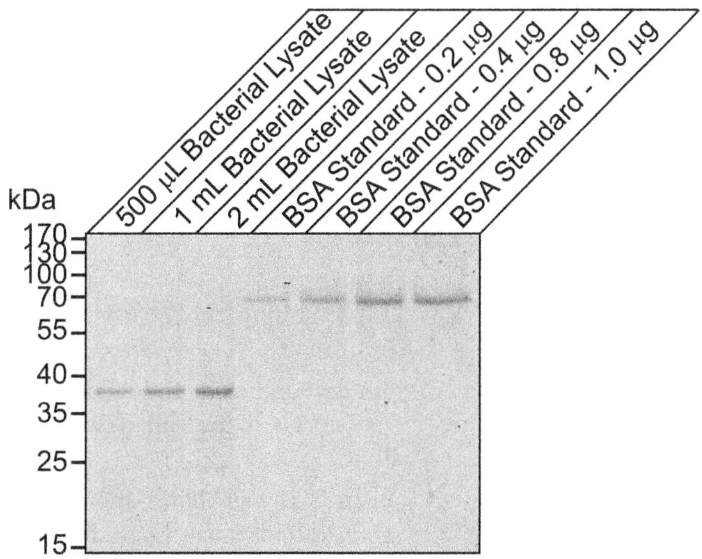

Fig. 2 An example for the estimation of the amount of GST-fused protein (GST-PSD95, residue 2-76) purified from *E. coli* extract. Different amounts of the extract of *E. coli* expressing GST-PSD95 (500 μL, 1 mL, and 2 mL) were applied onto 50 μL GSH sepharose beads. After washing the beads, GST-fused proteins were eluted with the GSH Elution Buffer. Eluates and BSA standards (0.2-, 0.4-, 0.8-, and 1.0 μg) were loaded for SDS-PAGE and staining with CBB. The gel image was taken using a LI-COR Odyssey Infrared Imaging System 9201

3.3 Immobilization of GST-Fused Proteins on GSH Sepharose Beads

In the example experiment to purify GST-PSD95, -SAP97, and -RIM2-binding proteins, seven columns were prepared. GST or GST-fused proteins immobilized on GSH beads are as listed below;

1. GST.
2. GST-PSD95.
3. GST-SAP97.
4. GST-RIM2.
5. GST-PSD95.
6. GST-SAP97.
7. GST-RIM2.

Mouse brain extract was applied to columns 1–4, while CHAPS Binding Buffer was applied to columns 5–7. Column 1 is a negative control to exclude nonspecific proteins that bind to GSH beads or to the GST tag. Columns 5–7 are negative controls to rule out proteins that are derived from the *E. coli* suspension. These two sets of negative controls are absolute requirements to identify specific binding partners in this current protocol.

1. Apply the amount of bacterial suspension estimated in Subheading 3.2 to 50 μL of GSH sepharose beads in the custom-made column to immobilize 60 μg of GST-fused protein.
2. Wash beads three times with 250 μL of Washing Buffer to remove the proteins bound to beads nonspecifically.
3. To remove bacterial proteins bound to GST proteins (e.g., heat hock proteins), wash the beads three times with 250 μL of High Salt Buffer.
4. Equilibrate the beads three times with 250 μL of CHAPS Binding Buffer to change the buffer condition.

3.4 Preparation of Whole Mouse Brain Extract

1. Homogenize two whole mouse brains in 4 mL of Sucrose Solution using the Teflon Homogenizer and Potter rotor at 900 rpm at 4 °C (*see* **Note 7**).
2. Spin down nuclear and debris fractions at 900 × *g* for 10 min at 4 °C. Harvest the supernatant as the postnuclear supernatant (PNS).
3. Quantify the protein concentration of the PNS using a standard Protein Assay (e.g., Bradford or BCA method). Adjust the protein concentration of the PNS to 10 mg/mL with Sucrose Solution.
4. Mix 10 mL of the PNS with 10 mL of 2× CHAPS Binding Buffer (*see* **Note 8**). Incubate the PNS for 30 min at 4 °C to extract membrane proteins. Avoid forming bubbles.
5. Ultracentrifuge the PNS in a fixed-angle rotor at 100,000 × *g* for 1 h at 4 °C to remove the insoluble fraction. Harvest the supernatant as the PNS CHAPS Extract.

6. Incubate 25 mL of the PNS CHAPS Extract with 0.5 mL of GSH sepharose beads equilibrated with CHAPS Binding Buffer for 1 h at 4 °C on a rotator set to 10 rpm (*see* **Note 9**). After incubation, spin down beads at 800 ×*g* for 1 min at 4 °C and harvest the supernatant. This preclearing step reduces the endogenous GST and proteins with a high affinity to GSH sepharose beads in the PNS CHAPS Extract.
7. Filtrate the PNS CHAPS Extract with a 0.22 μm-pore filter unit.

3.5 Loading Brain Extracts onto Columns and Elution of Binding Partners

1. Cover the tip of a 20 mL syringe with five layers of parafilm (Fig. 1c) to connect the custom-made column (Fig. 1d).
2. Load 20 mL of the PNS CHAPS Extract or CHAPS Binding Buffer to the beads and let columns run by gravity flow. To avoid air bubbles in the flow, use a 2 mL syringe with a 20G needle to load the PNS CHAPS Extract as shown in Fig. 1d.
3. After the first flow, load the flow-through to the GSH sepharose beads again. After the second flow, remove the columns from the syringe and place them onto microcentrifuge tubes, as shown in Fig. 1b.
4. Wash the beads with 250 μL of CHAPS Binding Buffer. Repeat this twice.
5. Set the columns into new microcentrifuge tubes as shown in Fig. 1b and elute the bound proteins with 250 μL of NaCl Elution Buffer.
6. Set the columns into new microcentrifuge tubes and elute bound proteins together with GST-fused proteins with 250 μL of GSH Elution Buffer.

   Eluates from the **steps 5** and **6** will be analyzed by SDS-PAGE and CBB staining in Subheading 3.6.

3.6 Identification of Specific Binding Partners

In order to avoid contamination with human keratin, handle all samples, buffers, tubes, and glassware with gloves.

1. Add 40 μL of 3× Laemmli Buffer (with DTT) to 80 μL of each eluate (NaCl and GSH Eluates from Subheading 3.5, **steps 5** and **6**).
2. Incubate samples at 65 °C for 20 min with gentle agitation.
3. Load 37.5 μL of each sample into an individual lane of a 4–12 % Bis-Tris polyacrylamide gel.
4. Run SDS-PAGE in MOPS-SDS Running Buffer applying a current of 100–200 V until the dye front reaches the bottom of the gel.
5. After SDS-PAGE, perform all subsequent staining procedures on a desktop shaker at room temperature. Use glass containers

to avoid any contamination by polymers and plasticizers that might disturb analysis by mass spectrometry.

6. Incubate the gel with water briefly and subsequently with the Gel Fixation Solution in a glass petri dish for at least 60 min. The Gel Fixation Solution should completely cover the gel.
7. Wash the gel with water for 10 min two times. The water should cover the gel completely.
8. Incubate the gel in Colloidal Coomassie Working Solution overnight. Colloidal Coomassie Working Solution should cover the gel completely.
9. Remove the Colloidal Coomassie Working Solution and then wash the gels with the Destaining Solution until Coomassie particles are removed and the gel background is clear. Change Destaining Solution several times during this procedure.
10. Take an image of the gel with a conventional flat bed scanner or with a near infrared imaging system (e.g., Odyssey System, LI-COR; *see* above). Bands present in the samples with GST-fused proteins loaded with the PNS CHAPS extract but not in the negative control samples are considered as specific binding partners and are manually excised to be subjected to mass spectrometric protein identification (Fig. 3a, b). It is recommended to excise bands from the corresponding positions within the GST control lane as well, the parallel processing of which will reveal useful information on the background proteome binding to GST-beads alone. Proteins in bands highlighted in Fig. 3b were identified by mass spectrometry, as previously described [14]. Note that they appear to include reported binding partners (Table 1), demonstrating that this protocol is a reliable method for screening for proteins that bind proteins in the CAZ and PSD.

4 Notes

1. Keep the cuvettes and competent cells on ice before transformation and avoid heating up the competent cells with your fingers. Avoid pipetting the competent cells up and down or forming bubbles while transferring competent cells into the cuvette. These considerations are crucial to achieve a high transformation efficiency.
2. Incubation time and temperature for bacterial growth should be optimized to achieve a good yield and recovery of GST-fused proteins. We recommend incubation of *E. coli* cultures at 16–20 °C.

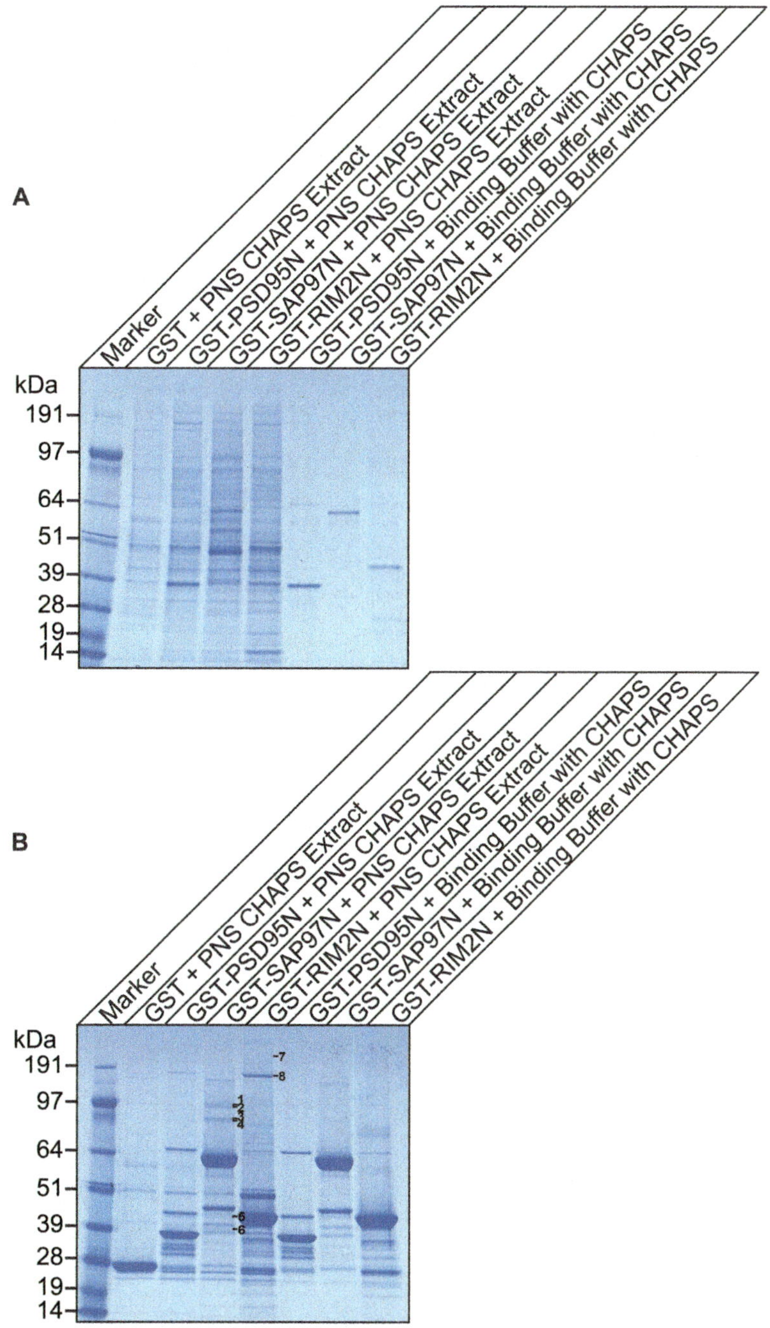

Fig. 3 Affinity purification of Binding Proteins of GST-fused PSD95, SAP97, and RIM2. 60 μg GST, GST-fused PSD95 (residue 2–76), SAP97 (residue 2–407), or RIM2 (2–155) were immobilized on GSH sepharose beads, where PNS CHAPS Extract of the mouse whole brain or CHAPS Binding Buffer was loaded. After washing the beads, bound proteins were eluted with 1 M NaCl (**a**) and 40 mM GSH (**b**) subsequently. Protein bands enriched in the eluate of GST-fused PSD95, SAP97, or RIM2 were analyzed by mass spectrometry. The results of mass spectrometric protein identification are listed in Table 1 for eight marked protein bands in Fig. 3b

Table 1
Identification of Binding Partners of GST-SAP97 and GST-RIM2 (related to Fig. 3b). For LC-MS analysis, short linear gradient separations (5–40 % acetonitrile in 20 min) were combined with data-independent acquisition (MS^E) on a quadrupole time-of-flight hybrid instrument. A minimum of two peptides was required for the identification and the false discovery rate was limited to 1 %

	Number of Band in Fig. 3b	Protein name	Accession number (UniProtKB/Swiss-Prot)	Molecular mass	Sequence coverage [%]
SAP97	1	Neuronal cell adhesion molecule	Q810U4	138,434	17.3
	2	Glutamate receptor 2	P23819	98,599	38.1
	3	Glutamate receptor 2	P23819	98,599	49.7
		Glutamate receptor 3	Q9Z2W9	100,462	34.9
		Glutamate receptor 1	P23818	101,504	43.8
		Glutamate receptor 4	Q9Z2W8	100,782	37.8
		Neuroligin-3	Q8BYM5	91,104	16.4
		Neuroligin 4-like	B0F2B4	97,287	8.3
	4	Glutamate receptor 2	P23819	98,599	56.3
		Glutamate receptor 3	Q9Z2W9	100,462	43.2
		Glutamate receptor 1	P23818	101,504	50.6
		Glutamate receptor 4	Q9Z2W8	100,782	47.8
		Neuroligin-2	Q69ZK9	90,932	16.5
		Neuroligin-3	Q8BYM5	91,104	17.5
	5	Voltage-gated potassium channel subunit beta-1	P63143	44,694	68.3
	6	Voltage-gated potassium channel subunit beta-2	P62482	40,995	73.6
RIM2	7	Protein unc-13 homolog A	Q4KUS2	193,658	20.9
		Protein unc-13 homolog B	Q9Z1N9	181,696	9.2
	8	Protein unc-13 homolog A	Q4KUS2	193,658	71.3
		Protein unc-13 homolog B	Q9Z1N9	181,696	14
		Protein unc-13 homolog C	Q8K0T7	249,684	7.5

3. IPTG concentration (0.1–1.0 mM) and induction time (2–20 h) should be optimized to achieve a good yield and recovery of GST-fused proteins.
4. *E. coli* should be frozen in Resuspension Buffer to achieve good recovery.
5. Different types of detergents (e.g., Triton-X100, sodium cholate, or n-octyl glucoside) in Resuspension Buffer could result in better extraction and recovery of a GST-fused protein from the *E. coli* lysate.
6. Avoid air bubbles while packing the GSH sepharose beads into the column. This would interrupt the flow in the column.
7. Two mouse whole brains (6-week-old animals) yield protein amounts sufficient for one column in this protocol.
8. Including different types of detergents (e.g., Triton-X100, sodium cholate, or n-octyl glucoside) in the Binding Buffer may result in the identification of different collections of binding partners.
9. Avoid forming bubbles to prevent proteins from partial denaturation.

Acknowledgments

We thank J. Day for discussion and advice on the manuscript. Expression vectors for GST-PSD-95 and GST-SAP97 are kind gifts from O. Schlüter. We thank M. Uecker, D. Hesse, B. Hesse-Niessen, K.-P. Hellmann, I. Thanhäuser, D. Schwerdtfeger, C. Harenberg, and F. Benseler for excellent technical assistance. This work was supported by the Max Planck Society and by grants from the German Research Foundation (SPP1365/KA3423/1-1 and KA3423/3-1; to H.K.), and the Fritz Thyssen Foundation (to H.K.).

References

1. Palade GE, Palay LS (1954) Electron microscope observations of interneuronal and neuromuscular synapses. Anat Rec 118:335–336
2. Gray EG (1959) Axo-somatic and axo-dendritic synapses of the cerebral cortex: an electron microscope study. J Anat 93:420–433
3. Couteaux R (1963) The differentiation of synaptic areas. Proc R Soc Lond B Biol Sci 158:457–480
4. Vaughn JE (1989) Fine structure of synaptogenesis in the vertebrate central nervous system. Synapse 3:255–285
5. Harlow ML, Ress D, Stoschek A, Marshall RM, McMahan UJ (2001) The architecture of active zone material at the frog's neuromuscular junction. Nature 409:479–484
6. Jin Y, Garner CC (2008) Molecular mechanism of presynaptic differentiation. Annu Rev Cell Dev Biol 24:237–262
7. Carlin RK, Grab DJ, Cohen RS, Siekevitz P (1980) Isolation and characterization of postsynaptic densities from various brain regions: enrichment of different types of postsynaptic densities. J Cell Biol 86:831–845

8. Boeckers T, Liedtke T, Spilker C, Dresbach T, Bockmann J, Kreutz M, Gundelfinger E (2005) C-terminal synaptic targeting elements for postsynaptic density proteins ProSAP1/shank2 and ProSAP2/shank3. J Neurochem 92:519–524
9. Inoue A, Okabe S (2003) The dynamic organization of postsynaptic proteins: translocating molecules regulate synaptic function. Curr Opin Neurobiol 13:332–340
10. Cingolani L, Goda Y (2008) Actin in action: the interplay between the actin cytoskeleton and synaptic efficacy. Nat Rev Neurosci 9:344–356
11. Kawabe H, Hata Y, Takeuchi M, Ide N, Mizoguchi A, Takai Y (1999) nArgBP2, a novel neural member of ponsin/ArgBP2/vinexin family that interacts with synapse-associated protein 90/postsynaptic density-95-associated protein (SAPAP). J Biol Chem 274:30914–30918
12. Takeuchi M, Hata Y, Hirao K, Toyoda A, Irie M, Takai Y (1997) A family of Psd-95/Sap90-associated proteins localized at postsynaptic density. J Biol Chem 272:11943–11951
13. Persaud A, Alberts P, Amsen EM, Xiong X, Wasmuth J, Saadon Z, Fladd C, Parkinson J, Rotin D (2009) Comparison of substrate specificity of the ubiquitin ligases Nedd4 and Nedd4-2 using proteome arrays. Mol Syst Biol 5:333
14. Ott C, Martens H, Hassouna I, Oliveira B, Erck C, Zafeiriou MP, Peteri UK, Hesse D, Gerhart S, Altas B, Kolbow T, Stadler H, Kawabe H, Zimmermann WH, Nave KA, Schulz-Schaeffer W, Jahn O, Ehrenreich H (2015) Widespread expression of erythropoietin receptor in brain and its induction by injury. Mol Med 21:803–815

Chapter 7

In Situ Protein Binding Assay Using Fc-Fusion Proteins

Nirmala Padmanabhan and Tabrez J. Siddiqui

Abstract

This protocol describes an *in situ* protein-protein interaction assay between tagged recombinant proteins and cell-surface expressed synaptic proteins. The assay is arguably more sensitive than other traditional protein binding assays such as co-immunoprecipitation and pull-downs and provides a visual readout for binding. This assay has been widely used to determine the dissociation constant of binding of trans-synaptic adhesion proteins. The step-wise description in the protocol should facilitate the adoption of this method in other laboratories.

Key words Fc-fusion protein, In situ protein binding assay, Protein-protein interaction

1 Introduction

Identification and characterization of interactions among synaptic proteins lead to the unraveling of previously unknown pathways in the development and function of synapses [1–4]. Once a new protein-protein interaction is detected by any of the in vitro biochemical approaches such as pull-down or co-immunoprecipitation (Co-IP) followed by mass-spectrometry, it is essential to validate the interaction in the physiological milieu [3–6]. Here, we describe a method to validate cell-surface protein-protein interactions in situ. This method has been widely used in assessing direct binding of synaptic adhesion proteins with their recombinant ligands. A protein of interest is expressed and purified from mammalian cells as a recombinant Fc-fusion protein and incubated with COS7 cells (*see* **Note 1**) that surface-express their putative interacting partners tagged with an epitope such as myc (*see* **Note 2**). The cells are subsequently fixed and detection is done by fluorescent conjugated antibodies that recognize the appropriate tags. This method is quantitative and weak interactions that may not be readily detectable by other methods such as Co-IP or pulldown may also be observed.

Alexandros Poulopoulos (ed.), *Synapse Development: Methods and Protocols*, Methods in Molecular Biology, vol. 1538, DOI 10.1007/978-1-4939-6688-2_7, © Springer Science+Business Media LLC 2017

2 Materials

2.1 Cell Lines

1. COS7.
2. Human Embryonic Kidney 293T.

2.2 Cell Culture and Transfection

1. DMEM: Dulbecco's Modified Eagle Medium.
2. DMEM-5: DMEM supplemented with 5 % Fetal Bovine Serum (FBS).
3. DMEM-10: DMEM supplemented with 10% FBS.
4. AIM V® Medium (Life technologies).
5. Transfection Reagent: FuGENE (Roche), LT1 (Mirus), or equivalent.
6. Trypsin Solution: 0.25% Trypsin-EDTA mix for tissue culture.
7. Zeocin Medium: 0.5 mg/ml Zeocin (e.g., Invitrogen), 10% FBS in DMEM.
8. T300 or T150 cell culture flasks.
9. Cell culture dishes, 100 and 60 mm.
10. 12-well cell culture plate.
11. Fine tipped forceps.
12. Haemocytometer.
13. 18 mm glass coverslips.

2.3 Protein Purification

1. PMSF Solution: 2% w/v Phenyl Methyl Sulphonyl Fluoride (PMSF) in isopropanol.
2. Protease Inhibitors: Protease Inhibitor Cocktail EDTA-free COMPLETE Mini protease inhibitor tablet (Roche).
3. Concentrating filter units with 30 kDa cutoff, for large and small volumes (e.g., Centricon Plus-70 and Centricon Plus-20, Millipore. *See* **Note 3**).
4. Protein G columns: Protein G sepharose beads packed in polypropylene columns (e.g., Poly-Prep columns, #7311550, Bio-rad).
5. Binding Buffer: 20 mM sodium phosphate, pH 7.0 with HCl.
6. Elution Buffer: 100 mM Glycine, pH 2.7 with HCl.
7. Tris–Cl: 1 M Tris base, pH 8.5 with HCl.

2.4 Protein Binding Assay

1. EGB buffer: 168 mM NaCl, 2.6 mM KCl, 10 mM HEPES, 2 mM CaCl2, 2 mM MgCl2, 10 mM d-glucose, 100 μg/ml BSA, pH 7.2 with HCl.
2. Antibodies: Rabbit anti-myc antibody, alexa-568 conjugated anti rabbit antibody, FITC-conjugated anti human IgG 3 (*see* **Note 4**).

3. Cell Fixing Solution: 4% formaldehyde, 4% sucrose in PBS.
4. Blocking Solution: 10% BSA in PBS.
5. Antibody Incubation Solution: 3% BSA in PBS.
6. Mounting agent (e.g., elvanol: Tris–HCl, glycerol, polyvinyl alcohol with 2% 1,4-diazabicyclo[2,2,2] octane).

2.5 Imaging and Analysis

1. Fluorescence microscope (e.g., Zeiss Axioplan 2) with 63×/1.4 NA, 40×/1.3 NA and 25×/1.25NA objectives (*see* **Note 5**).
2. Cooled CCD Camera (e.g., Photometrics sensys cooled camera).
3. Image analysis software (e.g., Metamorph, Molecular Devices).
4. Custom filters.

3 Methods

3.1 Protein Production from Mammalian Cells

Transiently expressing cells for sample protein production

1. Grow human embryonic kidney 293T cells (HEK293) in DMEM-10.
2. Split HEK293 cells intoT300 flasks (or T150 flask) containing 40 ml DMEM-10. Seed approx 2×10^7 cells per flask (*see* **Note 6**).
3. The following day change medium to DMEM-5 (*see* **Note 7**).
4. To transfect cells, incubate 30 μg of DNA encoding the Fc-fusion protein in 3 ml of (serum free) DMEM for 5 min (*see* **Note 8**).
5. Add 100 μl of transfection reagent (optimized for FuGENE, *see* **Note 9**) to the preincubated DNA/DMEM mix.
6. Vortex lightly and incubate the transfection complex mixture for 20 min at room temperature (RT).
7. Distribute 3 ml transfection complex drop-wise to each T300 flask and mix gently ensuring that the cells are not disturbed.
8. Let the transfection complex incubate with the cells overnight.
9. Change the medium to 40 ml AIM V medium for protein production and collect after 48 h of incubation.

Stably expressing cells for batch protein production

1. Alternatively, to generate a stable cell line for protein production, grow HEK cells in a 10 cm dish and transfect with an Fc-fusion protein construct containing zeocin resistance. cassette.
2. Select cells with Zeocin Medium for 28 days. Change to fresh Zeocin Medium every third day. Passage cells when confluence is reached.
3. Subsequent to Zeocin selection, plate cells in a T300 flask.

4. Change the medium to 40 ml AIM V medium.
5. Collect the medium every alternate day for 4 weeks for a total of about 400–500 ml medium.

3.2 Protein Purification and Validation

1. To concentrate protein, collect conditioned medium in conical centrifugation tubes.
2. Add 50 μl PMSF Solution per 50 ml conditioned medium.
3. To remove cell debris, centrifuge at 9000 × *g* for 10 min at 4 °C.
4. Transfer the supernatant to a new tube and repeat centrifugation at 9000 × *g* at 4 °C for 10 min.
5. Centrifuge the supernatant through concentrating filter unit at 3200 × *g* for 10–30 min at 4 °C.
6. Repeat with additional medium using the same concentrator unit (*see* **Note 10**).
7. Invert the unit and collect the concentrate by centrifuging at 1000 × *g* for 3 min at 4 °C.
8. Adjust the volume of the concentrate to about 10 ml.
9. Add protease inhibitor cocktail as per the manufacturer's recommendation. For EDTA-free COMPLETE Mini protease inhibitors (Roche) add one tablet per 10 ml concentrate.
10. Store at 4 °C until purification (*see* **Note 11**).
11. Wash Protein G columns with two column volumes of binding buffer before applying the sample.
12. Add Binding Buffer to the concentrate at 1:1 volume, i.e,, 10 ml binding buffer to 10 ml concentrate.
13. Run the sample through the column and save the flow-through for analysis.
14. Wash the column with 5–10 volumes of Binding Buffer.
15. Elute with 5 ml of Elution Buffer in a tube containing 1 ml of Tris–Cl. Mix lightly during elution.
16. Add 10–15 ml of cold sterile PBS to the eluted fraction.
17. Pass through a small concentrating filter unit for 20–30 min.
18. When volume reduces to less than 1 ml, refill with PBS and centrifuge for 20–30 min.
19. Concentrate to a final volume of 500–1000 μl.
20. To clean column, pass 12 ml of Elution Buffer through it, followed by 5 ml distilled water containing 0.02 % sodium azide.
21. Store at 4 °C in 3 ml water containing 0.02 % sodium azide (*see* **Note 12**).
22. Analyze eluted protein by 6–8 % SDS-PAGE (*see* **Note 13**). Load 10 μl each of eluate and flow through and stain the gel with Sypro Ruby protein gel stain.

23. For quantitation, run a BSA standard 0.05 μg through 3 μg and three different volumes of the eluate in an SDS-PAGE gel and stain with Sypro Ruby protein gel stain.
24. Use densitometry analysis to measure protein concentration by comparing to BSA standard.

3.3 Cell Culture and Transfection for Binding Assay

1. Take a confluent 100 mm dish of COS7 cells.
2. Wash 1× with PBS.
3. Aspirate and add 2 ml of pre-warmed Trypsin Solution.
4. Place the dish in a cell culture incubator and monitor detachment of cells (~3–5 min).
5. Do not proceed if the detachment takes longer than 5 min.
6. Add 8 ml of DMEM-10 and thoroughly dislodge cells. The serum components will inactivate the trypsin.
7. Centrifuge the medium containing the detached cells at 1500 × *g* for 5 min.
8. Aspirate the supernatant and resuspend the pelleted cells in 10 ml of complete DMEM.
9. Count the cells using haemocytometer.
10. For passaging, plate 6×10^5 cells per 100 mm tissue culture dish. Plated at this density, the COS7 cells will be ready to be split in about 3 days.
11. Prepare coverslips (18 mm) before trypsinizing the COS7 cells. To sterilize, grab the coverslips using fine-tipped forceps, dip in ethanol, dab off excess ethanol, and pass them through flame. Flip over the coverslip by rotating your hand to limit the heat generated, thereby preventing the formation of cracks. Place the flame-sterilized coverslips in a well of a 12-well plate (*see* **Note 14**).
12. Plate the COS7 cells at a density of 6×10^4 cells per well of a 12-well plate in DMEM-10.
13. The next day, exchange the medium of the 12-well plate with fresh DMEM-10 (1 ml per well).
14. The cells are now ready for transfection.
15. Use 500 ng of plasmid DNA encoding the putative myc-tagged binding partner of the recombinant Fc-fusion protein produced.
16. Aliquot the appropriate amount of plasmid into a microfuge tube.
17. Add transfection reagent (e.g., FuGENE or Mirus LT1) to serum-free DMEM (the standard ratio of FuGENE:DNA is 3:1, see manufacturer's instructions for optimum ratios).
18. Vortex briefly at low speed.
19. Incubate at RT in the cell culture cabinet for at least 20 min.

20. Add the transfection complex drop-wise to the well of the 12-well plate.
21. Shake the plate gently in multiple directions to disperse the complexes.
22. Allow transfection to take place overnight.
23. There is no need to wash the cells if using FuGENE or Mirus LT1 as these reagents are generally not toxic.
24. Expect a transfection efficiency of at least 40%.

3.4 In Situ Binding Assay

1. Wash the transfected COS7 cells on coverslips with EGB buffer.
2. Apply the Fc-fusion protein at the appropriate concentration (*see* **Note 15**) in EGB up to a total volume of 100 μl onto the coverslip with the transfected cells. Include the primary antibody such as anti-myc in this solution. Apply double the amount of antibody used in regular immunostaining.
3. Incubate for 45 min at 4 °C (*see* **Note 16**).
4. Wash 3× with EGB allowing a 5 min incubation with the buffer each time.
5. Fix cells in pre-warmed PFA solution.
6. Incubate in Blocking Solution for 30 min at RT.
7. Incubate with the appropriate dilutions of FITC-conjugated antihuman IgG to detect bound Fc-fusion protein and with Alexa-568-conjugated anti-mouse or anti-rabbit antibody to detect surface myc for 30 min at 4 °C in Antibody Incubation Solution.
8. Wash coverslips with PBS, aspirate and mount using mounting medium of choice (e.g., elvanol).
9. Let dry and proceed with imaging.

3.5 Imaging and Quantitation

1. Acquire images using a fluorescence microscope hooked up to a cooled CCD camera, image acquisition software and custom filters.
2. For quantitation, fix and stain experimental sets simultaneously and image with identical settings.
3. Acquire images in gray scale and analyze using an image analysis software such as Metamorph or Image J.
4. Draw regions around the perimeter of each COS7 cell and measure the average intensity values of bound protein and expressed protein within the region. Correct for average off-cell background.
5. Complete analysis using a statistical software program such as Graphpad Prism.

4 Notes

1. Other cell types such as Human Embryonic Kidney 293 and L cells may also be used.
2. Here, myc tag is shown as an example (Fig. 1). Any other fluorescent or nonfluorescent tag can be used in place of the myc tag.
3. The KDa cutoff is dependent on the putative size of the fusion protein.
4. Or another primary antibody if the epitope is different. Also, other secondary antibodies may be used, ensuring that the secondary antibodies do not have much spectral overlap.
5. 63× or 40× images may be used for quantitative analysis. 25× images may be used to demonstrate recombinant Fc protein binds to cells expressing the binding partner.

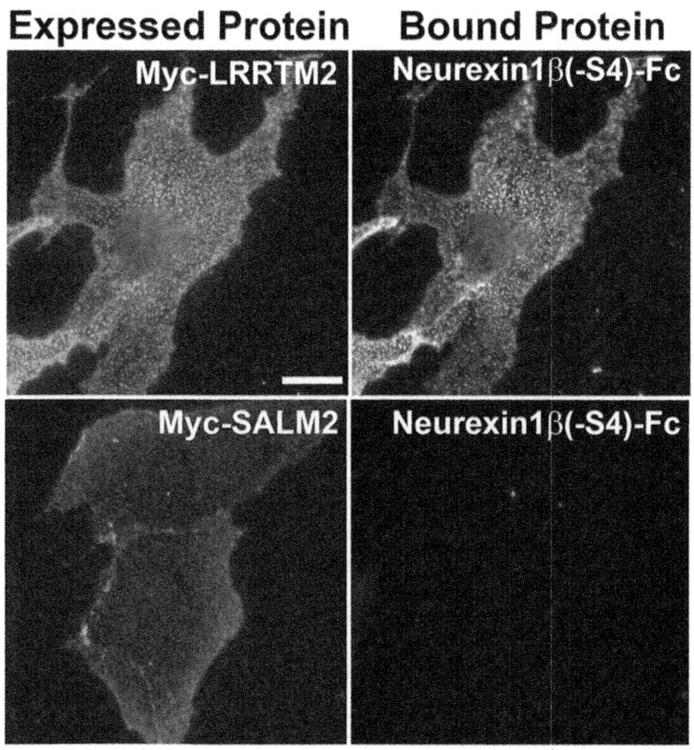

Fig. 1 An example of in situ binding of a recombinant protein to a tagged cell-surface expressed protein. Recombinant neurexin1β(−S4)-Fc was incubated with COS7 cells transfected with Myc-LRRTM2 or Myc-SALM2. Myc-LRRTM2, detected by surface staining for myc, bound neurexin1β(−S4)-Fc which was detected by FITC-labeled anti-human IgG. Myc-SALM2 was the negative control for binding. *Scale bar*, 10 μm

6. Typically, three T300 flasks provide sufficient medium for producing a small quantity of recombinant protein for several experiments. The number of flasks may be scaled up accordingly if more protein is required or if the yield of the protein is low.
7. The ideal confluence level of the cells should be between 75 and 85 % for maximum transfection efficiency.
8. This is the amount of DNA for one T300 flask.
9. Different transfection reagents have different optimum volumes and DNA:reagent ratio. It is best to take into account manufacturer's recommendations and optimize with one's own plasmid DNA.
10. Typically for a 500 ml volume, two centricon-70 units are sufficient.
11. Do not store at 4 °C for longer than 48 h.
12. Columns may be reused several times.
13. The gradient range of the SDS-PAGE for protein analysis may be varied depending upon the size of the fusion protein.
14. Alternatively, the coverslips may be autoclaved.
15. The amount of Fc-fusion protein will depend on the concentration and the K_d of binding. If a new interaction is being tested, it is worth testing a range of concentrations from 1 nM to 500 nM. Typical trans-synaptic adhesion protein interactions have been reported to have a K_d ranging from 3 nM to 30 nM [2–5, 7–9]. However, lower affinity interactions (K_d ~ 300 nM) have also been reported [10].
16. Length of incubation (30 min–1 h) and temperature (4–37 °C) can be varied.

References

1. de Wit J et al (2013) Unbiased discovery of glypican as a receptor for LRRTM4 in regulating excitatory synapse development. Neuron 79(4):696–711
2. Pettem KL et al (2013) The specific alpha-neurexin interactor calsyntenin-3 promotes excitatory and inhibitory synapse development. Neuron 80(1):113–128
3. Siddiqui TJ et al (2013) An LRRTM4-HSPG complex mediates excitatory synapse development on dentate gyrus granule cells. Neuron 79(4):680–695
4. Ko J, Fuccillo MV, Malenka RC, Sudhof TC (2009) LRRTM2 functions as a neurexin ligand in promoting excitatory synapse formation. Neuron 64(6):791–798
5. Siddiqui TJ, Pancaroglu R, Kang Y, Rooyakkers A, Craig AM (2010) LRRTMs and neuroligins bind neurexins with a differential code to cooperate in glutamate synapse development. J Neurosci 30(22):7495–7506
6. Linhoff MW et al (2009) An unbiased expression screen for synaptogenic proteins identifies the LRRTM protein family as synaptic organizers. Neuron 61(5):734–749
7. Gale NW et al (1996) Eph receptors and ligands comprise two major specificity subclasses and are reciprocally compartmentalized during embryogenesis. Neuron 17(1):9–19
8. Takahashi H et al (2011) Postsynaptic TrkC and presynaptic PTPsigma function as a bidirectional excitatory synaptic organizing complex. Neuron 69(2):287–303

9. Takahashi H et al (2012) Selective control of inhibitory synapse development by Slitrk3-PTPdelta trans-synaptic interaction. Nat Neurosci 15(3):389–398, S381-382

10. Comoletti D et al (2003) Characterization of the interaction of a recombinant soluble neuroligin-1 with neurexin-1beta. J Biol Chem 278(50):50497–50505

Chapter 8

Reconstitution of Synaptic SNAREs into Large Liposomes with Reduced Curvature Stress

Javier M. Hernandez

Abstract

Liposomes constitute a convenient biochemical model system to investigate mechanistic aspects of the membrane fusion of synaptic vesicles. The proteins responsible for mediating fusion are the SNAREs that belong to a highly conserved family of transmembrane proteins. Reconstituting SNAREs into liposomes using detergents has become a common approach not only to understand how SNAREs work, but also how fusion is regulated by the vast array of accessory proteins present at the presynapse. However, a concern has been that the high curvature stress of the small liposomes (diameters of ~40 nm) frequently used in many studies renders them prone to spontaneous fusion, bringing into question whether the measurements obtained faithfully represent SNARE-mediated fusion. By systematically varying the detergent concentration and characterizing the SNARE-liposome size distributions by light scattering, we describe a detailed procedure to reconstitute SNAREs into large liposomes with considerably reduced curvature stress.

Key words SNARE, Membrane fusion, Liposome, Exocytosis, Neurotransmitter release

1 Introduction

The engines behind membrane fusion of synaptic vesicles (SVs) with the plasma membrane leading to neurotransmitter release are the proteins belonging to the highly conserved family known as Soluble N-ethylmaleimide–sensitive factor Attachment protein Receptors, or SNAREs [1, 2]. The SNAREs involved in the fusion of SVs are vesicular synaptobrevin-2 (syb) and the plasma membrane proteins SNAP-25A (SN25) and syntaxin-1A (syx). According to the zipper hypothesis, the N-termini of syb (residing on the vesicle) and the binary syx:SN25 acceptor complex (residing on the plasma membrane) nucleate to form a four-helix bundle called the SNARE complex that brings opposing membranes together (also referred to as docking) and culminates in membrane fusion as assembly proceeds toward the C-termini [2] (*see* Fig. 1a).

A vast wealth of information on how synaptic SNAREs operate has been obtained by reconstituting these proteins into liposomes

Alexandros Poulopoulos (ed.), *Synapse Development: Methods and Protocols*, Methods in Molecular Biology, vol. 1538, DOI 10.1007/978-1-4939-6688-2_8,

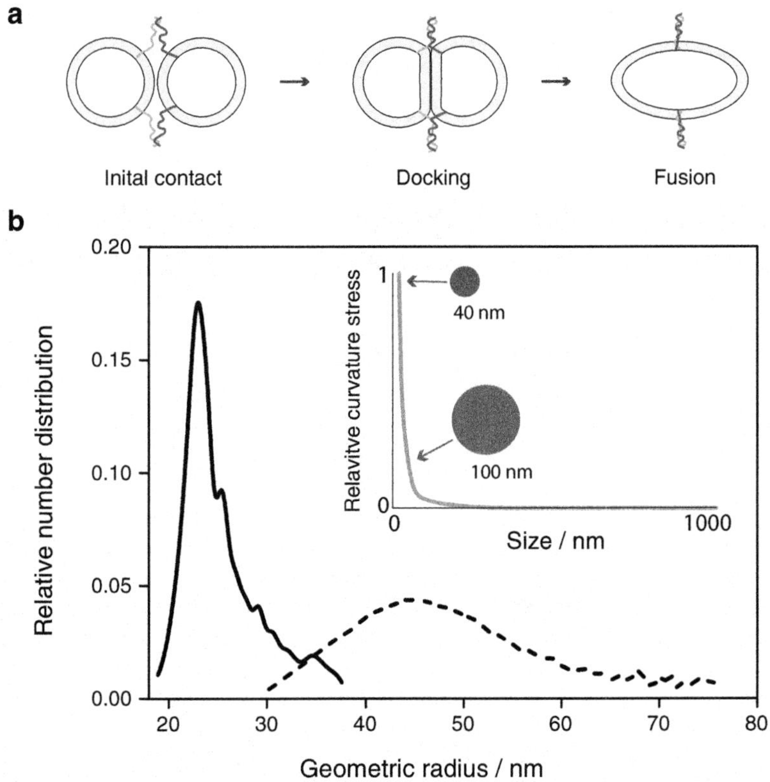

Fig. 1 Use of liposomes as a tool to investigate SNARE-mediated fusion. (**a**) Schematic depiction of liposome fusion as mediated by SNAREs. Initiation of SNARE complex assembly (conceptually depicted as coiled curves) brings liposomes to initial contact, while further progression of the SNARE assembly effectively docks membranes onto each other. The force generated by the formation of the SNARE complex eventually causes membranes to merge and fuse, a process that can be measured by lipid-mixing using the fluorescent probes NBD-PE/RHO-PE as mentioned in the methods section. (**b**) Relative number size distribution of a representative preparation of a small syb-liposome (*solid line*) compared to a large syb-liposome (*dash line*) measured by field-flow fractionation coupled to multi-angle laser light scattering (FFF-MALLS). Although the size distribution of the large liposomes appears much broader, when the size range is normalized to the peak size (the polydispersity index is a measure of this) the broadness of both distributions is comparable. *Inset*: a schematic depicting how curvature stress (measured as elastic bending energy of the membrane [14]) is sharply reduced relative to a small 40 nm liposome

and using fluorescence-based methods to measure a fusion-related property such as lipid or content mixing [3]. These SNARE-liposomes have the advantage of allowing researchers to study SNARE-mediated membrane fusion in well-defined biochemical conditions, thus simplifying an already complex process and allowing to probe new hypotheses. Recent examples of how this minimalist model system has been useful to gain new insights include knowledge acquired on the function of chaperone proteins Munc13 and Munc18 [4] and the Ca^{2+}-sensor synaptotagmin 1 [5].

While liposome fusion assays are today widely employed, various different protocols are available in the literature on how to reconstitute SNAREs into liposomes. An important parameter to consider is the size of liposomes which is directly related to the curvature stress of the membrane. A concern has been that the high curvature stress of the more commonly used 40 nm liposomes prepared by co-micellization makes liposome more prone to non-SNARE-mediated spontaneous fusion and is thus not suitable for mechanistic studies involving SNAREs [6]. A solution is to increase the liposome size 2–3 fold which substantially reduces the curvature stress of the membrane [7] (Fig. 1b). The use of these larger liposomes has been imperative, for instance, for slowing down the conversion from docking to fusion to uncover putative fusion intermediates [7], or for demonstrating the variable cooperativity in SNARE-mediated fusion [8].

To prepare large SNARE-liposomes with optimal reconstitution properties and diameters of ~100 nm, we performed a systematic study based on the three-stage model of Rigaud and Levy on the detergent-assisted reconstitution of transmembrane proteins [9] (Fig. 2). In particular, we attempted to find conditions for direct reconstitution of SNAREs (also known as the "step-by-step" reconstitution) into preformed large liposomes exposed to a critical amount of detergent and a micellar solution of SNAREs [7]. We reported the quantitative insertion of syb into liposomes of 80–100 nm diameter. However, the random orientation of syb on these SNARE-liposomes suggested that the mechanism of insertion is more in accord with a hybrid mechanism where the size of the liposomes is mostly preserved during reconstitution but some minor elements pertaining to the co-micellization method are also evident.

We next provide a detailed description of the reconstitution of SNAREs into large liposomes with reduced curvature stress. Keeping in mind that much of what we know about the reconstitution of transmembrane proteins is still very much empirically determined, we will make brief comments about factors or observations related to the mechanism of reconstitution which may be relevant to consider when making alterations to better fit the reader's particular needs (e.g., when preparing liposomes with a different lipid composition, or when using SNAREs with truncations or mutations). In this case, we strongly recommend checking parameters such as incorporation efficiency, orientation, and size distributions of the resulting SNARE-liposomes [7].

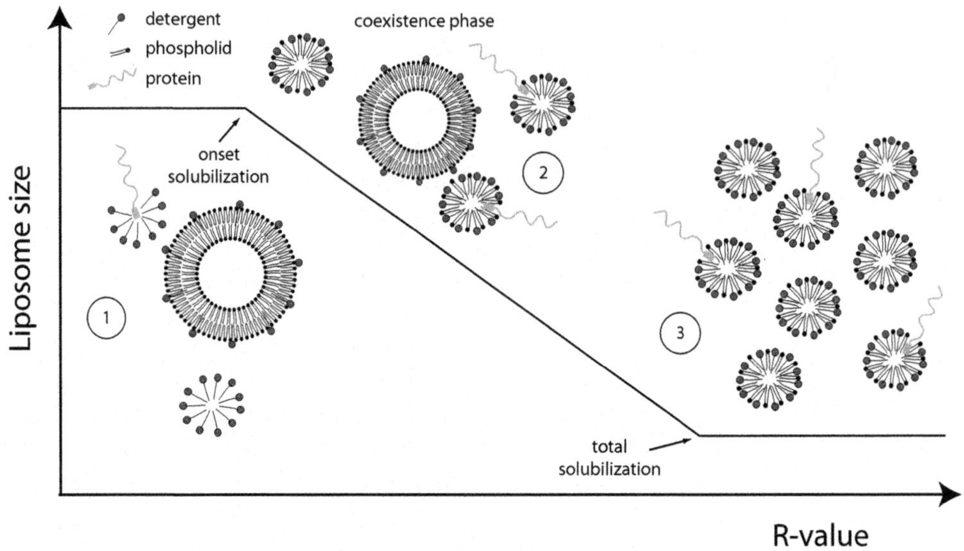

Fig. 2 Empirical relation between liposome size and detergent concentration for the reconstitution of membrane proteins. The mechanisms by which transmembrane proteins are reconstituted and how this affects liposome size have been rationalized as part of a 3-stage model where the critical parameter is the excess amount of detergent above the critical micelle concentration (CMC) relative to the total lipid content of the liposomes. This ratio, denoted as the *R*-value, is formalized by the expression [9]:

$$\frac{[D_{\text{total}}]-[D_{\text{water}}]}{[\text{lipid}]} = R \quad \text{value}$$

where $[D_{\text{total}}]$ and $[D_{\text{water}}]$ are the total detergent concentration and the detergent concentration in the water phase, respectively. In detergent/liposome mixtures, $[D_{\text{water}}]$ is the number with most uncertainty, but is approximated by $[D_{\text{cmc}}]$. At high *R*-values where co-micellization takes place, liposomes are completely solubilized resulting in the loss of size information of the liposome after detergent removal. By adding non-solubilizing amounts of detergent (i.e., low *R*-values at or very near the onset of solubilization [15]), the goal is to directly incorporate transmembrane proteins into liposomes, thus allowing better control of the size. At detergent levels in between those required for the onset and total solubilization, detergent micelles will partially solubilize liposomes and give rise to proteoliposomes with hybrid characteristics

2 Materials

2.1 Reagents

1. Chloroform:methanol mix: stock of 2:1 (v/v) chloroform: methanol stored in the dark.
2. Diethyl ether (HPLC grade).
3. Brain porcine purified lipids: phosphatidylcholine (PC), phosphatidylethanolamine (PE), and phosphatidylserine (PS) (powder form).
4. Cholesterol from ovine wool (powder form).
5. Ultra-high purity grade Nitrogen gas.

6. Liposome Reconstitution Buffer: 20 mM HEPES, 150 mM KCl, 1 mM DTT, pH 7.4; 0.5 mL.
7. RHO-PE Solution: 1 mg/mL RHO-PE (1,2-dioleoyl-*sn*-glycero-3-phosphoethanolamine-N-(lissamine rhodamine B sulfonyl) in chloroform.
8. NBD-PE Solution: 1 mg/mL NBD-PE (1,2-dioleoyl-*sn*-glycero-3-phosphoethanolamine-N-(7-nitro-2-1,3-benzoxadiazol-4-yl), in chloroform.
9. OG Solution: 300 mM n-octyl-β-D-glucoside in Liposome Reconstitution Buffer that can be kept at 4 °C for 1–2 months.
10. Bio-Beads SM-2 (Bio Rad).

2.2 Instrumentation

1. Standard rotary evaporator connected to vacuum pump with pressure controller.
2. 100 mL pear-shaped flask (or 50 mL if unavailable).
3. Hamilton syringes for handling different volumes of lipids.
4. Tip sonicator (e.g., Branson Sonifier).
5. Mini-Extruder Kit (e.g., Avanti Polar Lipids) with 1 mL gas-tight syringes.
6. Polycarbonate membrane (19 mm) of pore size 0.1 and 0.4 μm and filter supports for Mini-Extruder.
7. Dialysis cassette (0.5 mL, MW cutoff 2 kDa, e.g., Slide-A-Lyzer) with float buoy.
8. Dialysis cassette syringes (1 mL) with 18-gauge, 1-in. beveled needles.

3 Methods

The protocol is divided into three parts: (1) preparation of 1 mL multi-lamellar liposomes via reverse phase evaporation with a total lipid concentration of 8 mM, (2) extrusion to produce unilamellar (large) liposomes with nominal diameter of 100 nm, and then (3) incubation and exposure of large liposomes to a micellar solution of SNAREs followed by dialysis to form large SNARE-liposomes. Because of lipid loss occurring mainly during extrusion, the actual total lipid concentration of the large liposome suspension in **step 2** is reduced to ~6 mM if the protocol is followed as described. A fraction of these liposomes will then be used to prepare large SNARE-liposomes using a micellar solution of syb as an example (0.5 mL). The protocol can be upscaled or downscaled accordingly, but because of potential losses of sample volume during extrusion and evaporation, we recommend preparing at least a 0.5 mL suspension of large liposomes and no less than 0.2 mL of SNARE-liposomes.

3.1 Preparation of Large Liposomes by Reverse Phase Evaporation

1. Prepare stock solutions (25 or 10 mg/mL) of each lipid (PC, PE, PS, and Chol) in 2:1 (v/v) chloroform:methanol and store in sealed glass vials purged with N_2 at −20 °C until further use (*see* **Note 1**). For the first preparation of fluorophore-labeled liposomes, transfer chloroform solutions of NBD-PE or RHO-PE as shipped from the manufacturer directly into sealed glass vials.
2. Using Hamilton syringes, mix appropriate volumes of each lipid so that PC:PE:PS:Chol are in the molar ratios 5:2:2:1 for unlabeled liposomes (Table 1) into a pear-shaped flask previously purged with N_2 (*see* **Note 2**). If using NBD/RHO-labeled liposomes for lipid-mixing experiments, mix PC:PE:PS:Chol:NBD-PE:RHO-PE in ratios of 5:1.7:2:1:0.15:0.15 (Table 2).
3. Connect pear-shaped flask into a rotary evaporator and evaporate solvent by decreasing the pressure step-wise from 500 to 20 mbar done over the course of ~40 min (this can be done manually or automated depending on the availability of a controller). We recommend a rotation speed of ~200 rpm. The resulting lipid film should be homogeneously deposited throughout the bottom of the pear-shaped flask. Purge flask with N_2 before removing it from the rotary evaporator.
4. Using a Hamilton syringe, thoroughly redissolve lipid film in the pear-shaped flask with diethyl ether (1.5 mL). Immediately close flask to minimize exposure to air.
5. Add 0.5 mL of Liposome Reconstitution Buffer, noting the separation of the aqueous/organic phases.
6. Using a thin sonicator tip that easily reaches into the bottom of the flask, disperse the two-phase mixture with low-energy pulses (3×45 s, 50% duty cycle) with ~1 min ice-cooling periods (*see* **Note 3**). The resulting dispersion should be cloudy and stable (i.e., should not partition again after a few minutes).

Table 1
Example of volumes with corresponding concentrations used for preparing a lipid mix for making 1 mL of an unlabeled liposome suspension with nominal total lipid concentration of 8 mM

mol % (n/n)	Lipid	MW (g/mol)	Lipid stock (mg/mL)	Lipid stock (mM)	Lipid amount (μmol)	Stock volume (μL)
50	PC	760.1	25	32.9	4.00	121.6
20	PE	746.1	10	13.4	1.60	119.4
20	PS	812.1	10	12.3	1.60	129.9
10	Chol	386.7	10	25.9	0.80	30.9
0	NBD-PE	924.2	1	1.1	0.00	110.9
0	RHD-PE	1301.7	1	0.8	0.00	156.2

Table 2
Example of volumes with corresponding concentrations used for preparing a lipid mix for making 1 mL of an NBD/RHO-labeled liposome suspension with a nominal total lipid concentration of 8 mM

mol % (n/n)	Lipid	MW (g/mol)	Lipid stock (mg/mL)	Lipid stock (mM)	Lipid amount (μmol)	Stock volume (μL)
50	PC	760.1	25	32.9	4.00	121.6
17	PE	746.1	10	13.4	1.36	101.5
20	PS	812.1	10	12.3	1.60	129.9
10	Chol	386.7	10	25.9	0.80	30.9
1.5	NBD-PE	924.2	1	1.1	0.12	110.9
1.5	RHD-PE	1301.7	1	0.8	0.12	156.2

7. Connect flask back to the rotary evaporator. Immediately decrease the pressure to 500 mbar and leave for ~10 min. In increments of ~50–100 mbar, continue to gradually decrease the pressure step-wise during a ~40 min time span to 150 mbar. During evaporation, a gel phase is formed which coalesces into an aqueous suspension (*see* **Note 4**).
8. Once a clearly less turbid aqueous suspension is visible indicating the end of the gel phase, directly decrease pressure to 20–30 mbar to remove residual amounts of diethyl ether (10–15 min).
9. Using a Gilson-type pipette, transfer suspension to a 15 mL conical tube and make note of amount of water lost during evaporation (usually 50–100 μL). Make up volume to 0.5 mL with ultra-pure water and add another 0.5 mL to make a total suspension of 1 mL. The nominal total lipid concentration is now 8 mM. Store the liposome suspension temporarily at room temperature.

3.2 Extrusion of Aqueous Suspension to Form 100 nm Unilamellar Liposomes

1. Assemble Mini-Extruder according to the manufacturer's instructions using a polycarbonate membrane of 0.4 μm pore size and one filter support on each side of the membrane.
2. Clean and rinse 1 mL Hamilton gas-tight syringes with distilled water followed by Liposome Reconstitution Buffer. Using one syringe, fill the extruder chamber from one side with liposome reconstitution buffer and leave syringe inserted in the chamber.
3. Insert the liposome suspension with the remaining syringe into the chamber and extrude with 25 passes.
4. Remove both syringes keeping the liposome suspension inside one of the syringes (*see* **Note 5**).

5. Disassemble chamber and reassemble with a polycarbonate membrane of 0.1 μm pore size. Reuse the filter supports used in the first extrusion and using the empty syringe fill the chamber with liposome reconstitution buffer.
6. Insert the suspension-containing syringe on the opposite side of the chamber and extrude with 25 passes.
7. Remove liposomes and store in an Eppendorf-type tube at room temperature until used in the next step. Liposomes prepared in this way and stored at 4 °C retain their size of 90–120 nm for at least 3 weeks, although it is recommended to use them for the reconstitution of SNAREs within 2–3 days (*see* **Note 6**).

3.3 Reconstitution of SNAREs into Large Liposomes

The exact amount of reagents will depend on the desired final volume of SNARE-liposomes, the SNARE density on the surface of the liposome (the lipid/protein mole ratio, or l/p), and the concentration of the SNARE stock available, which varies from purification to purification. In this specific example, the preparation of 500 μL of syb liposomes with a l/p = 500:1 is described using a stock syb concentration of 77 μM. A 1 mL syb stock at this concentration allows ~19 liposome preparations as described below.

1. Express recombinant neuronal His-6-tagged SNAREs (from rat) in *E. coli* and purify using nickel-nitrilotriacetic acid (Ni-NTA) agarose chromatography followed by ion-exchange chromatography, as previously described in detail [10, 11]. We recommend that the final concentration of the protein stock be at least 60 μM. For the case of the preassembled syx/SN-25 binary complex (with or without a stabilizing C-terminal syb49-96 peptide [12]), the complex can be either coexpressed [13] or assembled from individual SNAREs [12]. Regardless of the specific SNAREs and or binary complexes used, the final ion-exchange purification step must be done in the presence of 34 mM n-octyl-β-D-glucoside (OG).
2. Amounts of liposomes, OG, syb, and extra buffer were calculated so that when all four components are mixed the final total lipid concentration is 4 mM (assuming total lipid concentration of liposome stock is 6 mM, *see* **Note 6**). Input variables used for the mixture calculation are shown in Table 3, while calculations of volumes of the four components are depicted in Table 4. Based on this, large liposomes (333 μL) are added to an Eppendorf-type tube followed by Liposome Reconstitution Buffer (78 μL).
3. To calculate the amount of OG and syb to add to the liposomes, the final molar ratio between the excess OG *above* the critical micelle concentration (CMC = 17 mM) and the total lipid concentration is set to 2 (the R-value as defined in Fig. 2, *see* **Note 7**). The detergent concentration already present in the SNARE/OG solution was included as part of the total OG

Table 3
Example of input parameters required for calculating the detergent/lipid/protein mixture required for the reconstitution of syb into large liposomes (total volume of 500 μL). Variables that can be modified according to the user's specific needs are shaded in gray, whereas we strongly recommend the remaining variables be left unchanged

	Input values for preparation of syb large liposomes	
	Constants used for calculations	
A	Critical micelle concentration (CMC) of OG (mM)	17.0
B	*R*-value = (Total detergent concentration—CMC)/total lipid concentration	2.0
	Stock concentrations and target volumes	
C	Total lipid concentration of 100 nm liposome stock (mM)	6.0
D	Concentration of OG stock (mM)	300
E	Concentration of syb stock (μM)	77
F	Concentration of OG in protein stock (mM)	34
G	Target volume of SNARE-liposome (μL)	500
H	Target molar lipid:protein ratio (l/p)	500
I	Target total lipid concentration of liposomes (mM)	4.0

Table 4
Calculation of volumes of syb, liposomes, OG, and buffer required for the reconstitution of syb into large liposomes based on variables defined in Table 3. Relevant equations used to calculate values are shown on the left column, while volumes to be used for the final mixture are shaded gray

	Output values for OG/SNARE/lipid mixture for preparation of large syb liposomes	
J = G*I/C	Volume of liposomes to be added (μL)	333
K = I/(H*1000)	Final concentration of syb in mixture (μM)	8.0
L = (K/E)*G	Volume of protein to be added (μL)	52
M = I*B + A	Final concentration of OG in mixture (mM)	25.0
N = M*G/1000	Final OG amount in mixture (μmol)	12.5
O = E*L/1000	OG amount already included in protein stock (μmol)	1.77
P = [(N-O)/D]*1000	Volume of OG stock to add (μL)	36
Q = G-J-L-P	Extra buffer to add to adjust total volume (μL)	79

amount. Using these criteria for the calculations (*see* Tables 3 and 4 for a breakdown of the calculation), OG (36 μL) and syb (52 μL) were mixed in a separate Eppendorf-type tube.

4. Gently add the syb/OG mixture to the liposome suspension taking care to minimize production of air bubbles.
5. Following the manufacturer's instructions (Thermo Scientific), collect mixture with a small syringe and inject into one of the ports of a dialysis cassette (0.5 mL, MW cut-off 2 kDa). Remove excess air from the cassette as the needle is being removed and mark the entry port used.
6. Dialyze mixture overnight at room temperature in Liposome Reconstitution Buffer (1 L) by attaching a float buoy to the cassette. Optional: add SM BioBeads (Bio Rad, final concentration 2 g/L) to the buffer to enhance detergent removal efficiency.
7. On the next day, perform a buffer exchange and continue dialyzing for another 2–3 h at room temperature (Fig. 3).
8. Using a different port, remove syb-liposomes from the dialysis cassette and store them in an Eppendorf-type tube. SNARE-liposomes prepared in this way should be used immediately or kept on ice up to a week.

4 Notes

1. It is important that the sealing is tight to minimize oxidation of lipids. However, we noticed that even if we were very cautious with preventing oxidation, the gel phase during reverse phase evaporation would not coalesce into an aqueous suspension when lipid stocks had been kept for too long. Based on our experience, we strongly recommend not to use lipids that have been stored for more than 3 months at −20 °C. For this reason, it is convenient to purchase small batches of lipids in powder form which can be requested from the supplier.
2. For a more precise preparation and to minimize cross-contamination of lipids, we recommend flushing out any lipid present in the needle's dead volume into the pear-shaped flask at least three times with chloroform:methanol, thus preventing that any lipid solution is lost during the washing step before moving onto the next lipid.
3. It is recommended that this step is first tried without a sample using only water and diethyl ether. Here, it is important that the intensity of the pulse is tested so that the dispersion does not splash into the flask's walls. Usually, the minimum operating intensity is used, but this will depend on the particular type of sonicator available.

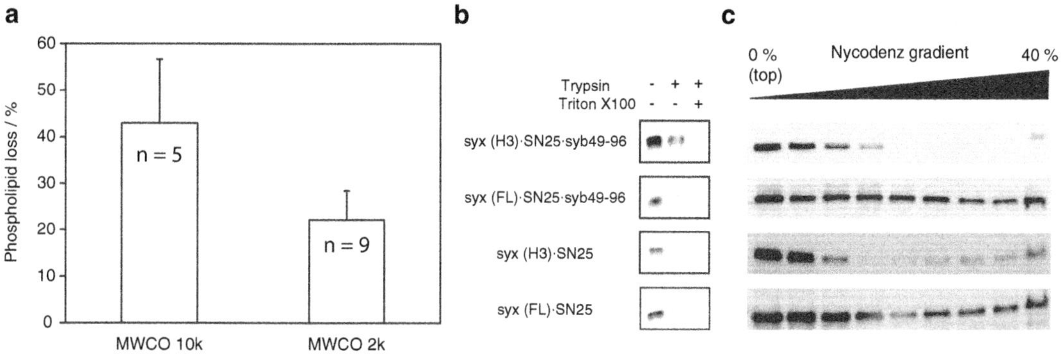

Fig. 3 Further biophysical characterization of large SNARE-liposomes. (**a**) We tested the lipid loss by measuring phosphate content occurring during the overnight dialysis of OG for the final step in the reconstitution of SNAREs into large liposomes. As can be appreciated, a more reproducible preparation with reduced lipid loss can be achieved by using a dialysis cassette with a molecular weight cutoff (MWCO) of 2 kDa. (**b**) Orientation assessment of the reconstitution of four different types of syx:SN25 binary complexes. Complexes are distinguished by two different biochemical features: (1) either full-length syx (FL) or a truncated version of syx (H3) without the N-terminal regulatory domain was used, and (2) a syb49-96 peptide included in the binary complex to stabilize it [12]. SNARE-liposomes were incubated with or without trypsin in the absence or presence of Triton X-100 and analyzed by Western blot and detection of SN25 (CL 71.1, Synaptic Systems). The assessment reveals that ~80–90 % of binary complexes are accessible to trypsin digestion which indicates a similar proportion of SNARES with inside-out orientation. *See* ref. 7 for further details on the procedure for carrying out the assessment. (**c**) Density centrifugation in a 0–40 % (w/v) Nycodenz gradient of large SNARE-liposomes containing the same syx:SN25 binary complexes as in (**b**). Liposomes were loaded at the bottom (40 % w/v) of the gradient; after centrifugation liposomes migrate to the top by virtue of the low density of lipids while unincorporated proteins remain at the bottom. Reconstituted binary complexes co-float with the liposomes at the top three to four fractions as evaluated by SDS-PAGE and Western blot analysis. Blots were immunostained for detection of SN25 (*see* **Note 8**)

4. During this critical step, a transient viscous gel forms that then transforms into an aqueous suspension of multi-lamellar liposomes. It is normal that bubbling is observed, but with practice this should be minimized by regulating the rate of the pressure decrease. For example, depending on the precision of the vacuum pump, it may be necessary to decrease the pressure in smaller incremental steps.
5. As the syringe containing the majority of the suspension is being withdrawn, slightly pull the syringe plunger to ensure removal of the residual amounts of liposomes that would otherwise be lost in the interior of the chamber.
6. Based on phosphate determination analysis of dozens of samples, the amount of lipid loss occurring during extrusion when performed as described is of the order of 20 %. Since the exact loss varies from preparation to preparation, we assumed a final total lipid concentration of 6 mM as a representative value after extrusion and was therefore used to calculate volumes in the next reconstitution step.

7. We observed that the chosen R-value during exposure of the liposomes to OG and SNAREs affects the final size distribution of the SNARE-liposomes (Fig. 4). When scanning R-values from 1 to 2.5, we noticed that none of the distributions of the SNARE-liposomes was identical to the size distribution of the liposomes before incubation with SNAREs (referred to as template in Fig. 4). This strongly argues against the idea that reconstitution occurs according to the direct reconstitution method in this system. Rather, it suggests a more complex mechanism of insertion may be occurring; for example, liposomes may partially solubilize resulting in a slight decrease in size, or alternatively, liposomes might increase in size due to spontaneous fusion between detergent-doped bilayers during exposure of liposomes to OG. Nevertheless, SNARE-liposomes still remain large and although the analysis reveals any R-value between 1.0 and 2.5

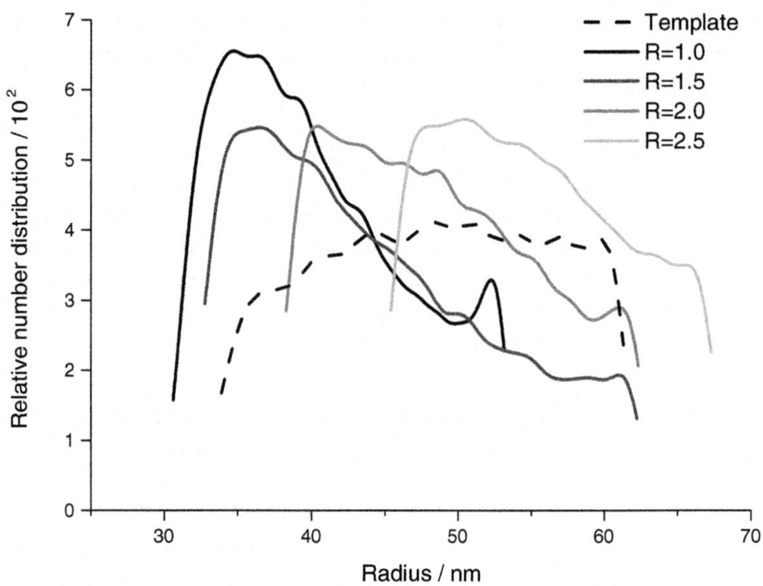

Fig. 4 Dependence of the size distribution of large syb-liposomes (l/p = 500:1) on the R-value used for reconstitution. Using FFF-MALLS to precisely characterize the relative number size distribution of the liposomes reconstituted with syb at the indicated *R*-values, we found that all distributions were different from the size distribution of the template liposomes (without SNAREs before exposure to OG), indicating that partial solubilization was occurring. Interestingly, the more OG was employed the larger the size of the resulting SNARE-liposomes. At *R* = 2.5 we observe that the size of the liposomes was even slightly larger than the template liposomes. One explanation for these observations is that there might be a balance between partial solubilization and spontaneous fusion of liposomes during OG exposure. Nevertheless, again these results stress the importance of carefully characterizing the preparation of SNARE-liposomes

would suffice to prepare large SNARE-liposomes, an R-value of 2.0 was chosen as a desirable OG concentration to more closely resemble the size of the template liposomes.

8. For reasons that are not yet understood, the analysis shows that binary complexes containing full length syx (FL) do not quantitatively incorporate into large liposomes (60–70% incorporation efficiency), highlighting the importance of characterizing the properties of SNARE-liposomes when variations are made. In this case, it would be recommended to perform an additional purification step (e.g., gel filtration) to remove unincorporated complexes from the preparation. Details for assessing incorporation efficiency can be found in reference [7].

Acknowledgments

I thank Reinhard Jahn and Petra Schwille for their constant support and encouragement, and Ioanna Bethani for critical comments. This work was supported by a grant from the Deutsche Forschungsgemeinschaft (SFB 803).

References

1. Sudhof TC, Rothman JE (2009) Membrane fusion: grappling with SNARE and SM proteins. Science 323(5913):474–477
2. Jahn R, Fasshauer D (2012) Molecular machines governing exocytosis of synaptic vesicles. Nature 490(7419):201–207. doi:10.1038/nature11320
3. Weber T, Zemelman BV, McNew JA et al (1998) SNAREpins: minimal machinery for membrane fusion. Cell 92(6):759–772
4. Ma C, Su L, Seven AB et al (2013) Reconstitution of the vital functions of Munc18 and Munc13 in neurotransmitter release. Science 339(6118):421–425. doi:10.1126/science.1230473
5. Lai Y, Diao JJ, Liu YX et al (2013) Fusion pore formation and expansion induced by Ca2+ and synaptotagmin 1. Proc Natl Acad Sci U S A 110(4):1333–1338. doi:10.1073/Pnas.1218818110
6. Rizo J, Chen XC, Arac D (2006) Unraveling the mechanisms of synaptotagmin and SNARE function in neurotransmitter release. Trends Cell Biol 16(7):339–350
7. Hernandez JM, Stein A, Behrmann E et al (2012) Membrane fusion intermediates via directional and full assembly of the SNARE complex. Science 336(6088):1581–1584. doi:10.1126/science.1221976
8. Hernandez JM, Kreutzberger AJ, Kiessling V et al (2014) Variable cooperativity in SNARE-mediated membrane fusion. Proc Natl Acad Sci U S A 111:12037–12042. doi:10.1073/pnas.1407435111
9. Rigaud JL, Levy D (2003) Reconstitution of membrane proteins into liposomes. Methods Enzymol 372:65–86
10. Fasshauer D, Antonin W, Margittai M et al (1999) Mixed and non-cognate SNARE complexes - characterization of assembly and biophysical properties. J Biol Chem 274(22):15440–15446
11. Fasshauer D, Margittai M (2004) A transient N-terminal interaction of SNAP-25 and syntaxin nucleates SNARE assembly. J Biol Chem 279(9):7613–7621
12. Pobbati AV, Stein A, Fasshauer D (2006) N- to C-terminal SNARE complex assembly promotes rapid membrane fusion. Science 313(5787):673–676
13. Stein A, Radhakrishnan A, Riedel D et al (2007) Synaptotagmin activates membrane fusion through a Ca2+-dependent trans interaction with phospholipids. Nat Struct Mol Biol 14:904–911. doi:10.1038/nsmb1305
14. Malinin VS, Lentz BR (2004) Energetics of vesicle fusion intermediates: comparison of calculations with observed effects of osmotic and curvature stresses. Biophys J 86(5):2951–2964
15. Lichtenberg D, Robson RJ, Dennis EA (1983) Solubilization of phospholipids by detergents - structural and kinetic aspects. Biochim Biophys Acta 737(2):285–304

Chapter 9

Isolation of Synaptosomes, Synaptic Plasma Membranes, and Synaptic Junctional Complexes

Mary L. Michaelis, Lei Jiang, and Elias K. Michaelis

Abstract

Isolation of synaptic nerve terminals or synaptosomes provides an opportunity to study the process of neurotransmission at many levels and with a variety of approaches. For example, structural features of the synaptic terminals and the organelles within them, such as synaptic vesicles and mitochondria, have been elucidated with electron microscopy. The postsynaptic membranes are joined to the presynaptic "active zone" of transmitter release through cell adhesion molecules and remain attached throughout the isolation of synaptosomes. These "post synaptic densities" or "PSDs" contain the receptors for the transmitters released from the nerve terminals and can easily be seen with electron microscopy. Biochemical and cell biological studies with synaptosomes have revealed which proteins and lipids are most actively involved in synaptic release of neurotransmitters. The functional properties of the nerve terminals, such as responses to depolarization and the uptake or release of signaling molecules, have also been characterized through the use of fluorescent dyes, tagged transmitters, and transporter substrates. In addition, isolated synaptosomes can serve as the starting material for the isolation of relatively pure synaptic plasma membranes (SPMs) that are devoid of organelles from the internal environment of the nerve terminal, such as mitochondria and synaptic vesicles. The isolated SPMs can reseal and form vesicular structures in which transport of ions such as sodium and calcium, as well as solutes such as neurotransmitters can be studied. The PSDs also remain associated with the presynaptic membranes during isolation of SPM fractions, making it possible to isolate the synaptic junctional complexes (SJCs) devoid of the rest of the plasma membranes of the nerve terminals and postsynaptic membrane components. Isolated SJCs can be used to identify the proteins that constitute this highly specialized region of neurons. In this chapter, we describe the steps involved in isolating synaptosomes, SPMs, and SJCs from brain so that these preparations can be used with new technological advances to address many as yet unanswered questions about the synapse and its remarkable activities in neuronal cell communication.

Key words Synaptosomes, Synaptic plasma membranes, Post synaptic densities, Synaptic junctional complexes, Ultracentrifugation, Synaptic membrane proteins

1 Introduction

Development of subcellular fractionation techniques with brain tissue has made it possible to obtain highly enriched preparations of synaptic terminals [1–13] and, in conjunction with numerous other technological advances, to answer many

Alexandros Poulopoulos (ed.), *Synapse Development: Methods and Protocols*, Methods in Molecular Biology, vol. 1538, DOI 10.1007/978-1-4939-6688-2_9, © Springer Science+Business Media LLC 2017

important questions about the cellular and molecular basis of neurotransmission. During homogenization of brain tissue in an iso-osmotic medium, the axon terminals, "synaptic boutons," are sheared off from the axons (Fig. 1). These nerve-ending particles or "synaptosomes" reseal into small vesicular structures containing all of the proteins and organelles that normally mediate the response to signals, such as depolarization, to bring about release of neurotransmitters stored in the synaptic vesicles [14–24]. Detailed studies of the composition of synaptosomes, including the capacity for protein synthesis [25] and identification of numerous synaptic proteins [26], have been performed.

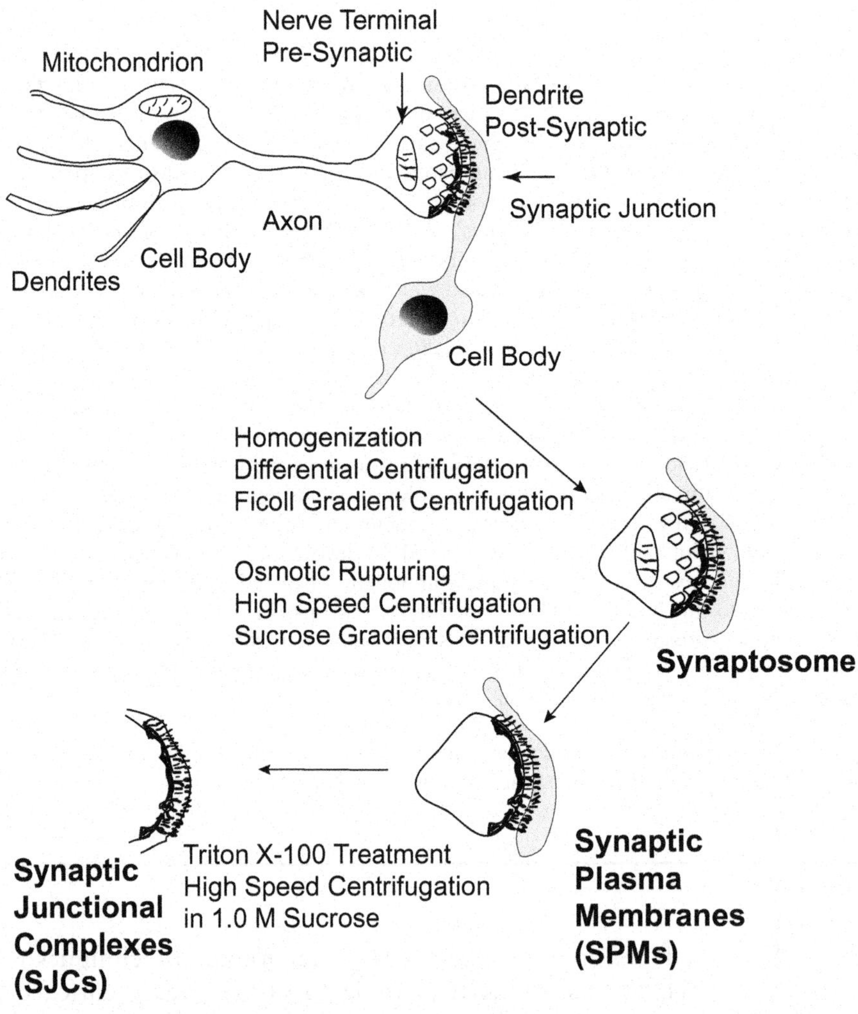

Fig. 1 Schematic representation of the origin of intact synaptosomes that can be isolated from the brain through a series of differential centrifugation steps. The synaptosomes can be osmotically ruptured and the synaptic plasma membranes recovered through density gradient centrifugation. And mild detergent treatment can be applied to solubilize the membranes on either side of the synaptic junction and the junctional complexes can be isolated with high speed centrifugation

It is estimated that a synaptosome has an average volume of ~0.4 μm^3 and contains ~380 synaptic vesicles and ~300,000 proteins, most of them involved in synaptic vesicle loading, exocytosis, and retrieval [26].

Synaptosomes can be lysed in hypotonic medium that leads to the release of most of the membranous intracellular organelles, such as mitochondria, synaptic vesicles, and endoplasmic reticulum. Differential centrifugation of the synaptosomal lysate yields a partially purified preparation of synaptic plasma membranes (SPMs). The SPMs consist of the plasma membranes of the presynaptic nerve terminals and of a fragment of the postsynaptic membranes that remains attached to the presynaptic "active zone" (Fig. 1). This partially purified preparation of SPMs can be recovered by centrifugation. The purity of the SPMs can be significantly increased by lysing the synaptosomes and subjecting the preparation to sucrose density gradient centrifugation to remove the remaining myelin, synaptic vesicle, endoplasmic reticulum, and mitochondrial components present in the partially purified fraction [1, 4, 5, 9, 13]. Importantly, the SPMs prepared in this manner retain the synaptic junctional complexes (SJCs), consisting of the presynaptic membrane at the "active zone" of transmitter release and the postsynaptic membrane and postsynaptic density (PSD) (Fig. 1). The copurification of pre- and postsynaptic plasma membranes is possible because postsynaptic membranes are tightly linked to presynaptic membranes through cell adhesion molecules, such as neurexins and neuroligins [27] that bridge the synaptic cleft. This renders the junction of pre- and postsynaptic membranes, including the PSDs, resistant to treatment with low concentrations of detergents such as Triton X-100 [1, 5–7, 11]. The postsynaptic plasma membrane and the PSD typically contain receptors for the released neurotransmitters, and tightly attached scaffolding proteins that hold the receptors in place or participate in signal transduction upon activation of receptors [5, 28–32].

The most widely used protocols for isolating synaptosomes, SPMs, and SJCs were developed by De Robertis, Whittaker, Cotman, Matthews, Salvaterra, Gurd, and Mahler [1, 5, 7–9, 13]. Based on these protocols, the overall strategy we typically follow in preparing these subfractions from brain is shown in Fig. 2 [33]. The major instruments required and the solutions that must be prepared are listed first, followed by the methods for isolation of synaptosomes, SPMs, and SJCs. The methods describe the subfractionation of tissue from 6 to 8 rat brains, but these can easily be scaled up for the preparation of synaptosomes from more rats or reduced for a smaller animal such as a mouse. The proportions of tissue-to-buffer volumes will remain essentially the same.

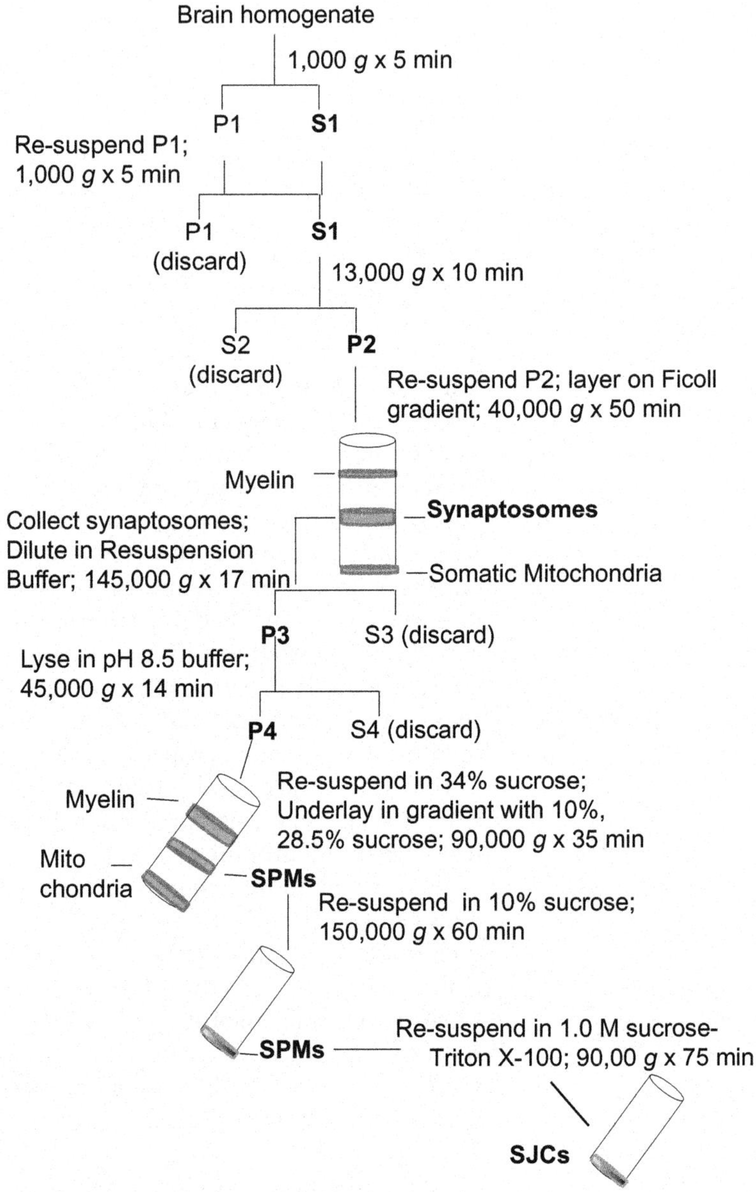

Fig. 2 Purification scheme for the various subcellular fractions that can be recovered from rat brain homogenates. Depending on the type of experiments to be done with the synaptic fractions, it is possible to stop with the recovery of the intact synaptosomes, with the synaptic plasma membranes, or go on to isolate the synaptic junctional complexes. Clearly, the amount of protein in the subfractions will decrease with the progression to the more purified components of the synaptic terminals

2 Materials

Subcellular fractionation of brain requires that specific types of centrifuges, rotors, and centrifuge tubes be available. These include a refrigerated preparative centrifuge that can create a centrifugal force of 40–50,000 × *g*'s, and a rotor that can accommodate tubes, that hold 20 mL or larger volumes (e.g., Beckman-Coulter Avanti JXN-26 centrifuge with a JA-20 rotor). An ultracentrifuge that can create 65,000–500,000 × *g*'s with a swinging bucket rotor for the Ficoll density gradients is needed to isolate intact synaptosomes (e.g., Beckman-Coulter Optima XE ultracentrifuge with an SW28 rotor, or Thermo-Sorvall WX ultracentrifuge with the same rotor). Isolation of SPMs and SJCs can be done in the same ultracentrifuge using a fixed angle rotor (e.g., Beckman-Coulter Type 70 Ti) that can create up to 500,000 × *g*'s.

All solutions used should be prepared in distilled/deionized water within 24 h of their use and maintained at 0–4 °C throughout all of the procedures.

1. Dissection instruments: small animal guillotine with sharp blade for decapitation, sharp, straight edge surgical scissors to cut through fur, small bone cutter to cut through bone on each side of head, curved forceps or tweezers with teeth to peel the skull back from the brain, curved forceps to lift the brain out of the skull, and small sharp scissors to mince the brain in a beaker.
2. Potter-Elvehjem glass homogenizer tubes of various volumes with Teflon pestles (0.13 mm clearance) driven by a motor with adjustable speeds between 1000 and 2000 rpm.
3. Homogenization Buffer: 0.32 M sucrose, 0.5 mM $MgSO_4$, 0.1 mM EGTA, 10 mM HEPES. Adjust pH to 7.4. Just before use, add benzamide, benzamidine hydrochloride (HCl), and 4-(2-aminoethyl)benzenesulfonyl fluoride-HCl (AEBSF) to the buffer to reach a final concentration of 0.1 mM for each. Also add a commercial protease inhibitor cocktail that contains pepstatin, bestatin, leupeptin, and aprotinin, as per the manufacturer's instructions.
4. Resuspension Buffer: 0.32 M sucrose, 10 μM EGTA. Adjust pH to 7.5 with 1 M Tris base. Just before use, add benzamide, benzamidine-HCl, and AEBSF to the buffer to reach a final concentration of 10 μM. Also add the protease inhibitor cocktail per instructions of the manufacturer.
5. Ficoll Density Gradient Solution: Dissolve Ficoll in Resuspension Buffer with all protease inhibitors. Prepare sufficient volumes of an 8 % w/v and a 14 % w/v Ficoll solution to have 10 mL of each solution for each gradient centrifuge tube.

To Proceed with Isolation of Synaptic Plasma Membranes:

6. Lysis Buffer: 3 mM Tris–HCl, pH 8.5. Add benzamide, benzamidine-HCl, AEBSF, and protease inhibitor cocktail at the same concentrations as in the Resuspension Buffer.
7. Sucrose Density Gradient Buffer: Prepare buffer containing 150 mM NaCl, 50 mM Tris–HCl, pH 7.5, and add sucrose to prepare solutions of 10, 28.5, and 34% w/v to be used in the discontinuous sucrose density gradients.
8. SPM Storage Buffer: 0.32 M sucrose, 50 μM $MgCl_2$, 10 mM Tris–HCl, pH 7.4. Add the same amount of benzamide, benzamidine-HCl, AEBSF, and protease inhibitor cocktail as added to the Ruspension Buffer.

To Proceed with Isolation of Synaptic Junctional Complexes:

9. SPM Resuspension Buffer for SJC Preparation: 1.0 M sucrose, 5 mM Tris–HCl, pH 7.4, plus the protease inhibitors at the same concentration as above.
10. SPM Detergent-Treatment Buffer: 1.0 M Sucrose, 5 mM Tris–HCl, protease inhibitors, plus Triton X-100 at a final amount equal to 2 mg Triton X-100 per mg of protein of resuspended SPMs. Triton X-100 is a viscous detergent with a density of 1.07 g/mL. To 10 mL of the 1.0 M Sucrose—Tris–HCl buffer solution described above, 1.0 mL of Triton X-100 should be added. This will be used as a stock solution for the treatment of SPMs prior to isolation of SJCs. Triton X-100 in this stock solution will have a density of 0.1 g/mL, or 0.1 mg/μL. Therefore, 20 μL of this solution should be added per mg of SPM protein in the SPM Resuspension Buffer for the preparation of SJCs.

3 Methods

3.1 Isolation of Synaptosomes (Fig. 2)

Animals should be handled according to the protocol approved by the institutional animal care and use committee (IACUC). The animals should be fully anesthetized by CO_2 inhalation, and the guillotine used for decapitation. Blood is rinsed away with ice-cold homogenization, buffer, and straight edge scissors are used to cut under the scalp at the posterior end of the head and deflect the scalp in the direction of the eyes. Using the tips of the surgical bone cutter, the bone surrounding the point of insertion of the spinal cord to the brain is cut laterally moving toward the side of each ear lobe. This allows for the skull to be opened and, using a forceps or tweezers with teeth, the skull is lifted in the direction of the nose so that the brain is revealed. The brain is rapidly scooped out of the skull by using a curved clamp and transferred to a beaker containing homogenization buffer and is kept on ice. The brains are usually obtained from six to eight rats for each preparation.

1. Using a small sharp scissors, mince the brain tissue in a small volume of the Homogenization Buffer in a beaker on ice. Transfer the tissue to a cold glass Potter-Elvehjem homogenization tube with a Teflon pestle (0.13 mm clearance) (*see* **Note 1**).
2. Homogenize the tissue in the Homogenization Buffer at a ratio of 10 mL Homogenization Buffer per gram of tissue weight. A rat brain weighs ~2 g, and thus ~20 mL of the Homogenization Buffer is added to the tube for 1 brain. Use 8–10 strokes up and down with the pestle driven at 1000–1500 rpm as the sheering action leads to complete homogenization. Intermittently put the homogenization tube into ice to prevent heating of the vessel as this can denature proteins. When homogenization is complete, measure and record the volume of the homogenate and remove a small aliquot for protein determination later on (*see* **Note 2**).
3. Transfer the homogenates to pre-cooled centrifuge tubes and centrifuge at 1000 × *g* for 5 min at 4 °C in a preparative centrifuge (e.g., Beckman-Coulter Avanti JXN-26) using a rotor such as the Beckman-Coulter rotor JA-20 to remove the nuclei and cell debris. Collect the supernatant from each tube and transfer to new centrifuge tubes pre-cooled at 0–4 °C.
4. Wash the pellet with ~2 mL of Homogenization Buffer by vortexing to resuspend the pellet. Centrifuge again at 1000 × *g* for 5 min at 4 °C and combine the supernatants from the two centrifugations.
5. Centrifuge the supernatants at 13,000 × *g* for 10 min at 4 °C using a preparative centrifuge and rotor as in **step 3** to obtain a crude mitochondrial fraction that contains free mitochondria and nerve ending particles, i.e., synaptosomes. Discard the supernatant.
6. Add 2 mL of the Resuspension Buffer to each pellet. Vortex and gently resuspend the pellet by using a Pasteur pipette to triturate the pellet of synaptosomes and free mitochondria (*see* **Note 3**). Measure and record the volume and save a small aliquot for protein determination. Transfer the resuspension to graduated cylinders and add more Resuspension Buffer to bring the suspensions to a volume of 17.5 mL per rat brain.
7. Prepare the Ficoll gradients (*see* **Note 4**). Add 10 mL of the 14 % Ficoll solution to the bottom of a polyallomer ultracentrifuge tube (38.5 mL capacity). Layer 10 mL of the 8 % Ficoll solution on the top of the 14 % Ficoll. Layer 17.5 mL of the resuspended crude mitochondrial sample on the top of the 8 % Ficoll. Add Resuspension Buffer as needed to balance the weight of each tube and to nearly fill the tubes for the centrifugation.

8. Perform discontinuous Ficoll density gradient centrifugation in an ultracentrifuge (Thermo Sorvall WX ultracentrifuge, or Beckman-Coulter Optima XE with a swinging bucket rotor such as SW28). The following parameters of centrifugation should be used: 63,000 × *g* for 50 min. Synaptosomes form a band at the interface of 8 and 14% Ficoll (Fig. 2).
9. Use a Pasteur pipette to aspirate the upper band and the liquid above it. Then collect the synaptosome band by introducing the tip of a Pasteur pipette near the wall of the tube at the interface of the two Ficoll layers, and transfer the synaptosomes to a tube for a fixed angle rotor (e.g., Beckman-Coulter Type 70 Ti).
10. Dilute the synaptosome fractions with 4 volumes of the Resuspension Buffer.
11. Centrifuge at 145,000 × *g* for 17 min at 4 °C (Beckman-Coulter Optima XE or Thermo Sorvall WX ultracentrifuge) to remove the Ficoll and obtain the isolated synaptosomes as the pellet.
12. Discard the supernatant and retain the pellet. If the goal is to obtain intact synaptosomes for specific types of assays, resuspend each pellet in 2 mL of the appropriate assay buffer and remove a small aliquot for protein determination. Use the fresh synaptosomes without freezing them.
13. If intact synaptosomes are not needed and the goal is to prepare SPMs, follow the methods outlined below (*see* **Note 5**).

3.2 Isolation of the SPMs

1. For the isolation of the partially purified SPMs, add ~0.5 mL of the Resuspension Buffer to either the fresh synaptosomes or to the frozen aliquots that have been rapidly thawed in a water bath at 37 °C. Then add 5 volumes of the hypotonic Lysis Buffer to each aliquot and transfer to a centrifuge tube. Invert the tubes several times to mix well. Allow the tubes to sit on ice for 20 min with intermittent gentle vortexing to allow for complete lysis of the synaptosomes.
2. Centrifuge the samples at 45,000 × *g* for 14 min at 4 °C in a fixed angle rotor (e.g., Type 70 Ti). Discard the supernatant and resuspend the pellet in a buffer appropriate for the type of studies you plan to perform with this partially purified SPM preparation.
3. In order to recover a more highly purified SPM fraction, resuspend each of the pellets from the above centrifugation step in 13 mL of the 34% sucrose buffer. This represents the resuspended, partially purified SPM fraction in 34% sucrose solution.
4. Prepare the discontinuous sucrose density gradient by first introducing 12 mL of the 10% sucrose solution at the bottom

of each centrifuge tube (38.5 mL capacity) for a fixed angle rotor (e.g., Beckman-Coulter Type 70 Ti). Using a long 18-gauge needle and syringe that reaches the bottom of the tube, slowly introduce 12 mL of the 28.5 % sucrose solution to the bottom of the tube under the 10 % sucrose to form a discontinuous density gradient of 10 and 28.5 % sucrose.

5. To complete formation of the sucrose density gradient, use a long needle to introduce the suspension of partially purified SPMs in 34 % sucrose to the bottom of the centrifuge tubes. This will generate the gradient of 10, 28.5, and 34 % sucrose.
6. Centrifuge the gradients at 90,000 ×*g* for 35 min in the Type 70 Ti rotor. In this flotation/sedimentation condition, myelin contaminants float to the interface between 10 and 28.5 % sucrose, SPMs float to the interface between 28.5 and 34 % sucrose, and the mitochondria sediment in the 34 % sucrose layer as a pellet at the bottom of the tube.
7. Using an 18-gauge needle and a syringe, remove the 10 % sucrose and the myelin band at the interface of 10 and 28.5 % sucrose, and then collect the band of SPMs at the interface of the 28.5 and 34 % sucrose.
8. Transfer the SPM sample to a clean tube for the same rotor and dilute the harvested membrane fraction with distilled/deionized H_2O to a final sucrose concentration of approximately 10 %. Subject the diluted SPM sample to centrifugation at 150,000 ×*g* for 60 min using the Type 70 Ti rotor.
9. If the SPM subfraction is to be used as a freshly prepared sample for the planned studies, then resuspend the pelleted, purified SPMs in the appropriate buffer for the conduct of such studies or in the SPM Storage Buffer. Remove and save a small aliquot for protein determination.
10. If the SPM subfraction is to be stored for later use, the SPM pellet should be resuspended in the SPM Storage Buffer at a protein concentration of approximately 8–10 mg/mL, divided into aliquots of 0.25–0.5 mL, snap frozen in liquid N_2, and stored at -80 °C (*see* **Note 6**).
11. If the SPM subfraction is to be used for the preparation of SJCs, then follow the procedure outlined below.

3.3 Isolation of Synaptic Junctional Complexes (SJCs)

1. Resuspend the SPM subfraction obtained as a pellet (according to methods described above) in a small volume (0.5 mL per centrifuged pellet) of the SPM Resuspension Buffer. The resuspended SPMs from all centrifuge tubes are combined and transferred to a beaker, and the protein concentration measured by standard protein measurement techniques (e.g., bicinchoninic acid [BCA] or microBCA procedure).

2. Based on the estimated total protein amount of SPMs that will be used for the preparation of SJCs, add the SPM Resuspension Buffer to dilute the protein to 0.1–0.2 mg/mL. To this diluted preparation of SPMs, add sufficient volume of SPM Detergent-Treatment Buffer to achieve a solution that has 1 mg of total SPM protein per 20 μL of the buffer. The addition should be made slowly while stirring the SPM suspension gently.
3. Continue stirring the diluted SPM preparation containing the SPM Detergent-Treatment Buffer gently while avoiding any foaming. Stir for 15 min at 4 °C.
4. Transfer the suspension of SPMs into centrifuge tubes for the ultracentrifuge rotor (e.g., Beckman-Coulter Type 70 Ti or Type 70.1 Ti if smaller total volumes are used). Top-off the membrane suspension with the SPM Resuspension Buffer. Fill the tubes to near the top and balance them.
5. Subject all membrane samples to centrifugation at 90,000 × *g* for 75 min. Carefully aspirate the supernatant. The pellet contains the SJCs. Resuspend the pellet at the bottom of each centrifuge tube using 0.5 mL of SPM Storage Buffer if the SJCs are to be stored, or of the buffer that will be needed for the planned studies if the membranes are to be used as fresh preparations. Remove a small aliquot of the SJC suspension for protein determination.

4 Notes

1. The brain tissue should be kept at 0–4 °C throughout the subfractionation procedure. Rotors, tubes, and centrifuges should be pre-cooled before recovering the brains. Keeping the tissue cold and adding the several protease inhibitors listed effectively prevents activation of proteases that are abundant in the brain. If the fractions being isolated will be used to study the activity of kinases or the presence of phosphorylated proteins, it is advisable to include a phosphatase inhibitor cocktail in the preparation steps as well.
2. It is advisable at each step in the isolation of the subfractions to keep track of the total volume of each fraction and to remove a small aliquot of the sample before moving on to the next step. This makes it possible to determine the degree of enrichment in synaptosome, synaptic membrane, and synaptic junctional complex proteins that has occurred through the various steps of subfractionation. Measure the protein concentration of each of the samples (mg/mL) and multiply that by the total volume (mL) of each tissue fraction so that it is possible to construct a table showing the total amount of protein in the

homogenate (100%) and then monitor the percent of the total recovered at each step leading to the more purified synaptic fractions. This information makes it possible to select one or more markers known to be present in the final subfraction and track its enrichment through all of the steps. The most widely used marker of synaptosomes and SPMs is the Na^+/K^+ ATPase (the Na^+ pump). The ATP-hydrolyzing activity of the pump can be readily assayed, and the specific activity of the enzyme per mg of protein should increase as the fractionation proceeds from the brain homogenate to the SPMs. A prominent marker of the SJCs is the postsynaptic density protein PSD95. Its enrichment can be tracked by immunoblot analysis or enzyme-linked immunosorbent assays (ELISA), adding the same μg of protein from each fraction to a lane of the SDS-PAGE gel or to the wells of a 96-well plate for the ELISA.

3. Isolation of synaptosomes, SPMs, and SJCs involves several steps in which brain subfractions are precipitated under high centrifugal force, and the pellets must be resuspended in fresh medium. The fractions must ultimately be homogeneous, with no chunks or particles. We have found that gently dislodging the pellet from the centrifuge tube by adding a small amount of the medium and mildly vortexing the tube makes it possible to transfer the sample to a small glass homogenizing tube. A Teflon homogenizer can then be used to hand homogenize the sample gently with the appropriate amount of medium on ice and ensure a homogeneous resuspension for the next step in the protocol. It is also possible to use a Pasteur pipette and gently triturate the sample up and down until it becomes homogeneous.
4. Ficoll rather than sucrose is used for preparing the discontinuous density gradient to isolate the synaptosomes because it preserves their structure. This is due to the fact that synaptosomes are three-dimensional particles filled with organelles and intracellular fluid. Ficoll has a much higher molecular mass (~400,000) than sucrose (~342). Therefore, to achieve the same density, a much lower molar concentration of Ficoll is needed than for sucrose. The large number of sucrose molecules tends to pull water from the synaptosomes and cause them to shrink. The resulting loss of osmotic pressure prevents the synaptosomes from functioning optimally for certain types of assays. An alternative procedure uses density gradients formed by Percoll rather than Ficoll to the same effect [12].
5. After isolating the synaptosomes it is possible to resuspend them in a small volume of SPM Storage Buffer, flash freeze them in liquid nitrogen, and store them at −80 °C. The samples can then be thawed later and used for the preparation of the SPMs. If intact synaptosomes are being prepared for functional

assays such as identifying modulators of neurotransmitter release [34], freshly prepared rather than previously frozen synaptosomes would be required.

6. Isolated SPMs, can be frozen as noted above and thawed for measuring the uptake of transmitters or of ions into resealed SPM vesicles [22, 35]. In order to promote the resealing of synaptic membrane vesicles for measuring transport activity, the frozen SPMs must be thawed very rapidly at 37 °C for 10 min in 6 volumes of the internal medium needed for the transport assay.

References

1. De Robertis E et al (1961) On the isolation of nerve endings and synaptic vesicles. J Biophys Biochem Cytol 9:229–235
2. De Robertis E (1967) Ultrastructure and cytochemistry of the synaptic region. The macromolecular components involved in nerve transmission are being studied. Science 156(3777):907–914
3. Lapetina EG, Soto EF, de Robertis E (1967) Gangliosides and acetylcholinesterase in isolated membranes of the rat-brain cortex. Biochim Biophys Acta 135(1):33–43
4. Cotman CW, Matthews DA (1971) Synaptic plasma membranes from rat brain synaptosomes: isolation and partial characterization. Biochim Biophys Acta 249(2):380–394
5. Salvaterra PM, Matthews DA (1980) Isolation of rat brain subcellular fraction enriched in putative neurotransmitter receptors and synaptic junctions. Neurochem Res 5(2):181–195
6. Cotman CW et al (1974) Isolation of postsynaptic densities from rat brain. J Cell Biol 63(2 Pt 1):441–455
7. Cotman CW et al (1971) An ultrastructural and chemical analysis of the effect of triton X-100 on synaptic plasma membranes. Biochim Biophys Acta 249(2):406–418
8. Cotman CW, Taylor D (1972) Isolation and structural studies on synaptic complexes from rat brain. J Cell Biol 55(3):696–711
9. Gurd JW et al (1974) Isolation and partial characterization of rat brain synaptic plasma membranes. J Neurochem 22(2):281–290
10. Therien H, Mushynski W (1976) Isolation of synaptic junctional complexes of high structural integrity from rat brain. J Cell Biol 71(3):807–822
11. Carlin RK et al (1980) Isolation and characterization of postsynaptic densities from various brain regions: enrichment of different types of postsynaptic densities. J Cell Biol 86(3): 831–845
12. Dunkley PR et al (1986) A rapid method for isolation of synaptosomes on percoll gradients. Brain Res 372(1):115–129
13. Gray E, Whittaker V (1962) The isolation of nerve endings from brain: an electron microscopic study of cell fragments derived by homogenization and centrifugation. J Anat 96(Pt 1):79
14. De Lores Arnaiz GR, Alberici M, De Robertis E (1967) Ultrastructural and enzymic studies of cholinergic and non-cholinergic synaptic membranes isolated from brain cortex*. J Neurochem 14(2):215–225
15. Azcurra JM, De Robertis E (1967) Binding of dimethyl-C^{14}-d-tubocurarine, methyl-C^{14}-hexamethonium, and H^3-alloferine by isolated synaptic membranes of brain cortex. Int J Neuropharmacol 6(1):15–IN1
16. De Robertis E, Fiszer S, Soto EF (1967) Cholinergic binding capacity of proteolipids from isolated nerve-ending membranes. Science 158(3803):928–929
17. Söllner T et al (1993) A protein assembly-disassembly pathway in vitro that may correspond to sequential steps of synaptic vesicle docking, activation, and fusion. Cell 75(3):409–418
18. Delorenzo RJ et al (1979) Stimulation of Ca^{2+}-dependent neurotransmitter release and presynaptic nerve terminal protein phosphorylation by calmodulin and a calmodulin-like protein isolated from synaptic vesicles. Proc Natl Acad Sci 76(4):1838–1842
19. Young AB, Snyder SH (1974) Gamma-aminobutyric acid binding to receptor sites in the rat central nervous system. Proc Natl Acad Sci 71(12):4802–4807
20. Nicholls DG (1989) Release of glutamate, aspartate, and γ-aminobutyric acid from isolated nerve terminals. J Neurochem 52(2):331–341
21. Enna S, Snyder SH (1975) Properties of γ-aminobutyric acid (GABA) receptor binding

in rat brain synaptic membrane fractions. Brain Res 100(1):81–97

22. Kanner BI, Sharon I (1978) Active transport of L-glutamate by membrane vesicles isolated from rat brain. Biochemistry 17(19):3949–3953
23. Danbolt NC, Storm-Mathisen J, Kanner BI (1992) An [Na^+ + K^+]coupledl-glutamate transporter purified from rat brain is located in glial cell processes. Neuroscience 51(2): 295–310
24. Wang Y-J, Mahler HR (1976) Topography of the synaptosomal membrane. J Cell Biol 71(2):639–658
25. Cotman CW, Taylor DA (1971) Autoradiographic analysis of protein synthesis in synaptosomal fractions. Brain Res 29(2):366–372
26. Wilhelm BG et al (2014) Composition of isolated synaptic boutons reveals the amounts of vesicle trafficking proteins. Science 344(6187):1023–1028
27. Tsetsenis T et al (2014) Direct visualization of trans-synaptic neurexin-neuroligin interactions during synapse formation. J Neurosci 34(45):15083–15096
28. Kornau H-C et al (1995) Domain interaction between NMDA receptor subunits and the postsynaptic density protein PSD-95. Science 269(5231):1737–1740
29. Naisbitt S et al (1999) Shank, a novel family of postsynaptic density proteins that binds to the NMDA receptor/PSD-95/GKAP complex and cortactin. Neuron 23(3):569–582
30. Ehrlich I, Malinow R (2004) Postsynaptic density 95 controls AMPA receptor incorporation during long-term potentiation and experience-driven synaptic plasticity. J Neurosci 24(4):916–927
31. Srivastava S et al (1998) Novel anchorage of GluR2/3 to the postsynaptic density by the AMPA receptor–binding protein ABP. Neuron 21(3):581–591
32. Kennedy MB (2000) Signal-processing machines at the postsynaptic density. Science 290(5492):750–754
33. Michaelis EK et al (1983) High affinity Ca^{2+}-stimulated Mg^{2+}-dependent ATPase in rat brain synaptosomes, synaptic membranes, and microsomes. J Biol Chem 258(10):6101–6108
34. Michaelis ML, Michaelis EK, Myers SL (1979) Adenosine modulation of synaptosomal dopamine release. Life Sci 24(22):2083–2092
35. Michaelis ML, Michaelis EK (1981) Ca^{++} fluxes in resealed synaptic plasma membrane vesicles. Life Sci 28(1):37–45

Chapter 10

Purification of Synaptosome Populations Using Fluorescence-Activated Synaptosome Sorting

Elisa Luquet*, Christoph Biesemann*, Annie Munier, and Etienne Herzog

Abstract

For several decades, neurobiologists have used subcellular fractionation methods to analyze the molecular structure and some functional features of the cells in the central nervous system. Indeed, brain tissue contains a complex intermingled network of neuronal, glial, and vascular cells. To reduce this complexity biochemists have optimized fractionation protocols that enrich in specific compartments such as synapses (called "synaptosomes") and synaptic vesicles, for example. However, recently, these approaches suffered from a lack of specificity and purity. In a recent effort, we extended the conventional synaptosome preparation to purify fluorescent synaptosomes on a cell sorter. We could prove that our method allows for the steep enrichment in fluorescent excitatory VGLUT1^{venus} synaptosomes containing the presynaptic element and the tip of the post-synaptic element and a strong depletion in neuronal and glial contaminants. Here, we propose a detailed procedure for the implementation of Fluorescence Activated Synaptosome Sorting.

Key words Synaptosome, Fluorescence activated synaptosome sorting, Synapse, Subcellular fractionation, Cytometry, Micro-particle sorting, VGLUT1, Glutamatergic neurotransmission

1 Introduction

The brain contains a complex meshwork of neurons, glial cells, and blood vessels. Within this meshwork, synapses are micro-structures (0.5–2 μm most of the time) formed between neurons. The daunting complexity of this tissue presents a challenge for research on synapse structure and function. Therefore, numerous subcellular fractionation approaches were developed to enrich isolated synaptic particles (called synaptosomes) and synaptosome-derived subsynaptic compartments [1–6]. Synaptosomes are functional synaptic particles consisting of a resealed presynaptic compartment and part of the postsynaptic element [7, 8], with which a wealth of knowledge about synapse structure, composition, and function has

*These authors contributed equally.

Alexandros Poulopoulos (ed.), *Synapse Development: Methods and Protocols*, Methods in Molecular Biology, vol. 1538,
DOI 10.1007/978-1-4939-6688-2_10, © Springer Science+Business Media LLC 2017

been generated [1, 5, 9]. Also, generic proteomes of synaptic structures were produced [5, 6, 10–12]. All these previous studies provide precious resources about the average composition of synapses and corresponding substructures from the brain.

A major limitation of the conventional synaptosome-based preparations used so far is that they contain a mixture of synapse types (49.1 % of all particles) and many neuronal and non-neuronal contaminations (50.9 % of all particles) [3, 13, 14]. This lack of specificity and the major contamination by non-synaptic compartments confound the interpretation of synaptosome-derived data and limit the identification of new synapse-specific proteins by proteomics. Current neurobiological research requires the highest possible specificity regarding functional synaptic subpopulations. To circumvent these issues effort has been made to dissect molecular scaffolds isolated through co-immunoprecipitation assays. Indeed, contaminations are partly removed in this manner but the output material is limited to medium sized (1000–5000 kD) macromolecular scaffolds [15], while the resolution of isolated complexes is limited by regional tissue dissections combining several synaptic origins within the explants [16]. Others combined synaptosome immunolabeling with flow cytometry; however, uncertainties regarding the efficiency of these methods remain, and they remain to be successfully implemented on unfixed samples [17–19].

To reach specificity and significantly higher purity of unfixed synaptosome samples, we established the first protocol for Fluorescence Activated Synaptosome Sorting (FASS) of subpopulations of synapses [20]. We targeted VGLUT1 expressing synapses using our knock-in mouse line that expresses a fully functional fluorescent VGLUT1^{VENUS} protein instead of wild type (WT) VGLUT1 [21]. Our FASS protocol represents a novel approach to purify specific synapses to near homogeneity. Purified material can be studied by immunofluorescence and electron microscopy, immuno-blotting, and proteomic techniques. Further, FASS allows high-resolution separation of synaptic and extra-synaptic components of neurons and the identification of novel components of VGLUT1-containing synapses [20].

While one can find the general principles, outline, and flowchart of the FASS approach in our original publication [20], we propose here to detail all elementary steps necessary to sort and collect a given synapse population (VGLUT1^{venus} from mouse forebrain as an example) including recent optimizations to the original protocol. Beyond the efficient sorting of VGLUT1 synapses, extensions of FASS to other populations of synapses and other types of labeling approaches can be envisioned. In addition, our basic approach may turn out to be useful to other microparticle/organelle sorting projects.

2 Materials

2.1 Synaptosome Preparation

1. Dissection tools, including curved thin tweezers and razor blades.
2. Drill type homogenizer (i.e., Heidolph RZR 2102 control).
3. Glass/Teflon potter with a capacity of 2 cm^3.
4. Plate shaker.
5. Centrifuge with fixed angle rotor (equivalent to Sorval SS-34 or Beckmann JA17).
6. Ultracentrifuge with swing-bucket rotor (i.e., Beckman-Coulter Optima LE80K ultracentrifuge with SW40Ti rotor or Thermo WX Ultra 90 ultracentrifuge with TH 641 rotor).
7. Ultra-clear ultracentrifuge tubes for your swinging bucket rotor (i.e., Beckman Coulter 344059 that fit in SW40Ti; *see* Fig. 1).
8. Isosmolar Buffer: 0.32 M sucrose 4 mM HEPES pH 7.4 (40 mL per prep).

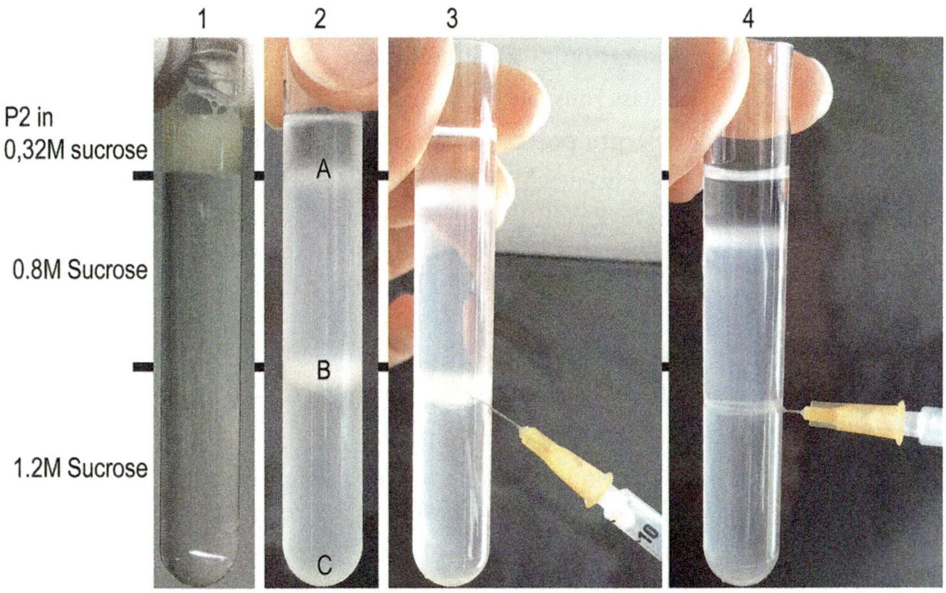

Fig. 1 Sucrose gradient handling for final synaptosome fractionation. (*1*): P2 resuspension in 0.32 M sucrose is layered on the sucrose gradient. (*2*): After ultracentrifugation, three fractions *A*, *B*, and *C* are clearly visible, the *A* fraction is prominently populated by myelin sheath fragments, the *B* fraction is the synaptosome-rich fraction, and the *C* pellet contains larger debris and free mitochondria as described earlier. (*3*): the *B* fraction is collected by inserting a 27-gauge needle armed 1 mL syringe from the side of the tube below the *B* fraction. During collection the level is descending and one should adjust the needle orientation to fully grasp the bulk of the *B* interface. (*4*): at the end of the procedure, the *B* interface is nearly cleared and no contamination has been added from the top myelin rich layer

9. 0.8 M Sucrose Solution: 0.8 M sucrose 4 mM HEPES pH 7.4 (25 mL per prep).
10. 1.2 M Sucrose Solution: 1.2 M sucrose 4 mM HEPES pH 7.4 (25 mL per prep).
11. FM4-64 from Interchim (FP-41109C) or any other provider.
12. Protease Inhibitors: Protease inhibitor cocktail 80 μM Aprotinin, 5 mM Bepstatin, 1.5 mM E-64, 2 mM Leupeptin, 1 mM Pepstatin A. If using commercial protease inhibitor cocktail, ensure it is compatible with downstream applications. Check with your proteomics facility for a cocktail of inhibitors that is compatible with mass spectrometry (i.e., without AEBSF, such as Halt Protease and Phosphatase Inhibitor Cocktail from Thermo Scientific).

2.2 FACS Analyses and Sorting Steps

1. 5 mL Polystyrene Round-Bottom Tube 12 × 75 mm style (i.e., Falcon #352058 by BD Labware).
2. 15 mL Centrifuge tubes (i.e., Sarstedt #62.554.502, *see* **Note 1**).
3. FACSAria cell sorter from Becton-Dickinson Inc. (*see* **Note 2**) with sample collection cooling unit.
4. Sheath fluid (i.e., BD FACSFlow ref: 342003).
5. Filtration of sorted samples.
6. Glass reusable filtration unit and rotary vane vacuum pump.
7. 0.1 μm polycarbonate filter (i.e., Isopore™ polycarbonate filter, Merck-Milipore VCTP02500).
8. SDS-PAGE loading buffer: 2.5 % SDS, 0.002 % Bromophenol Blue, 0.7135 M (5 %) β-mercaptoethanol, 10 % glycerol, 62.5 mM Tris with HCl to pH 6.8.

2.3 Synaptosome Fixation on Coverslips

1. coverslip holder.
2. 12 mm cover slips.
3. 24-well flat bottom plates.
4. Centrifuge with microplate swing rotor (i.e., Beckman-Coulter J-26 XP centrifuge with JS 5.3 rotor).
5. Chrome Alum Gelatin Slide Coating: 1 % Gelatin, 1 % Chromium potassium sulfate, in distilled water.
6. PFA Sucrose Buffer: 4 % paraformaldehyde, 4 % sucrose in PBS.

3 Methods

On the day preceding the experiment, prepare all materials and buffers. Buffers should be stored at 4 °C. If you are unsure of their cleanness, wash the required glass/Teflon potter(s) by rinsing three times with distilled water, brushing three times with distilled water, brushing three times with 70% ethanol, rinsing three times with 70% ethanol, and drying overnight. Next morning, brush the potter(s) three times with acetone, rinse three times with milliQ water, let it dry, store on ice until use, add protease inhibitors to Isosmolar Buffer and PBS (volume needed to do synaptosome dilution, typically 50 mL for two forebrain samples).

3.1 Brain Tissue Dissection

1. Organize four areas on the bench to sacrifice the mouse, extract the brain, dissect regions of interest, and homogenize tissue samples.
2. Sacrifice the first mouse by cervical dislocation (follow the local regulatory guidelines regarding ethics on animal handling). Decapitate and extract the brain from the skull.
3. Clean and cool down the brain by placing it in ice-cold PBS (without protease inhibitors). Cut a piece of mouse-tail to secure a genotyping biopsy backup and freeze it at −20 °C. While working in the cold room is advisable but not mandatory, all subsequent steps must be performed on ice.
4. Dissect out the regions of interest. For forebrain dissection proceed as follows. Using a sharp razor blade cut and discard olfactory bulbs, brain stem and cerebellum. Split the two brain hemispheres, flip them over with cortices facing down, and pinch out thalamus, colliculi, and white matter using curved tweezers. Cut forebrains in three or four pieces and store in a weighing dish on ice.
5. Repeat steps in Subheading 3.1 for each mouse. We routinely performed our FASS with one mouse forebrain (250 mg approximately), we recently used smaller brain regions and downscaled to roughly 70 mg for each sample (*see* **Note 3**).

3.2 Tissue Homogenization

1. Weigh each sample if you need to measure the yield of your fractionation. Introduce tissue samples into a 2 cm^3 clean ice-cold glass/Teflon potter with tweezers. Add 2 mL of ice cold Isosmolar Buffer into the potter, and connect to the homogenizer rotor. Homogenize tissue with 12 strokes at 900 rpm. Keep the potter ice cold as much as possible during homogenization.
2. Collect the homogenate in a 15 mL falcon tube. Add 2.5 mL of ice cold Isosmolar Buffer in the glass potter. Rinse potter with three strokes at 900 rpm and add the wash to the homogenate falcon tube. Further rinse the Teflon potter by running 3 mL ice cold Isosmolar Buffer and letting it collect in the glass tube and add it to the homogenate falcon tube. Mix the homogenate by inverting the tube three times. The final homogenate should be at least 5 volumes of the tissue sample (grams of tissue/mL buffer) for proper subsequent fractionation. Aliquot 150 μL of homogenate (H fraction) prior to proceeding. Repeat this step for each sample.

3.3 Synaptosome Fractionation

1. Equilibrate the weight of the homogenates in pairs to balance the centrifuge, then centrifuge at 1000 × *g* for 8 min at 4 °C (we use Awel AMF 20–8 rotor; adjust sedimentation time to your rotor according to standard formulas).
2. Transfer 6 mL of each supernatant (S1) to new 15 mL falcon tubes, mix by inverting each tube three times. Discard the left-over of S1.
3. Resuspend the pellet (P1) in 2.5 mL of ice cold Isosmolar Buffer with a 5 mL pipet and a vortex.
4. Aliquot 150 μl of S1 fraction and P1 fraction.
5. Equilibrate S1s weight in pairs, then centrifuge at 12,500 × *g* for 13 min at 4 °C (we use Awel AMF 20–8 rotor; adjust time of sedimentation to your rotor according to standard formulas).
6. During this centrifugation step lay discontinuous sucrose gradients in new and not reused ultracentrifuge tubes (we use 14 × 89 mm Beckman Coulter 344059, *see* **Note 4**). First add 5 mL of ice cold 1.2 M Sucrose Solution, then slowly lay on top 5 mL of 0.8 M Sucrose Solution with a mechanical pipette dispenser (*see* Fig. 1).
7. Aliquot 150 μl of the supernatant (S2) then discard S2.
8. Resuspend P2 pellets in 1 mL of ice cold Isosmolar Buffer with several up and down strokes with a p1000 micropipette. Aliquot 150 μl of P2 then layer the remaining P2 slowly on the top of sucrose gradients with a mechanical pipette dispenser.
9. Precisely equilibrate the weights of the gradients in the ultracentrifuge tubes.
10. Centrifuge at 50,000 × *g* for 1 h 10 min at 4 °C in the ultracentrifuge swinging rotor. Set the acceleration to 6 of 10, and the deceleration at 4 of 10, to preserve the gradient layering.
11. Collect synaptosomes (B fraction) after the centrifugation with a 1 mL syringe and a 25G 16 mm needle through the tube wall (pierce bellow the B fraction interface; *see* Fig. 1 and **Note 4**). Collect the fraction in a 2 mL micro-centrifuge tube; mix by inverting the tube three times (avoid vortexing).
12. Dilute 100 μl of synaptosomes in 20 mL ice cold 0.22 μm filtered PBS with protease inhibitors. Mix by inverting the tube three times (a 1/200 dilution factor was determined empirically for 1 forebrain of 250 mg; this pre-dilution prevents aggregation of particles *see* **Note 4**).
13. Protect the suspension from light by wrapping the tube with aluminum foil. These diluted samples correspond to the master samples for FACS analysis and sorting. The remaining concentrated samples should be kept for protein titration and western-blot analysis.

Table 1
Optical parameters for VGLUT1^{venus} FASS analysis

Channel	Filter and gain
FSC	ND1.0, 340 V
SSC	488/10 nm, 365 V
FITC (VGLUT1^{venus})	530/30 nm, 700 V
PerCP (FM4-64)	675/20 nm, 700 V

3.4 FACS Analysis and Sorting

1. The FACS should be set as follows: 70 μm Nozzle, sample shaking at 300 rpm, sample temperature at 4 °C, FSC neutral density filter 1.0, 488 nm Laser on, Area Scaling 1.18, Window Extension 0.0, FSC Area Scaling 0.96, Sort Precision: 0-16-0. Thresholding on FM4-64 with a threshold value of 800. Optical parameters should be set as appeared in Table 1.
2. All samples should be diluted in 0.22 μm filtered PBS with protease inhibitors until reaching approximately 20,000 events per second at a flow rate between 1 and 3 (*see* **Note 5**). Sample shaking should run at 300 rpm and temperature set to 4 °C.
3. The detection threshold is read on FM4-64 fluorescence. Stain each sample by adding 1 μg/mL of FM4-64 prior to loading in the cell sorter.
4. Noise baseline check with PBS: Load PBS + FM4-64. Record for 1 min. Noise events should be down to 200/min approximately. Should noise rise above 400–500 events/min (*see* **Note 1**), then run ultra pure water for 10 min at maximal flow rate. Run PBS again with a flow rate of 3. Repeat wash steps until the baseline falls below 400 events/min.

   Autofluorescence baseline check with WT sample and gating strategy: Add FM4-64 and mix by inverting the tube three times. Flow the sample for 1 min in the sorter before recording one million events at a flow rate of 3.0 (*see* **Note 5**). Slide gate P1 toward lower forward scatter channel (FSC) and side scatter channel (SSC) scores to include around 80% of total detected events in this gate (Fig. 2a). P1 represent the non-aggregated population of particles in the sucrose synaptosome sample. Create a second contour plot graph displaying venus fluorescence peak area in Y and FSC peak area in X. Restrict the display to P1 population in this new graph (*see* Fig. 2b). Center gate P2 on the WT autofluorescent population. Set gate P3 above to get no events in this gate but to have P3 as close as possible to P2 (Fig. 2b). Put gate P5 around the aggregating diagonal population to follow unwanted aggregation processes over-time (Fig. 2a). Additional graphs in the sheet may help in monitoring the sample but they are not critical for the gating strategy.

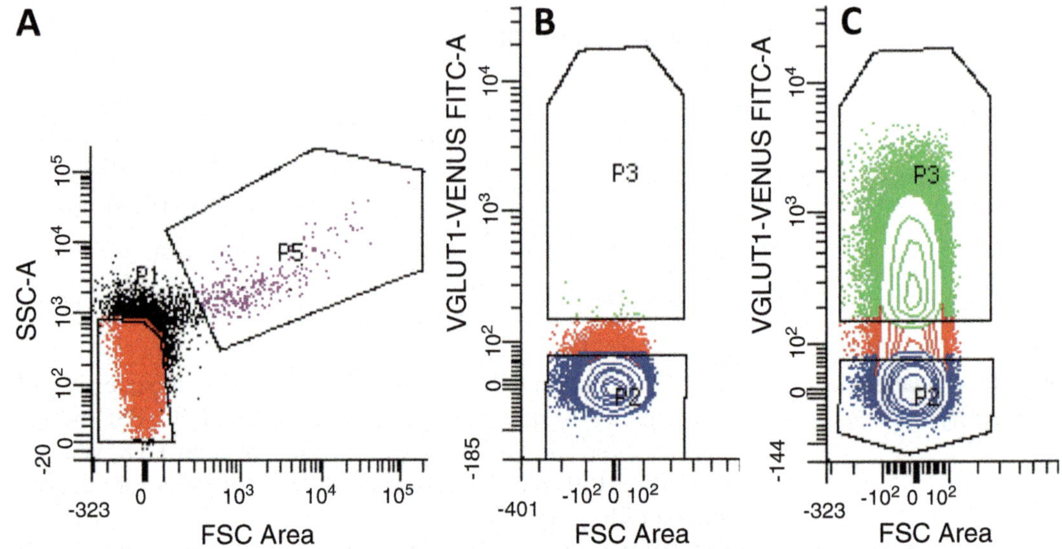

Fig. 2 Analysis and gating of VGLUT1^{venus} synaptosomes. (**a**) Light scattering of all particles detected through FM4-64 threshold. P1 gate represents single particles and generally accounts for 80 % of the total. P1 is shifted to the lower left side of the main population as we observed small aggregates in the top right part. P5 gate covers large aggregates and allows monitoring aggregation over time. It should remain below 10 %. Note that FACSAria cannot resolve size differences in the sucrose synaptosomes. The P1 gated population is then displayed in a new contour plot (*see* **b** and **c**). (**b**) Wild Type P1 population baseline VGLUT1^{venus} fluorescence. P2 gate is positioned to cover the whole wild type P1 population. (**c**) VGLUT1$^{v/v}$ samples provide a significant fluorescent population above the P2. P3 is designed to cover this population and exclude the WT baseline. All dot plots are displayed in bi-exponential mode as our samples display in the lowest range of values. Note that protocol optimizations included here allow a much better detection of VGLUT1$^{v/v}$ particles compared to our first report of the FASS method [20]. As a consequence, sort rates are sharply increased

5. Fluorescence analysis with VGLUT1^{venus} sample: Let the sample flow for 1 min in the sorter before recording one million events at a flow rate of 3.0 (*see* **Note 5**). Adjust P3 top gate to include all highly VGLUT1^{venus} fluorescent particles.
6. Sorting fluorescent synaptosomes: The number of events/s should remain stable between 18,000 and 25,000 events/s with a flow rate at 3.0 or below. Use the 15 mL Falcon tube collection device cooled to 4 °C. The left tube will receive P3 gated events (VGLUT1^{venus} fluorescent particles) while the right tube will collect non-fluorescent P2 gated events that will eventually be discarded (*see* **Note 6**). Perform a test sort for quality control of the streams and make sure left and right streams fall in their respective tubes. Place new 15 mL tubes. Open the sorting sheet. Set sorting precision to 0-16-0 (*see* **Note 6**), add P3 gate to the left side and P2 gate to the right side. Start sorting. During sorting keep the speed of analysis below 22,000 events/s roughly and the sorting efficiency above 80 % by modulating the flow-rate speed. The rate of particle aggregation seen in gate P5

should remain below 10% of total events. Otherwise, prepare a fresh VGLUT1^{venus} tube. On average, a new sample tube has to be prepared every hour. Check the volume of collected sample directly into the collection tube. On average, one million events generate 1 mL of collected sample. Do not sort more than 14 million events in one 15 mL Falcon tube.

7. Sort reanalysis: The main quality control of the sort is performed through the reanalysis of sorted samples in the sorter. Use at least 500 μl of the sorted sample to perform the reanalysis. Add FM4-64, mix and run this sample first 1 min at a flow rate of 11, then 1 min at a flow rate of 3.0 before recording for 1 min. This procedure ensures a thorough cleaning of the sample line from any contamination with the unsorted sample that was processed before, and to stabilize the flow of sorted events. The flow of events in reanalysis usually runs between 100 and 200 events per second (roughly 100-fold slower than during analysis and sorting). A good sort is indicated by more than 60% of events seen in P3 and less than 12% of events in P2 (*see* Fig. 3). Additionally, sorted synaptosomes should display a more compact population in FSC/SSC scatter plots (*see* Fig. 3c). Eventually one may monitor the global fluorescent population in the sorted sample (*see* the inset graph in Fig. 3c). Total fluorescence in a good sort represents more than 80% of the total events.

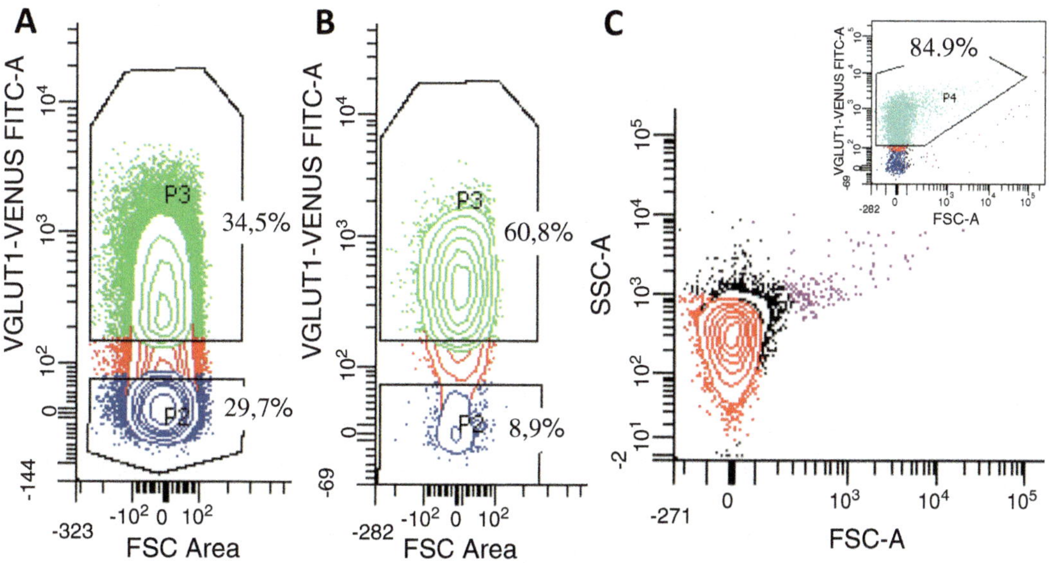

Fig. 3 Reanalysis of sorted samples. (**a**) Same graph as Fig. 2c. On a good day, around 35% of the total sample is in P3 and around 30% in P2. (**b**) Reanalysis of sorted P3 population. P3 population rises up to 55–65% of the total while P2 negative particles drop to less than 10%. (**c**) On a light scatter plot, the sample seems now more compact in the P1 gate, though aggregation still takes place after sorting. In the inset, we estimate the total fluorescence in the sorted sample as all particles originating from P3 sorting. P4 usually rises to 75–85%

3.5 FASS Sample Processing

1. FASS sample concentration for proteomics approaches: As soon as the quality control reanalysis has been validated, FASS synaptosomes collected in 15 mL tubes are concentrated by filtration on 25 mm diameter polycarbonate filters with 0.1 μm pores at 4 °C. This step may be very lengthy if the vacuum is not strong enough. We therefore recommend optimizing the filtration setup on sucrose samples prior to the first sorting day. Overlay all FASS samples of the same origin onto the filter over the afternoon of sorting. For a typical day of sort we collect around 80 million synaptosomes.
2. After completion of filtration, remove the filter with tweezers. Place the filter onto a small petri dish and cut it into 10–12 pieces with a blade and place them into a 0.5 mL microcentrifuge tube. To keep the 0.5 mL tube clean put it into a 50 mL falcon tube. Freeze it in liquid nitrogen and store it at −20 °C until use.
3. Thaw and resuspend proteins in 50–70 μl of SDS-PAGE loading buffer and incubate at 70 °C for 3 min (for full denaturation while preserving complex hydrophobic proteins like neurotransmitter transporters). Cool down on ice. Put the closed 0.5 mL tube top down with most of the sample in the cover. Pierce the bottom of the 0.5 mL tube with a 27-gauge needle and place in a clean 1.5 or 2 mL tube. Centrifuge for 1 min at full speed in a microcentrifuge to collect the sample in the larger tube.
4. FASS sample protein titration: Sucrose synaptosome proteins should be titrated using routine methods (Bradford, Lowry, or others). Then FASS samples are titrated using silver staining performed on SDS-PAGE gels against a standard curve of sucrose synaptosomes. Following titration, western blots are performed using around 0.5 μg of proteins. Optimizations allow to probe several proteins per lane (up to 6–8 using a striping procedure).

3.6 FASS Sample Recovery on Coverslips

1. Gelatinize coverslips: place 12 mm coverslips in a coverslip holder and immerse in Chrome Alum Gelatin Slide Coating for 4 min, remove and allow to air dry in a 60 °C oven for at least 2 h minimum.
2. Place gelatinized cover slips into a 24-well flat bottom microplate.
3. Add 1 mL of sorted synaptosomes per well.
4. Centrifuge microplate for 34 min at 6800 × *g* on a microplate swing-rotor centrifuge (we use JS 5.3 rotor using Beckman J-26XP centrifuge). Discard supernatant.
5. Fix with 500 μl of PFA Sucrose Buffer. Incubate at room temperature 20 min under slow shaking.

6. Eliminate PFA and wash three times for 5 min with PBS at room temperature under slow shaking. Store at 4 °C in PBS until use.
7. Upon immunofluorescence staining, coverslips may be mounted on glass slides using standard procedures for downstream imaging applications [20].

4 Notes

1. FM4-64 based thresholding of particle detection requires one to be extremely careful regarding noise issues. Indeed, FM4-64 will fluoresce in the presence of detergent or hydrophobic molecules. Any unwanted release of particles or molecules prone to provide a hydrophobic milieu to FM4-64 molecules will induce noise in the experiment. We faced this problem with certain plasticware and when the cell sorter is not fully rinsed with water after the cleaning procedure from Becton-Dickinson (BD). Hence, we suggest to be extremely cautious when changing/testing plasticware and to always check the noise baseline at the beginning of a sort. This type of noise will not be harmful for sample sorting at high event rates, but will be really damaging for reanalysis of sorted samples at low event rates.
2. Our experience of synaptosome sorting is successful with the FACSAria cell sorter from BD Bioscience which does not mean that it may not work with any other cell sorter. Indeed, one can see from the FSC/SSC scatter plots that the FACSAria is nearly blind in this size range of biological samples and much could probably be done to optimize microparticle detection and sorting with this equipment. However, we had the opportunity to work for several years with another cell sorter, the Moflo XDP from Beckmann-Coulter with very little success. One chief aspect for the success of a sorting day is the stability of the fluidics system of the sorter over-time. This criterion has never been met with the Moflo XDP even though some short sequences of sorting turn out promising in the reanalysis. Overall, a properly set FACSAria can be stable for a full day easily. Importantly, the quality of the objective lens and gel coupling to the analysis chamber should be checked by the maintenance team from the provider as optimal quality is critical with our small and dim particles. Upon lens replacement we gained a lot of signal on our current system.
3. The amount of input material in the cell sorter is clearly not a critical element for FASS success. Indeed, most of the VGLUT1^{venus} sample generated for one experiment is not reaching the sorter at the end of the day because the bottleneck is the maximum processing speed of 20,000 events per second. Thus

in principle a very small microdissection of brain should be suitable for FASS, provided that the fractionation methods are scaled down appropriately. A critical aspect though is to have a high level of fluorescent particles in the sample so that the time spent analyzing unwanted events is lower. Indeed, the larger the population in the P3 gate is, the faster the sorting will be. Along these lines, we did try to use post nuclear pellet 2 fractions in the sorter, but the quality and speed of sorts suffered from this. Another critical issue regarding the sample quality is to ensure specific labeling of the target compartment. We achieve this using VGLUT1^{venus} knockin mice as VGLUT1 is a protein highly confined to synapses in VGLUT1-positive excitatory neurons [21, 22]. One has to pay attention to this when implementing the method with another labeling approach.

4. Aggregation of particles is a critical parameter to control for the success of FASS. With VGLUT1^{venus} labeling, the main population of synapses is fluorescent; hence, most aggregates contain at least one fluorescent particle. Thereby, when particles aggregate, small aggregates that escape FSC resolution tend to be sorted with the single particles. Also, the increasing aggregated population (seen in P5) is removing events of interest in P3 and sort rates drop dramatically. Even though we do not fully understand which parameter is responsible for the massive aggregation process that occurs on conventional sucrose samples we suspect that myelin and bacteria/fungi derived trace contaminants to catalyze it. Through methodic comparisons, we identified several critical parameters to block the process almost completely. These include: a thorough cleaning procedure for the potters (*see* Subheading 3), the use of fresh filter sterilized buffers, the exclusive use of new tubes for the three centrifugation steps (in our case the first 2 centrifugations are using 15 mL falcon tubes), syringe based collection of the sucrose synaptosome B fraction (*see* Fig. 1), the vortex-free gentle mixing of suspensions using tube inversions, pre-dilution of synaptosome fractions right after the collection from the sucrose gradient, and finally the constant handling of samples at 4 °C (also mandatory for protein handling in general).

5. The optimal focusing of the sample flow in the sorter will positively influence the fluorescence detection. Indeed, the flow rate of sample entering the acceleration chamber can be set from 0 to 11 on the FACSAria. This tuning actually sets the pressure difference between the sample line and the flow analysis chamber. We observed that changes in the flow rate used on the instrument change the measurements on our sample. Indeed, at small flow rates, the sample is efficiently sheathed by the sheath fluid and optimally focused on the laser beams. While loose focusing may be harmless for the measurements on a 10–15 μm cell, it

becomes critical when dealing with 1 μm particles. Hence, we insist on using a flow rate of 3.0 or less for all analyses, sorts, and reanalyses in the method. Playing with dilutions will allow reaching the proper flow rate when higher values are needed. On a similar note, we observed that sample measurements/statistics are most of the times not stable before the sample has run for 1 min in the sorter. Hence, we always run the sample for 1 min before refreshing the graphs and recording a sample.

6. Sorting quality depends on the sorting precision/mask used and on the actual quality of the stream. In our hands, the best results were obtained upon using the 0-16-0 mask that selects events only when they are estimated to fall in the center of the drop of interest (and no contaminant is close by in the flow). In our hands, the simultaneous sorting of P2 non-fluorescent events on the opposite side allowed increasing the sort quality of the P3 fluorescent events. We believe that this occurs through the better focusing/lower spraying of the central trash flow and by the active removal of negative events from the central trash flow. We then tried to analyze these P2 gated samples but soon figured out that the silver gel profiles are too different from our sucrose synaptosome standard so that protein titration can be considered accurate. Also, in principle, this P2 gated population is a nonrelevant ensemble of particles of many different origins that only share the common feature of being negative to VGLUT1^{venus} fluorescence.

References

1. Whittaker VP (1993) Thirty years of synaptosome research. J Neurocytol 22:735–742
2. Huttner WB, Schiebler W, Greengard P et al (1983) Synapsin I (protein I), a nerve terminal-specific phosphoprotein. III. Its association with synaptic vesicles studied in a highly purified synaptic vesicle preparation. J Cell Biol 96: 1374–1388
3. Cotman CW, Matthews DA (1971) Synaptic plasma membranes from rat brain synaptosomes: isolation and partial characterization. Biochim Biophys Acta 249:380–394
4. Jones DH, Matus AI (1974) Isolation of synaptic plasma membrane from brain by combined flotation-sedimentation density gradient centrifugation. Biochim Biophys Acta 356:276–287
5. Boyken J, Grønborg M, Riedel D et al (2013) Molecular profiling of synaptic vesicle docking sites reveals novel proteins but few differences between glutamatergic and GABAergic synapses. Neuron 78:285–297
6. Weingarten J, Lassek M, Mueller BF et al (2014) The proteome of the presynaptic active zone from mouse brain. Mol Cell Neurosci 59:106–118
7. Gray EG, Whittaker VP (1962) The isolation of nerve endings from brain: an electron-microscopic study of cell fragments derived by homogenization and centrifugation. J Anat 96:79–88
8. Whittaker VP (1959) The isolation and characterization of acetylcholine-containing particles from brain. Biochem J 72:694–706
9. Bai F, Witzmann FA (2007) Synaptosome proteomics. Subcell Biochem 43:77–98
10. Takamori S, Holt M, Stenius K et al (2006) Molecular anatomy of a trafficking organelle. Cell 127:831–846
11. Burré J, Zimmermann H, Volknandt W (2007) Immunoisolation and subfractionation of synaptic vesicle proteins. Anal Biochem 362:172–181
12. Wilhelm BG, Mandad S, Truckenbrodt S et al (2014) Composition of isolated synaptic boutons reveals the amounts of vesicle trafficking proteins. Science 344:1023–1028

13. Henn FA, Anderson DJ, Rustad DG (1976) Glial contamination of synaptosomal fractions. Brain Res 101:341–344
14. Dodd P, Hardy JA, Oakley AE et al (1981) Synaptosomes prepared from fresh human cerebral cortex; morphology, respiration and release of transmitter amino acids. Brain Res 224:419–425
15. Schwenk J, Harmel N, Brechet A et al (2012) High-resolution proteomics unravel architecture and molecular diversity of native AMPA receptor complexes. Neuron 74:621–633
16. Schwenk J, Baehrens D, Haupt A et al (2014) Regional diversity and developmental dynamics of the AMPA-receptor proteome in the mammalian brain. Neuron 84:41–54
17. Wolf ME, Kapatos G (1989) Flow cytometric analysis and isolation of permeabilized dopamine nerve terminals from rat striatum. J Neurosci 9:106–114
18. Wolf ME, Kapatos G (1989) Flow cytometric analysis of rat striatal nerve terminals. J Neurosci 9:94–105
19. Gylys KH, Fein JA, Yang F et al (2004) Enrichment of presynaptic and postsynaptic markers by size-based gating analysis of synaptosome preparations from rat and human cortex. Cytometry A 60:90–96
20. Biesemann C, Grønborg M, Luquet E et al (2014) Proteomic screening of glutamatergic mouse brain synaptosomes isolated by fluorescence activated sorting. EMBO J 33:157–170
21. Herzog E, Nadrigny F, Silm K et al (2011) In vivo imaging of intersynaptic vesicle exchange using VGLUT1 venus knock-in mice. J Neurosci 31:15544–15559
22. Fremeau RT, Voglmaier S, Seal RP et al (2004) VGLUTs define subsets of excitatory neurons and suggest novel roles for glutamate. Trends Neurosci 27:98–103

Part III

Observing Synaptic Structure and Ultrastructure

Chapter 11

Optimized Protocol for Imaging Cleared Neural Tissues Using Light Microscopy

Yoh Isogai, Douglas S. Richardson, Catherine Dulac, and Joseph Bergan

Abstract

Understanding physical and chemical processes at an organismal scale is a fundamental goal in biology. While science is adept at explaining biological phenomena at both molecular and cellular levels, understanding how these processes translate to organismal functions remains a challenging problem. This issue is particularly significant for the nervous system where cell signaling and synaptic activities function in the context of broad neural networks. Recent progress in tissue clearing technologies lessens the barriers that previously prevented the study of large tissue samples while maintaining molecular and cellular resolution. While these new methods open vast opportunities and exciting new questions, the logistics of analyzing cellular processes in intact tissue have to be carefully considered. In this protocol, we outline a procedure to rapidly image intact brain tissue up to thousands of cubic millimeters. This experimental pipeline involves three steps: tissue clearing, tissue imaging, and data analysis. In an attempt to streamline the process for researchers entering this field, we address important considerations for each of these stages and describe an integrated solution to image intact biological tissues. Hopefully, this optimized protocol will lower the barrier of implementing high-resolution tissue imaging and facilitate the investigations of mesoscale questions at molecular and cellular resolution.

Key words Lightsheet, Confocal, CLARITY, OptiView, Tissue clearing, Whole brain imaging

1 Introduction

Mapping the structure and connectivity of neural circuits is essential to understand brain function in both healthy and diseased states. A powerful set of tools already exists to characterize neural circuits, processes, synapses, and proteins at multiple spatial resolutions. For example, classic retrograde and anterograde tracing techniques efficiently map the broad connections between brain regions [1–3], viral strategies allow entire neural circuits to be traced from genetically defined neural populations [4, 5], and histological techniques such as antibody staining and in situ hybridization allow the precise localization of individual biomolecules [6, 7]. Central to each of these techniques is the need to analyze the location, shape, and connectivity of individually labeled neurons, synapses, or proteins. In the past, processing histological brain tissue typically

Alexandros Poulopoulos (ed.), *Synapse Development: Methods and Protocols*, Methods in Molecular Biology, vol. 1538,
DOI 10.1007/978-1-4939-6688-2_11,

required sectioning samples into thin (~10–100 μm) sections to allow visualization of deep structures. The process of sectioning tissue can be time consuming, difficult to automate, prone to tissue imperfections (tearing and folding), and destroys many essential long-range connections. Recent advances in tissue clearing techniques, however, make mapping neural circuits within large volumes of intact tissue both feasible and practical [8–17].

Techniques for making biological samples transparent date back more than a century [18, 19], and the potential for these strategies to image large volumes of intact tissue at high resolution is finally being realized (Figs. 1, 2, and 3). Each tissue clearing protocol has both drawbacks and benefits, most of which have been discussed previously [20, 21]. Instead, our goal here is to: (1) Outline a cost-effective and reliable protocol in sufficient detail to rapidly produce high-quality tissue samples; (2) Indicate factors within this protocol that are particularly sensitive and provide guidance on the selection of these parameters; (3) Describe the process of preparing tissue for imaging; (4) Compare and contrast microscopy strategies for imaging large volumes of brain tissue; (5) Discuss important considerations for the analysis of the potentially massive datasets generated by this approach.

The procedures outlined below are intended as a reasonable entry point for imaging large volumes of intact tissue; however, some refining of these procedures will undoubtedly be required to optimize results with each unique preparation. Important parameters that may require tuning are indicated throughout the protocol.

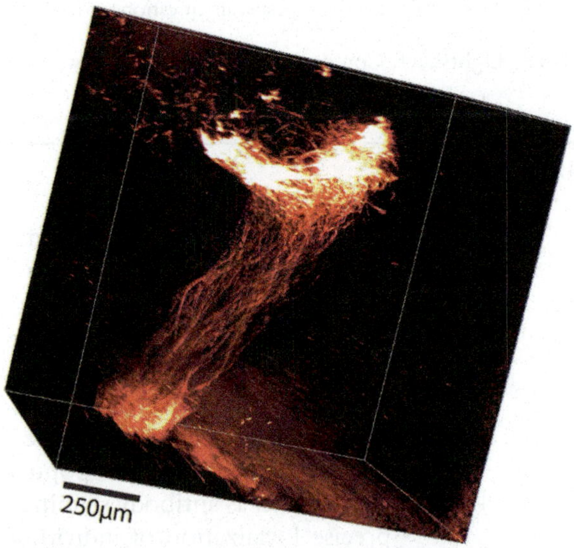

Fig. 1 Three dimensional reconstruction (1.2 mm × 1.2 mm × 0.4 mm) of vomeronasal sensory neuron axons terminating in the accessory olfactory bulb. Transgenically expressed tdTomato in specific vomeronasal receptor neurons is imaged using a Zeiss LSM710 confocal microscope with a 10× objective immersed in OptiView. Optical sections were 3D rendered using Imaris

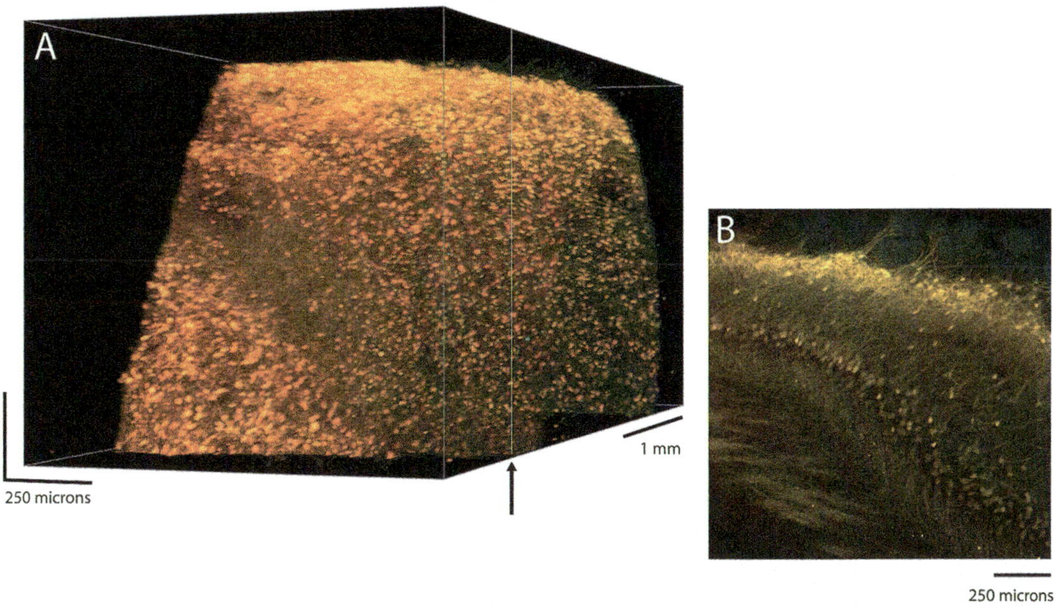

Fig. 2 Three-dimensional reconstruction of fluorescently labeled neurons in the olfactory bulb. (**a**) A 1.25 mm × 1.25 mm × 3.5 mm region of a CLARITY treated mouse olfactory bulb was placed in OptiView and imaged with a Zeiss Lightsheet Z.1 microscope. TdTomato was expressed in mitral and tufted cells by crossing a Tbet:Cre line [35] with the Ai9 tdTomato reporter line [36]. Initial image processing was done using FIJI, and the volume image was rendered using the maximum intensity projection rendering algorithm in Vaa3D. (**b**) An individual imaging plane acquired at a depth of 1 mm (*arrow panel A*) with lightsheet microscopy

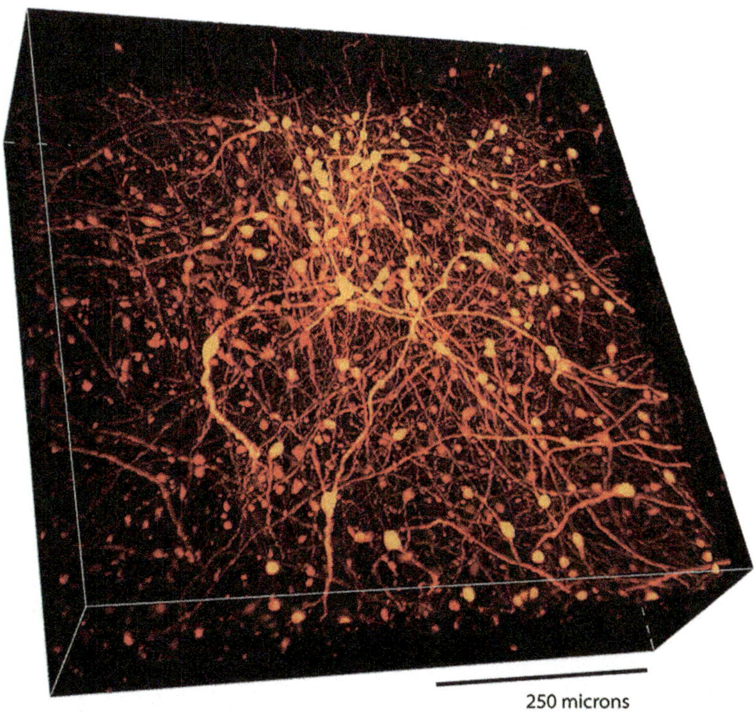

Fig. 3 Three-dimensional reconstruction of cell bodies and processes of fluorescently labeled neurons in the olfactory bulb. A 600 μm × 600 μm × 150 μm region of mouse olfactory bulb (same conditions as in Fig. 1) was imaged with a Zeiss Lightsheet Z.1 microscope. Initial image processing was done using FIJI, and the volume image was rendered using the maximum intensity projection rendering algorithm in Vaa3D. Axons, dendrites, and potential synaptic contacts are easily observed with this imaging strategy

2 Materials

2.1 Sample Considerations

Fluorophores of choice (*see* **Notes 1** and **2**).

2.2 Homemade CLARITY Electrophoresis Chamber

Parts list for buffer recirculating system for CLARITY electrophoretic clearing (*see* **Note 3**).

Vendor	Item description	Catalog no.
Grey beard pumps	The Niagara with 120-Volt transformer - designed for water	316
US plastic	3″ opening 5 gal carboy	A00357
KORAD	Precision variable adjustable 60 V 5A DC linear power supply	KA6005D
McMaster-Carr	Water filter with light-blocking blue bowl	4448K35
McMaster-Carr	Replacement filter cartridge	4422K61
McMaster-Carr	Patch cord—banana plug to alligator clip (12″)	6927K42
McMaster-Carr	1/4″ Tubing—5 ft	2085T5
McMaster-Carr	Single barbed tube fitting	5463K247
McMaster-Carr	Single barbed tube fitting (7/16″ hex for 1/4 tubing)	5463K245
McMaster-Carr	Durable nylon tight-seal barbed tube fitting	5463K247

2.3 Electrophoretic Clearing of Mouse Brains

1. Refractometer (e.g., Milwaukee MA871).
2. Hydrogel monomer solution: 4 % acrylamide, 0.0,25 % bisacrylamide, 0.25 % VA-044 initiator (Wako), 1× PBS, 4 % paraformaldehyde.
3. PBST: 0.1 % Tween-20 in PBS.
4. Clearing solution: 200 mM boric acid, pH8.5, 4 % SDS [8].
5. Superglue.
6. Imaging medium: FocusClear (CelExplorer Labs) or OptiView (172 mM sodium diatrizoate, 815 mM meglumine diatrizoate, 0.5 mM EDTA, 0.1 % Tween-20, pH 8, refractive index adjusted to 1.46 using a refractometer, filtered through 0.22 μm pore filter) (*see* **Note 4**).

2.4 Microscopy Equipment

Most imaging systems were not designed to image samples centimeters in size; however, with careful consideration of the working distance of available objectives, sample mounting technique, and refractive index of the final clearing solution a number of home-built and commercially available confocal, multi-photon, and light-sheet microscopes can be adapted for this task.

2.4.1 Objectives

The objectives available will determine the maximum depth in the sample can be imaged, what lateral and axial resolutions can be obtained, and which microscopy techniques are best suited to the experiment.

Working Distance

In a properly cleared, highly transparent tissue, imaging depth is most often limited by objective working distance. Routinely, only low numerical aperture (NA) objectives (less than 0.5) have a working distance greater than 2 mm. Unfortunately, if the NA is too low, 3D imaging is not worthwhile due to the poor axial resolution. Currently, researchers must choose between one of three options.

1. Recommended: a number of long working distance (above 5 mm), high NA (more than 0.9) objectives are now commercially available and can be used in conjunction with confocal, multi-photon, and lightsheet microscopes.
2. Alternative #1: a low/moderate NA (less than 0.6), long working distance objective can be used in a lightsheet configuration where the thinness of the lightsheet, and not the imaging objective, determines the axial resolution [14, 17, 22].
3. Alternative #2: an objective with a modest NA and long working distance (at least 3 mm) can be used in conjunction with a chamber that places the sample between two coverslips [8]. Here, approximately half of the sample is imaged from each side and the two 3D volumes are merged using image processing software.

Refractive Index

Most moderate to high NA, long working distance objectives are corrected to the refractive index (RI) of water (1.33). Many of the modern clearing solutions, including OptiView, have an RI close to 1.46. This refractive index mismatch (1.33 versus 1.46) will increase aberration, reducing imaging depth and resolution in all dimensions. Again, three options exist:

1. Recommended: use of objectives that are specifically matched to a refractive index of 1.45, or objectives with correction collars that can be tuned over this range. These are now commercially available.
2. Alternative #1: Use a moderate to low NA water immersion objective. Although, not perfect, the refractive index of water, 1.33, is a closer match than air. Additionally, the lower NA will be less susceptible to aberration and the objective will have a longer working distance.

3. Alternative #2: The introduction of adaptive optics to some imaging platforms can be used to correct for aberrations and improve imaging depth and resolution [23].

2.4.2 Imaging Modality

Two commercially available modalities exist that can be adapted to imaging large, cleared tissue samples: point-scanning confocal (*see* **Note 6**), and lightsheet microscopy.

Point-Scanning Confocal Microscopy

Most commercially available confocal microscopes can be used (Fig. 1); however, the following modifications are highly recommended.

1. Upright microscope (inverted microscopes are poorly suited to image large samples).
2. Large Z (axial) objective travel range.
3. Adjustable height stage.
4. Fully automated (x, y) mechanical stage.
5. Long working distance, RI-matched objective (*see* Subheading 2.4.1.2).
6. Full complement of yellow and red laser lines (561 nm, 594 nm, 633 nm, tunable IR).
7. 80/20 transmission/reflection dichroic mirror (allows imaging of reflected light, *see* Subheading 3.2 below).

Lightsheet Microscopy

Lightsheet imaging of large, cleared tissues has mainly been performed using two designs, both of which have now been commercialized:

1. Ultramicroscopy [9] uses a setup similar to a dissecting microscope. The sample is placed in the base of the microscope containing optics that produce a thin lightsheet that can be focused through the sample. A long working distance, low NA air objective or moderate NA dipping objective capable of capturing a large field of view sits above the sample. Here, the entire sample can be imaged in a single field of view, or zoom optics and a mechanical stage can be used to reduce the field of view and tile and stitch together a final image. Fluorescence captured by the imaging objective is projected onto a camera.
2. Selective plane illumination microscopy [24] suspends the sample in front of a standard compound microscope objective between two additional excitation objectives that project a lightsheet into the sample. In this system, the sample is raster or "tile" scanned in front of the imaging objective as it is too large to fit in a single field of view. The fluorescence captured by the imaging objective is delivered to a camera. By removing the need for laser scanning, an entire Z plane can be captured in as little as 10 milliseconds. This means an entire brain can be imaged in a few hours or less, depending on the field of view of the microscope (Figs. 2 and 3).

Recommended components for a lightsheet microscope:

1. Cylindrical lens to form the lightsheet.
2. Dual side excitation light paths.
3. Dual view light path and two sCMOS cameras for simultaneous two channel fast readout.
4. Adjustable mirror to ensure lightsheet is positioned at focal plane of imaging objective.
5. Resonance scanning mirror to provide a pivot scan to reduce line artifacts (*see* Subheading 3.4 below; [25].
6. Laser blocking filter for all but one excitation laser (allows imaging of scattered light, *see* Subheading 3.2 below).

2.5 High Performance Computer System with Large Bandwidth and Storage

See **Note 5** for recommended specifications.

2.6 Image Processing Software

Several open source and commercial software packages exist to facilitate processing large imaging datasets. The following freeware/open source options are recommended:

1. FIJI/ImageJ/BioFormats are useful to convert proprietary file formats to .tif, downsample data, and stitch tiles [26, 27]. Additionally, a large and growing number of plugins are available that enable a wide range of procedures (image filtering, cell counting, colocalization, and many more).
2. Vaa3d [28] allows fast and detailed 3D visualization of image stacks. While currently more limited in analysis capabilities, the array of plugins available for Vaa3d is rapidly growing (Fig. 3).
3. Commercial software suites such as Arivis Vision4D (Arivis AG) and Imaris (Bitplane) are also capable of handling TB-sized data sets, provide 3D rendering, and allow a diverse range of data analysis tools.

3 Methods

3.1 Electrophoretic Clearing of Mouse Brains

Caution: Hydrogel solution contains formaldehyde and acrylamide and is highly toxic. Perform perfusion strictly under a ventilation hood with proper protective gear and dispose of waste chemicals under institutional guidelines.

1. Perfuse animal with 50 ml PBS followed by 25 ml hydrogel monomer solution and transfer the brain tissue to a sealable tube containing ~25 ml cold hydrogel solution [8].

2. Flush excess oxygen from the tissue-containing tube with nitrogen gas.
3. Keep brains at 4 °C (with aluminum foil to shield them from light) for 2 days, inverting one to two times on the second day.
4. After 24 h (this may be extended), replace oxygen with nitrogen a second time. Because oxygen inhibits acrylamide polymerization, it is important to remove as much oxygen from the tissue containing tube as possible before initiating acrylamide polymerization (**step 5**). Alternating rounds of exerting a vacuum on the uncapped tissue-containing tube and replacing the removed gases with pure nitrogen can easily achieve this. Repeating this **step 3** times evenly spaced over 15 min is sufficient.
5. Bring a water bath to 38 °C and move the tube containing the brain directly from ice into this water bath (*see* **Note 7**).
6. Carefully remove the brain from the hydrogel, and gently roll the brain across a Kimwipe to take off any remaining hydrogel (*see* **Note 8**).
7. Wash the brain in PBST overnight at room temperature while rocking.
8. Wash the brain in Clearing solution overnight at room temperature while rocking (*see* **Note 9**).
9. Place brain into an electrophoresis chamber, tighten the cap, and cover with aluminum foil to prevent exposure to ambient light.
10. Turn on the pump and allow clearing solution to flow for several minutes before turning on the electrophoresis power supply.
11. Adjust the voltage to achieve an initial current of ~0.5 A. If using one of the designs that can be found at http://clarityresourcecenter.org/, this will likely require a voltage of ~30 V. It is important to monitor the chamber temperature to ensure that the temperature does not rise above 45 °C. The current source should be immediately shut down if the temperature exceeds 45 °C to prevent damage to the tissue and reduction of fluorescence.
12. The duration required to clear a tissue sample depends primarily on the size of the tissue sample and the length of time the tissue has been allowed to passively clear. Tissue samples less than 2 mm are likely to be cleared after ~24 h of electrophoresis. Larger samples may require up to 4 days. This can be done all at once or broken into parts to maintain appropriate temperature and voltage. The temperature should always stay between 37 and 45 °C.
13. Remove the brain and wash overnight at room temperature in PBST while rocking.
14. Place brains in 200 ml of OptiView solution for 3–5 days prior to imaging. Prior to imaging, move brain into a fresh 50 ml of OptiView solution for 1 day. Closely monitor pH (8.0) and

refractive index (1.46) of this solution, as it will slowly be diluted with PBST if reused. For optimal results replace solution when refractive index begins to drop. Gently invert several times once a day to ensure thorough mixing (*see* **Note 9**).

3.2 Imaging Cleared Brains

3.2.1 Sample Mounting

Determining the best way to mount samples requires consideration of both the imaging system and objective to be used. Coverslip-corrected objectives require a coverslip to be placed between the sample and the objective. Ensure that the coverslip is the correct thickness (usually #1.5, 0.17 mm) and that the coverslip rests perfectly flat. *It is absolutely essential to remove all air bubbles from the mounted sample, as it is impossible to image through air bubbles regardless of the specific imaging strategy (see below)*. If the working distance of the objective is insufficient to cover the entire sample, sandwich the sample between two coverslips and image from both sides. A suggested dual coverslip chamber design is presented in Chung et al., 2013. A more convenient option is to use a dipping objective on an upright microscope. In this configuration the sample can be glued to a petri dish, which is filled with the final clearing solution, assuming the objective has the required working distance. Finally, certain lightsheet designs require the sample to be suspended in a fluid filled chamber. One successful strategy is to glue a small magnet to the base of the tissue away from any regions of interest, and use this magnet to reversibly mount the tissue.

Whether using confocal or lightsheet imaging, it may be helpful to acquire an additional data channel of autofluorescence or reflected light. This is best done with a long wavelength (>600 nm) laser as it has the best ability to penetrate through the tissue. This dataset can serve multiple purposes including: (1) image registration and (2) subtraction of the tissue autofluorescence signal from fluorescently labeled channels.

3.2.2 Imaging with an Upright Point Scanning Confocal Microscope (See Note 10)

1. Superglue brain to bottom of a high-walled 60–100 mm Petri dish.
2. Fill petri dish with OptiView solution until brain is submerged.
3. Lower stage of upright microscope if required to allow for clearance of long working distance objective and large sample.
4. Submerge objective (if dipping) and acquire tiled Z-stacks over area of interest with 5–10% overlap between tiles and an optical zoom of 1–1.3 (to reduce tiling edge artifacts in final image).

3.2.3 Imaging by Lightsheet Microscopy

1. Orient mounted brains such that any areas of interest are >1 mm from the attachment site.
2. Fill the imaging chamber with OptiView, and lower the brain into the microscope chamber.

3. Allow the tissue sample to rest in the imaging chamber for several hours prior to imaging (optional, *see* **Note 9**). This allows time to equalize any refractive index differences that may have developed during the mounting procedure, air bubbles at the top of the imaging chamber to disperse, and temperature differences to normalize. During this step, it is important to cover the imaging chamber to prevent evaporation of the imaging medium.
4. Acquire tiled z-stacks of the desired regions with at least 5 % overlap between tiled views. 5 % overlap is important to ensure the images can be accurately stitched together into a single dataset. Often, it is helpful to image large tissue samples from multiple angles, as distinct regions of any tissue sample will be imaged most clearly when the light path distance is minimized.
5. After imaging, remove brain and place in PBST at 4 °C for storage.

3.3 Data Merging/Image Processing

If the sample is larger than a few mm in any direction, it is likely that multiple image stacks will need to be stitched together, and a variety of tools exist to make this process easier, more accurate, and efficient. Two options stand out as particularly effective for large datasets.

1. The grid/Collection stitching feature in FIJI [26] is easy to use and suitable for small datasets. As the program is memory intensive, large datasets must be processed in virtual stack mode, which increases processing time. An advantage of this method is that the output transformation can be applied to images of any size and/or color that share the same reference frame.
2. Terastitcher—Built as a standalone program, as well as a plugin for FIJI and Vaa3D. Terastitcher provides a fast and accurate means to assemble multiple tiled image stacks into a large 3D volume dataset and allows nonrigid stitching [29]. Terastitcher is more efficient with memory usage and suitable for larger datasets.

3.4 Reducing Artifacts

The most common artifacts seen during imaging are dark lateral (lightsheet) or axial (confocal) stripes. These are most often caused by the introduction of air bubbles into the sample or hydrogel, the accumulation of debris on the sample surface, or poorly cleared tissue regions. Lightsheet microscopes are more susceptible to casting shadows behind areas of strong staining due to the high absorption of excitation light in these regions. To reduce these artifacts:

1. Only handle the sample in a clean environment. Do not allow towels or blotting paper that will deposit lint to come in contact with the sample.
2. Avoid puncturing/damaging the hydrogel or sample with sharp forceps or other dissection tools.

3. Keep samples fully submerged in imaging medium at all times and transfer tissue samples (that are not enclosed by a cover-slip) to the imaging chamber as quickly as possible to reduce the introduction of air bubbles.
4. Prior to filling the sample holder/imaging chamber ensure it is completely dry and no aqueous solutions are present that could alter the RI of the imaging solution.
5. In a lightsheet microscope, illuminating the sample from both the left and right sides relative to the imaging objective can reduce artifacts, provided both lightsheets penetrate to the current field of view.
6. The introduction of a resonance scanning mirror that "pivots" the lightsheet allowing it to enter the sample from multiple angles can greatly reduce stripe artifacts [25]. However, care should be taken with this step as pivot scanning may introduce additional noise if the refractive index of the medium is not properly calibrated for the objectives.
7. Many image processing software packages have a line or stripe removal option. This often relies on identifying stripes in a Fourier transformation of the image and removing that data from the Fourier transform. *Caution*: this is not compatible with intensity-based analyses as required data may be removed.

3.5 Troubleshooting

1. Inconsistent clearing:

   Perhaps the most common cause for inconsistent clearing is that the tissue sample is highly myelinated. White matter is more resistant to the clearing process, and often requires longer durations of active or passive clearing. If the white matter remains opaque, one should extend the clearing time until better results are obtained. Extending the duration of passive clearing at 37 °C can also be effective. If extending the clearing procedure is not successful, it will likely be necessary to reduce the tissue size.
2. Cleared brains have brownish tint and/or burnt spots:

   One common problem of the electrophoretic step is that brains obtain a yellow or brown color. Some yellowing should be expected; however, excessive voltage, duration, or heat during clearing can potentiate this issue. It is recommended to lower the voltage and current and to monitor the temperature carefully. Another approach is to clear tissue in multiple short periods of time, for example, alternating hours of active and passive clearing. Tissue that comes into close, or direct, contact with electrodes may show burnt spots. Make sure that the electrophoresis chamber has enough room so that the brain does not come in direct contact with the electrodes.
3. Low fluorescence compared to freshly fixed tissues:

   There will be some fluorescence loss from the clearing procedure. However, further fluorescence loss can be minimized by avoiding excessive heat in the electrophoresis chamber and by avoiding over-clearing.

4. Yellowing of the clearing solution:

   The clearing solution can turn slightly yellow over time, but this does not seem to affect the clearing step or fluorescence intensity. Carefully follow the initial conditions as outlined in Subheading 3.1. Clearing solution should be renewed when, or preferably before, there is a change in electrical resistance—as evidenced by a noticeable difference in the current to voltage relationship or an increase in heat production.

5. Image stitching failure; tiles do not perfectly register:

   When imaging a large tissue sample at single cell resolution, it is extremely important to keep the sample steady throughout the acquisition. It is important to acquire multiple tiled stacks such that the order of tiles minimizes sample movement. A failure to keep the sample steady will result in altered three-dimensional geometry of the sample, and it can cause errors in stitching. Also, consider the working distance of the objectives. Frequently, the objective working distance will determine the maximum depth that can be imaged. Attempting to image deeper than appropriate for the given objective will result in distorted images that are inaccurate and will be difficult to stitch together.

6. Blurry images:

   Blurry images often result from a mismatch in refractive index of the sample and the imaging medium. We recommend: (1) measuring the precise refractive index of each batch of imaging medium; (2) maintaining the samples in the same solution for at least 2 days prior to imaging while periodically adjusting the refractive index to the desired level; (3) transferring samples with the same solution for imaging without the introduction of air bubbles; (4) ensuring that the imaging chamber is completely dry before filling with imaging medium; (5) adjusting the refractive index of the imaging medium to optimize the balance between tissue clarity, refractive index mismatch with the objectives, and distortion in the lightsheet angles.

4 Notes

1. Maximizing the strength of fluorescence and the transparency of each tissue sample are primary concerns. Longer wavelength fluorophores (e.g., tdTomato) should be chosen whenever possible as the excitation wavelengths required for excitation can penetrate deeper, even in cleared tissue, and will likely provide more consistent results. The following protocol has been successfully used to image tissues expressing mCherry, GFP, YFP, mCitrine, and tdTomato transgenically, as well as tissues

expressing these fluorescent proteins from viral constructs. Shorter wavelength fluorescent proteins (e.g., BFP) have been imaged in deep tissues with only limited success.

2. Secondary labeling of tissue, for example by antibody staining, has been successful. However, one should currently expect secondary labeling procedures to take several weeks (if not months) when performed on large tissue samples. This is an active area of research and it is likely the antibody staining procedures will become more efficient in the near future. As before, choosing a fluorophore that is shifted toward the longer wavelengths and also provides a strong signal is essential.
3. The success of clarifying biological tissues relies on generating a tissue sample with a uniform refractive index. The CLARITY procedure [8] achieves this by removing lipids from the tissue sample with electrophoresis. A low-cost buffer recirculation system for this electrophoretic step can be constructed using the parts listed in Materials. This setup can be constructed for under $1000 and the electrophoretic step can be performed optimally at 37 °C.
4. The selection of imaging medium is a critical step since optimized tissue clearing requires the refractive index of tissue and medium to be carefully matched [30, 31]. However, not all refractive index matched solutions are equivalent. For example, organic solvents such as methyl salicylate and BABB clear well [32], but contain toxic compounds that can also damage microscope components. Many organic solvents also shrink tissue, possibly compromising the underlying structure, and displace water molecules that are essential for maintaining fluorescence from many fluorophores, including nearly all fluorescent proteins. Glycerol-based media are cheap and allow precise refractive index matching, but in our experience, the achieved transparency tends to be unstable, making long imaging experiments and tissue storage difficult. FocusClear, a commercially available solution [30, 33], works well for CLARITY cleared brains [8]. However, the cost of proprietary reagents may be prohibitively expensive for some experiments.
5. Spinning disk confocal microscopy is not well suited for imaging thick samples as the pinhole size and spacing of most disks are optimized for high NA objectives with short working distances. The largest drawback to point-scanning confocal is its slow speed. Depending on the chosen field of view, and lateral and axial resolutions, imaging of an entire brain would likely take days. Therefore, point scanning confocal is ideal for generating high-resolution 3D images of smaller regions and not generally useful for imaging an entire organ. Green/red/IR lasers are recommended as blue and near-UV lasers have trouble penetrating deep into tissue even when cleared.

6. The lack of a low-cost imaging medium that maintains the original structure provides suitable transparency of samples, and does not quench fluorescence signals motivated us to develop a new transparentizing reagent. This reagent, OptiView, consists of defined chemical components that can be titrated to optimize the refractive index for different samples. For imaging of CLARITY-treated mouse brains, we found that a refractive index of 1.46

   is optimal. When imaging datasets that require the acquisition of many Z stacks (tiles), datasets can quickly become prohibitively large. Depending on the number of tiles required and number of channels imaged, file sizes may exceed several terabytes in size. Therefore, planning how to store and analyze the data prior to data collection is essential. In the case of lightsheet microscopy, data is acquired extremely rapidly requiring a computer with fast write speeds. Although solid state hard drives are likely not necessary, a RAID array comprised of multiple 7200 rpm hard disk drives is a must. In addition, if different workstations are being used for acquisition and storage, a high-speed 10 Gbit network connection is recommended. It is also recommended to have 100–200 GB of RAM for both acquisition (to buffer images if there is a slowdown in data writing) and processing as some data analysis steps can require large amounts of memory. It is possible to reduce the need for RAM with most image analysis software suites by using virtual image stacks; however, in many cases this will noticeably slow image processing. Access to cluster computing may also be advantageous as many image processing steps can be performed in parallel.

7. For best passive clearing results, add a stir bar to the bath to cause the brain to rock, and cover the tube containing the brain with aluminum foil.

8. We use the same clearing solution as indicated in Chung et al. 2013. During the electrophoretic process a dark residue that reduces the imaging quality may be deposited on the tissue surface. One solution to this problem is to remove most, but not all of the excess hydrogel prior to the electrophoretic step. Once the tissue has been processed through the electrophoretic step, the remaining hydrogel can be removed along with the unwanted residue.

   Additionally, passive clearing for several days at 37 °C can reduce the required duration of electrophoretic tissue clearing and minimize damage from the prolonged heating during electrophoresis. Passive clearing by extending the incubation with clearing solution also maintains the highest level of fluorescence and can therefore be used to increase the final signal-to-noise ratio of genetically encoded fluorescent proteins.

9. OptiView can transparentize a 1 mm slice of fixed mouse brain without electrophoretic clearing within 1–2 days. It can be pushed to clear 2–3 mm tissues taking about a week but white matter will remain cloudy. In all cases, better transparency is achieved with longer incubation times. If possible, it is helpful to incubate the tissue sample for as long as possible prior to imaging.
10. In highly transparent, well-cleared tissue, multi-photon excitation is often not needed as visible wavelength lasers using a single photon process can often penetrate the entire working distance of the objective. As a single photon excitation event is more probable than a multi-photon event, this can result in a higher signal-to-noise ratio and the ability to scan faster [34]. However, multi-photon offers two advantages over single photon: reduced out-of-focal-plane-bleaching and increased penetration for excitation of blue/green emitting dyes [34] (*see* Subheading 2.4.2.1).

Acknowledgments

We would like to thank the Howard Hughes Medical Institute and the National Institute of Health (NIH/NIDCD RO1 5R01DC013087-02; F32 DC10089) for funding. We also thank the Harvard Center for Biological Imaging for providing infrastructure and support.

References

1. Petrovich GD, Canteras NS, Swanson LW (2001) Combinatorial amygdalar inputs to hippocampal domains and hypothalamic behavior systems. Brain Res Brain Res Rev 38(1–2):247–289
2. Swanson LW (2000) Cerebral hemisphere regulation of motivated behavior. Brain Res 886(1–2):113–164
3. Graybiel AM, Ragsdale CW Jr (1979) Fiber connections of the basal ganglia. Prog Brain Res 51:237–283
4. Watabe-Uchida M, Zhu L, Ogawa SK, Vamanrao A, Uchida N (2012) Whole-brain mapping of direct inputs to midbrain dopamine neurons. Neuron 74(5):858–873. doi:10.1016/j.neuron.2012.03.017
5. Wickersham IR, Lyon DC, Barnard RJ, Mori T, Finke S, Conzelmann KK, Young JA, Callaway EM (2007) Monosynaptic restriction of transsynaptic tracing from single, genetically targeted neurons. Neuron 53(5):639–647. doi:10.1016/j.neuron.2007.01.033
6. Lein ES, Hawrylycz MJ, Ao N, Ayres M, Bensinger A, Bernard A, Boe AF, Boguski MS, Brockway KS, Byrnes EJ, Chen L, Chen L, Chen TM, Chin MC, Chong J, Crook BE, Czaplinska A, Dang CN, Datta S, Dee NR, Desaki AL, Desta T, Diep E, Dolbeare TA, Donelan MJ, Dong HW, Dougherty JG, Duncan BJ, Ebbert AJ, Eichele G, Estin LK, Faber C, Facer BA, Fields R, Fischer SR, Fliss TP, Frensley C, Gates SN, Glattfelder KJ, Halverson KR, Hart MR, Hohmann JG, Howell MP, Jeung DP, Johnson RA, Karr PT, Kawal R, Kidney JM, Knapik RH, Kuan CL, Lake JH, Laramee AR, Larsen KD, Lau C, Lemon TA, Liang AJ, Liu Y, Luong LT, Michaels J, Morgan JJ, Morgan RJ, Mortrud MT, Mosqueda NF, Ng LL, Ng R, Orta GJ, Overly CC, Pak TH, Parry SE, Pathak SD, Pearson OC, Puchalski RB, Riley ZL, Rockett HR, Rowland SA, Royall JJ, Ruiz MJ, Sarno NR, Schaffnit K, Shapovalova NV, Sivisay T, Slaughterbeck CR, Smith SC, Smith KA, Smith BI, Sodt AJ, Stewart NN, Stumpf KR, Sunkin

SM, Sutram M, Tam A, Teemer CD, Thaller C, Thompson CL, Varnam LR, Visel A, Whitlock RM, Wohnoutka PE, Wolkey CK, Wong VY, Wood M, Yaylaoglu MB, Young RC, Youngstrom BL, Yuan XF, Zhang B, Zwingman TA, Jones AR (2007) Genome-wide atlas of gene expression in the adult mouse brain. Nature 445(7124):168–176. doi:10.1038/nature05453

7. Oh SW, Harris JA, Ng L, Winslow B, Cain N, Mihalas S, Wang Q, Lau C, Kuan L, Henry AM, Mortrud MT, Ouellette B, Nguyen TN, Sorensen SA, Slaughterbeck CR, Wakeman W, Li Y, Feng D, Ho A, Nicholas E, Hirokawa KE, Bohn P, Joines KM, Peng H, Hawrylycz MJ, Phillips JW, Hohmann JG, Wohnoutka P, Gerfen CR, Koch C, Bernard A, Dang C, Jones AR, Zeng H (2014) A mesoscale connectome of the mouse brain. Nature 508(7495):207–214. doi:10.1038/nature13186
8. Chung K, Wallace J, Kim SY, Kalyanasundaram S, Andalman AS, Davidson TJ, Mirzabekov JJ, Zalocusky KA, Mattis J, Denisin AK, Pak S, Bernstein H, Ramakrishnan C, Grosenick L, Gradinaru V, Deisseroth K (2013) Structural and molecular interrogation of intact biological systems. Nature 497(7449):332–337. doi:10.1038/nature12107
9. Dodt HU, Leischner U, Schierloh A, Jahrling N, Mauch CP, Deininger K, Deussing JM, Eder M, Zieglgansberger W, Becker K (2007) Ultramicroscopy: three-dimensional visualization of neuronal networks in the whole mouse brain. Nat Methods 4(4):331–336. doi:10.1038/nmeth1036
10. Erturk A, Becker K, Jahrling N, Mauch CP, Hojer CD, Egen JG, Hellal F, Bradke F, Sheng M, Dodt HU (2012) Three-dimensional imaging of solvent-cleared organs using 3DISCO. Nat Protoc 7(11):1983–1995. doi:10.1038/nprot.2012.119
11. Hama H, Kurokawa H, Kawano H, Ando R, Shimogori T, Noda H, Fukami K, Sakaue-Sawano A, Miyawaki A (2011) Scale: a chemical approach for fluorescence imaging and reconstruction of transparent mouse brain. Nat Neurosci 14(11):1481–1488. doi:10.1038/nn.2928
12. Ke MT, Fujimoto S, Imai T (2013) SeeDB: a simple and morphology-preserving optical clearing agent for neuronal circuit reconstruction. Nat Neurosci 16(8):1154–1161. doi:10.1038/nn.3447
13. Kuwajima T, Sitko AA, Bhansali P, Jurgens C, Guido W, Mason C (2013) ClearT: a detergent- and solvent-free clearing method for neuronal and non-neuronal tissue. Development 140(6):1364–1368. doi:10.1242/dev.091844
14. Susaki EA, Tainaka K, Perrin D, Kishino F, Tawara T, Watanabe TM, Yokoyama C, Onoe H, Eguchi M, Yamaguchi S, Abe T, Kiyonari H, Shimizu Y, Miyawaki A, Yokota H, Ueda HR (2014) Whole-brain imaging with single-cell resolution using chemical cocktails and computational analysis. Cell 157(3):726–739. doi:10.1016/j.cell.2014.03.042
15. Tainaka K, Kubota SI, Suyama TQ, Susaki EA, Perrin D, Ukai-Tadenuma M, Ukai H, Ueda HR (2014) Whole-body imaging with single-cell resolution by tissue decolorization. Cell 159(4):911–924
16. Yang B, Treweek JB, Kulkarni RP, Deverman BE, Chen CK, Lubeck E, Shah S, Cai L, Gradinaru V (2014) Single-cell phenotyping within transparent intact tissue through whole-body clearing. Cell 158(4):945–958. doi:10.1016/j.cell.2014.07.017
17. Renier N, Wu Z, Simon DJ, Yang J, Ariel P, Tessier-Lavigne M (2014) iDISCO: a simple, rapid method to immunolabel large tissue samples for volume imaging. Cell 159(4):896–910
18. Schultze O (1897) Über Herstellung and conservirung durchsichtigen embryonen zum stadium der skeletbildung. Anat Anz 13:3–5
19. Spalteholz W (1914) Über das Durchsichtigmachen von menschlichen und tierischen Präparaten und seine theoretischen Bedingungen. Leipzig, S. Hirzel
20. Chung K, Deisseroth K (2013) CLARITY for mapping the nervous system. Nat Methods 10(6):508–513. doi:10.1038/nmeth.2481
21. Kim SY, Chung K, Deisseroth K (2013) Light microscopy mapping of connections in the intact brain. Trends Cogn Sci 17(12):596–599. doi:10.1016/j.tics.2013.10.005
22. Tomer R, Ye L, Hsueh B, Deisseroth K (2014) Advanced CLARITY for rapid and high-resolution imaging of intact tissues. Nat Protoc 9(7):1682–1697. doi:10.1038/nprot.2014.123
23. Wang K, Milkie DE, Saxena A, Engerer P, Misgeld T, Bronner ME, Mumm J, Betzig E (2014) Rapid adaptive optical recovery of optimal resolution over large volumes. Nat Methods 11(6):625–628. doi:10.1038/nmeth.2925
24. Huisken J, Swoger J, Del Bene F, Wittbrodt J, Stelzer EH (2004) Optical sectioning deep inside live embryos by selective plane illumination microscopy. Science 305(5686):1007–1009. doi:10.1126/science.1100035
25. Huisken J, Stainier DY (2007) Even fluorescence excitation by multidirectional selective plane illumination microscopy (mSPIM). Opt Lett 32(17):2608–2610
26. Preibisch S, Saalfeld S, Tomancak P (2009) Globally optimal stitching of tiled 3D microscopic image acquisitions. Bioinformatics 25(11):1463–1465. doi:10.1093/bioinformatics/btp184

27. Schindelin J, Arganda-Carreras I, Frise E, Kaynig V, Longair M, Pietzsch T, Preibisch S, Rueden C, Saalfeld S, Schmid B, Tinevez JY, White DJ, Hartenstein V, Eliceiri K, Tomancak P, Cardona A (2012) Fiji: an open-source platform for biological-image analysis. Nat Methods 9(7):676–682. doi:10.1038/nmeth.2019
28. Peng H, Bria A, Zhou Z, Iannello G, Long F (2014) Extensible visualization and analysis for multidimensional images using Vaa3D. Nat Protoc 9(1):193–208. doi:10.1038/nprot.2014.011
29. Bria A, Iannello G (2012) TeraStitcher - a tool for fast automatic 3D-stitching of teravoxel-sized microscopy images. BMC Bioinformatics 13:316. doi:10.1186/1471-2105-13-316
30. Zhu D, Larin KV, Luo Q, Tuchin VV (2013) Recent progress in tissue optical clearing. Laser Photon Rev 7(5):732–757. doi:10.1002/lpor.201200056
31. Efimova OI, Anokhin KV (2009) Enhancement of optical transmission capacity of isolated structures in the brain of mature mice. Bull Exp Biol Med 147(1):3–6
32. Oldham M, Sakhalkar H, Oliver T, Allan Johnson G, Dewhirst M (2008) Optical clearing of unsectioned specimens for three-dimensional imaging via optical transmission and emission tomography. J Biomed Opt 13(2):021113. doi:10.1117/1.2907968
33. Genina EA, Bashkatov AN, Tuchin VV (2010) Tissue optical immersion clearing. Expert Rev Med Devices 7(6):825–842. doi:10.1586/erd.10.50
34. Denk W, Strickler JH, Webb WW (1990) Two-photon laser scanning fluorescence microscopy. Science 248(4951):73–76
35. Haddad R, Lanjuin A, Madisen L, Zeng H, Murthy VN, Uchida N (2013) Olfactory cortical neurons read out a relative time code in the olfactory bulb. Nat Neurosci 16(7):949–957. doi:10.1038/nn.3407
36. Madisen L, Zwingman TA, Sunkin SM, Oh SW, Zariwala HA, Gu H, Ng LL, Palmiter RD, Hawrylycz MJ, Jones AR, Lein ES, Zeng H (2010) A robust and high-throughput Cre reporting and characterization system for the whole mouse brain. Nat Neurosci 13(1):133–140. doi:10.1038/nn.2467

Chapter 12

Structured Illumination Microscopy for the Investigation of Synaptic Structure and Function

Soyon Hong*, Daniel K. Wilton*, Beth Stevens, and Douglas S. Richardson

Abstract

The neuronal synapse is a primary building block of the nervous system to which alterations in structure or function can result in numerous pathologies. Studying its formation and elimination is the key to understanding how brains are wired during development, maintained throughout adulthood plasticity, and disrupted during disease. However, due to its diffraction-limited size, investigations of the synaptic junction at the structural level have primarily relied on labor-intensive electron microscopy or ultra-thin section array tomography. Recent advances in the field of super-resolution light microscopy now allow researchers to image synapses and associated molecules with high-spatial resolution, while taking advantage of the key characteristics of light microscopy, such as easy sample preparation and the ability to detect multiple targets with molecular specificity. One such super-resolution technique, Structured Illumination Microscopy (SIM), has emerged as an attractive method to examine synapse structure and function. SIM requires little change in standard light microscopy sample preparation steps, but results in a twofold improvement in both lateral and axial resolutions compared to widefield microscopy. The following protocol outlines a method for imaging synaptic structures at resolutions capable of resolving the intricacies of these neuronal connections.

Key words Synapse, Neuron, Structured Illumination Microscopy, Colocalization, Immuno fluorescence, Channel alignment, Refractive index matching

1 Introduction

Synapses form key communication sites of the nervous system and as such, development and maintenance of healthy synapses are critical for the formation of functional neural networks. Loss of synaptic integrity and proper function has been linked to many developmental and neurodegenerative diseases including autism, schizophrenia, Alzheimer's disease, and Huntington's disease [1–4]. Careful analysis of synapse biology, in particular in intact brain tissue, has been limited with conventional light microscopy tools due to synapses' small size, high density, and complex matrix. Recent advances in the field of super-resolution light microscopy

*These authors contributed equally.

Alexandros Poulopoulos (ed.), *Synapse Development: Methods and Protocols*, Methods in Molecular Biology, vol. 1538, DOI 10.1007/978-1-4939-6688-2_12, © Springer Science+Business Media LLC 2017

now allow for the investigation of synapse structure and function at a more detailed level, opening a new method to study potential underlying mechanisms of synapse biology in both normal and diseased brains.

Region-specific synapse dysfunction and loss has long been considered a hallmark of Alzheimer's disease [5–8]; however, during the disease pathogenesis, how or when synapse loss occurs remains elusive. One approach is to examine the integrity of synapse structure in different brain regions of diseased versus control tissue by analyzing the close apposition of various pre- and postsynaptic markers. This has been challenging with conventional microscopy due to the diffraction limiting nature of light and the small size of synaptic structures. Electron microscopy studies show a typical synapse in the hippocampus to have a postsynaptic density ranging in area from 0.07 μm^2 to 0.42 μm^2 [9]. However, super-resolution microscopy techniques now provide the ability to visualize synapses in intact brain tissue and to resolve the distribution of proteins on their surface with a light microscope [10, 11]. These techniques also enable quantification of different populations of synapses within brain regions, something previously only possible with labor-intensive techniques such as immuno-electron microscopy or ultra-thin section array tomography.

A number of super-resolution techniques have been developed in recent years. In general, they can be classified under three umbrella terms: Structured Illumination Microscopy (SIM), Single Molecule Localization (SML), and REversible Saturable OpticaL Fluorescence Transitions (RESOLFT) [12]. All three of these techniques have been implemented commercially and are becoming widely accessible in core facilities and individual labs around the world. Although SIM does not attain the same level of resolution improvement that SML or RESOLFT do, SIM is appealing as it does not require large changes to standard light microscopy sample preparation, is relatively fast, and produces limited photo-damage and bleaching. In comparison to ultra-thin section array tomography, SIM provides comparable axial resolution, improved lateral resolution, simpler sample preparation, and eliminates artifacts caused by torn, folded, or incorrectly registered tissue sections.

A Structured Illumination Microscope uses constructive and deconstructive interference of excitation light at the focal plane of an objective to illuminate a sample with a series of sinusoidal stripes of high spatial frequency. These stripes are shifted laterally to illuminate the entire sample, and rotated to effect resolution enhancement in all lateral directions and an image is taken at each position [13, 14]. If the sample possesses high-frequency spatial information, the stripes produce moiré fringes that are observed in Fourier space and can be used to recover super-resolution data that is normally lost as light diffracts through the optics of the microscope. In an ideal sample, SIM produces a twofold improvement in resolution.

SIM can also be implemented to form a structured illumination pattern in the axial dimension, thereby producing a twofold resolution improvement in *x*, *y*, and *z* postprocessing [13, 15].

2 Materials

2.1 Solutions

All solutions should be prepared with ultrapure water and analytical-grade reagents.

1. TBS (tris-buffered saline): 7.86 g Tris–HCl, 8.67 g NaCl, Add H_2O to 1 L, Adjust pH to 7.4 with HCl.
2. PFA Solution: 4% Paraformaldehyde in PBS, freshly made, EM grade.
3. Sucrose Solution: 30% sucrose in PBS.
4. OCT: Cryosectioning specimen matrix (e.g., Tissue-Tek® O.C.T. compound; Sakura® Finetek USA).
5. Blocking Buffer: 20% goat serum (v/v) in TBS.
6. Antibody Dilution Buffer: 0.3% Triton X-100 (v/v), 10% normal goat serum (v/v) in TBS.
7. TDE Mounting Medium: 97% v/v 2,2′-Thiodiethanol (TDE), 2.4% v/v 1,4-Diazabicycle[2.2.2]octane (DABCO), 0.6% v/v PBS, Refractive Index = 1.51.

2.2 Microscope Setup

SIM has now been implemented by a number of commercial manufacturers. This protocol is routinely carried out on a Zeiss ELYRA PS1 microscope; however, it should be widely applicable to all other manufacturers.

Recommended features:

1. Multiple diffraction gratings or a tunable spatial light modulator matched to the wavelength of all laser lines to be used.
2. Optics to collect −1, +1, and 0 diffraction orders (0 order needed for 3D SIM).
3. Axial piezo stage with high accuracy and reproducibility. Step sizes of ~100 nm are needed to ensure Nyquist sampling in the axial dimension.
4. High sensitivity, low read noise, high bit depth camera. Both EMCCD and sCMOS cameras can be used. Pixel spacing should be <50 nm to ensure Nyquist sampling.
5. High numerical aperture (NA) objective lens (≥1.4).
6. Software to perform multichannel alignment (if imaging more than one channel).
7. Software to perform structured illumination calculations.
8. Analysis software for synapse quantification (e.g., Imaris, MatLab).

2.3 Color Alignment

Multicolor imaging requires multiple optical components (diffraction gratings, filter cubes) to be moved in and out of the light path. Each of these components can introduce slight shifts between color channels that can affect colocalization analysis (Fig. 1). Subdiffraction multicolor fluorescent spheres (Tetra-spec beads, 100–200 nm, Life Technologies) can be imaged to develop a transformation matrix that is applied to the final SIM image for color channel alignment.

2.4 Mounting Medium

A number of commercial mounting media are available for purchase. However, many super-resolution modalities are very sensitive to index of refraction mismatches. A slight RI mismatch can reduce axial resolution and imaging depth (as the structured illumination pattern quickly degrades) and produce artifacts during SIM processing (Figs. 2 and 3). Therefore, it is prudent to use mounting media

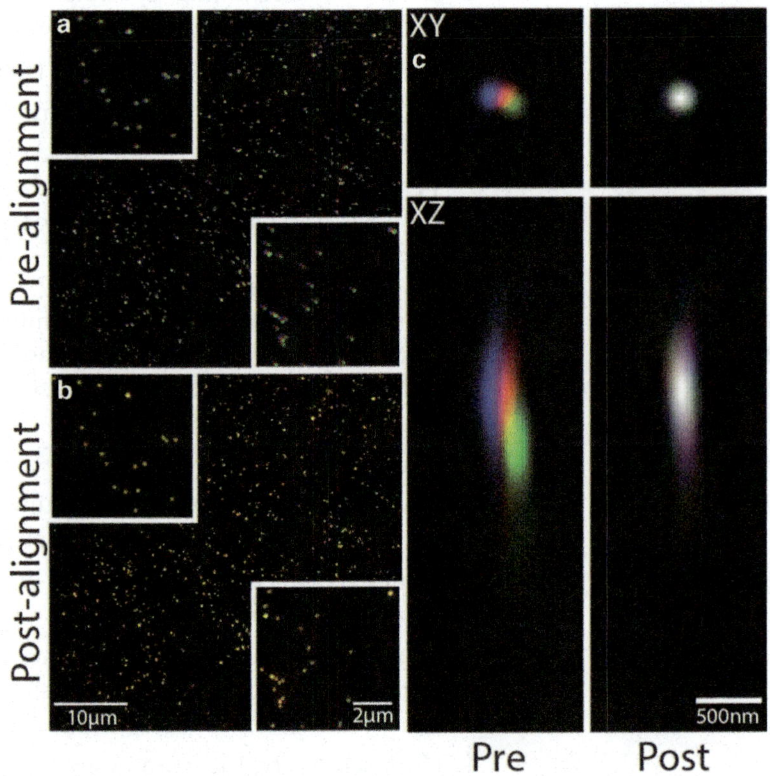

Fig. 1 Adjusting for chromatic aberration in SIM. (**a–c**) 200 nm beads imaged under the following conditions: 488 nm excitation, 495–575 nm emission collection (*green*); 561 nm excitation, 570–650 nm emission collection (*red*); 642 nm excitation, 655+ nm emission collection (*blue*). (**a**) SIM processed image of 200 nm beads. Enlarged regions in corners identify differing levels of chromatic shift within the same image. (**b**) Same image as in (**a**) after color channel alignment. (**c**) Lateral and axial projections of a single bead before and after color channel alignment

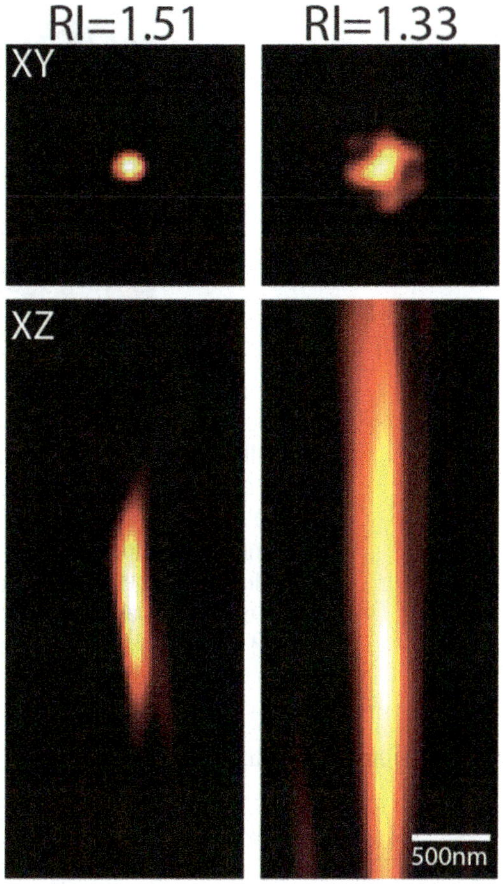

Fig. 2 Poor refractive index matching results in aberration. 100 nm red fluorescent beads were imaged (561 nm excitation, 570–650 nm emission) and processed using SIM. Beads were mounted in Immersol 518F (RI = 1.518, Carl Zeiss Microscopy) or Immersol W 2010 (RI = 1.33, Carl Zeiss Microscopy) under #1.5 thickness coverslips. Point-spread functions were measured on ten beads, averaged, and presented as lateral (*XY*) and axial (*XZ*) projections

matched as closely as possible to the immersion medium of your microscope objective. As most high NA objectives use oil immersion with an RI of approximately 1.51, glycerol-based mounting media are often not ideal (RI = 1.40–1.49; Table 1). If these must be used, it is recommended to allow them to cure for more than 72 h as the curing process will increase the RI. A number of solvent-based mounting media can be used that very closely match the RI of immersion oil and also have tissue clearing properties [16, 17]. As brain tissue is highly scattering, clearing properties are desirable. Caution must be used as some dyes are quenched in solvents, while others will increase in brightness [16]. Aqueous tissue samples will need to be moved through a dilution series of increasing solvent concentration prior to final coverslip mounting [16]. New, high RI aqueous mountants are now becoming available as well [18].

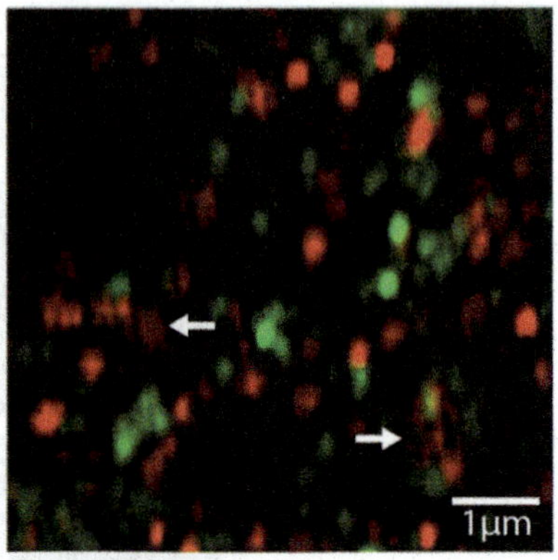

Fig. 3 Processing induced artifacts. 14 μm section of the dLGN of a p5 mouse stained with presynaptic marker VGlut2 and postsynaptic marker homer showing examples of grid line and noise artifacts that can arise during SIM processing. This can occur as a result of a poor signal-to-noise ratio in the sample, incomplete filtering of the grid frequency prior to inverse Fourier transformation, or intensity fluctuations between the raw phase and rotation images at the same *Z* plane (for example when photobleaching occurs). White arrows indicate regions where SIM processing artifacts were introduced

Table 1
Refractive indices of common mounting media

Mounting media	Refractive index	Manufacturer
Fluoromount G	1.40	SouthernBiotech
75 % Glycerol	1.44	Sigma
ProLong Gold	Fresh: 1.39 Cured 24 h: 1.40 Cured 4 days: 1.44	Life Technologies
Vectashield	Cured: 1.46	Vector Labs
Mowiol	Cured: 1.49	Sigma
SeeDB2	1.51	–
97 % TDE	1.51	Sigma
100 % DBE	1.56	Sigma
Wintergreen Oil	1.53	Sigma

2.5 Immersion Oil

The RI of immersion oil is measured at or near 23 °C. Differing from this temperature will change the RI of the oil. Depending on environmental conditions, ventilation around the microscope, how long the microscope has been running, and whether or not live-cell environmental stage incubators are present, the temperature near the sample may be far from 23 °C. A number of temperature-corrected immersion oils are commercially available; therefore, it is recommended that multiple oils be compared empirically (*see* **Note 1**).

2.6 Coverslips

Most microscope objectives are corrected for #1.5, or 170 μm, thick coverslips (this is indicated on the side of the objective lens; *see* **Note 2**).

3 Methods

3.1 Tissue Preparation

1. Deeply anesthetize a mouse with 2.5 % Avertin.
2. Perfuse the mouse with ice-cold 20–30 ml phosphate buffered saline (PBS), and then subsequently with 20–30 ml freshly made PFA Solution. Carefully remove the brain and place it in PFA Solution for 2 h on ice. Alternatively, the PFA perfusion step can be skipped. Perfuse mouse with PBS and drop fix the brain in PFA Solution for 2 h.
3. Rinse the brain 3× with PBS to remove any remaining PFA Solution and immediately cryoprotect by placing it in Sucrose Solution. Leave the brain in sucrose at 4 °C until the tissue sinks to the bottom of the tube (24–48 h).
4. Once the tissue sinks, prepare the brain for sectioning. The following steps are performed in the preparation for cryostat sectioning:
   (a) Incubate the tissue for 1 h at RT in 2:1 mixture of Sucrose Solution:OCT in a small tissue mold.
   (b) Freeze OCT-embedded brain by placing on dry ice. OCT-embedded brain can be stored at −80 °C for several weeks until ready to be sectioned.

3.2 Staining

1. Section the tissue using a cryostat (14 μm sections) ideally within 2 weeks of preparation and collect on superfrost slides.
2. Warm slides on a 37 °C heat block for 30 min.
3. Wash slides 2× with TBS.
4. Incubate slides with Blocking Buffer for 2 h at room temperature (RT).
5. Incubate slides with 1° antibody (diluted in Antibody Dilution Buffer to a working concentration that was previously empirically determined for the specific antibody) overnight at 4 °C.
6. Wash slides 3× for 10 min with TBS.
7. Incubate slides with 2° antibody (diluted in Antibody Dilution Buffer to a working concentration that was previously empirically determined) for 2 h at RT.

8. Wash slides 3× for 30 min with TBS.
9. Perform dehydration series if using solvent-based mounting medium.
10. Apply a small drop of desired mounting medium to each section and mount a coverslip on the top (22 × 50 mm, No. 1.5).
11. Seal edges of slides with nail polish and store at 4 °C until imaging session.

3.3 Imaging

1. Prepare multicolor bead slide for SIM calibration and channel alignment:
   (a) Dilute 100 nm Tetraspec beads (Life Technologies) 1:500 (v/v) in ethanol.
   (b) Sonicate for 3 min in a bench top water bath sonicator.
   (c) Place two drops of bead solution on a slide, spread across slide with a razor blade, and allow to dry.
   (d) Place a drop of mounting media on a coverslip and place over dried beads avoiding the introduction of air bubbles.
2. Image bead slide:
   (a) Acquire a full field of view 3D SIM image stack centered on the beads. *Z*-step = 90 nm, *Z*-range = 3.5 μm for each color channel present in the sample of interest.
   (b) Process the images using the same SIM algorithms to be used for the sample of interest (Fig. 1).
   (c) Measure full width half maximum (FWHM) of beads to assess performance of the microscope (if beads are subdiffraction limit in size).
   (d) Calculate a transformation matrix to align each color channel in *x*, *y*, and *z*. You will need an affine transformation; translation alone will not be sufficient (Fig. 1).
3. Image sample:
   (a) Individually adjust laser power, camera exposure time, and camera gain (if applicable) to fill a significant portion of the dynamic range of the camera. Avoid saturating any pixels. Avoid amplifying noise, nonspecific staining, autofluorescence, etc. if possible.
   (b) Depending on the number of grating rotations and channels imaged, acquisition times can be significant. Ensure all fluorophores used are stable and laser powers are not high enough to induce photobleaching resulting in decreasing levels of fluorescence at later rotations or *Z* planes as this can introduce artifacts (Fig. 3).

(c) If color channels are imaged sequentially, image higher wavelength channels first to reduce bleaching of yet to be imaged fluorophores.

(d) Acquire a 3D SIM image stack over desired range. Ensure Nyquist sampling in *x*, *y*, and *z* based on the expected resolution after SIM processing.

4. SIM processing:

(a) Process each color channel using SIM algorithms available in commercial or custom software (Fig. 4).

(b) Align each color channel based on the transformation matrix calculated in Subheading 3.3, **step 2**.

(c) Assess image for artifacts: background noise and/or incomplete removal of grid lines. If artifacts are seen, adjust parameters of SIM processing algorithm and reprocess (Fig. 3).

3.4 Synapse Quantification Using Imaris and Matlab

1. Open SIM-processed file in Imaris. Make sure that the Volume icon is selected in the top left menu.
2. Confirm the pixel values. Go to the Edit menu and select Image Properties. In this window, the *x*, *y*, and *z* voxel sizes are displayed. These values depend on the microscope, objective, camera, and *z*-step acquisition and can be found in the software used to acquire the image.
3. In the Display Adjustment window, maximize visualization of all immunoreactive synaptic puncta by increasing brightness for each channel (Fig. 5a).
4. Create spots for each channel:

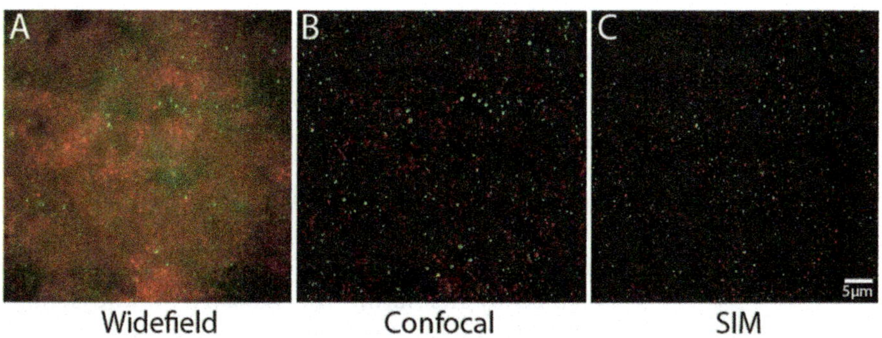

Fig. 4 SIM imaging of mouse brain sections. Widefield (**a**), point scanning confocal (**b**), and SIM (**c**) images of a 14 μm section of an adult (3 months) mouse brain stained with presynaptic marker synaptotagmin (*green*) and postsynaptic homer (*red*). Images were processed identically using ImageJ: a smoothing filter was applied to all channels, and a 0.5 gamma adjustment was applied to all green channels (to allow visualization of dim synapses, without saturating larger puncta)

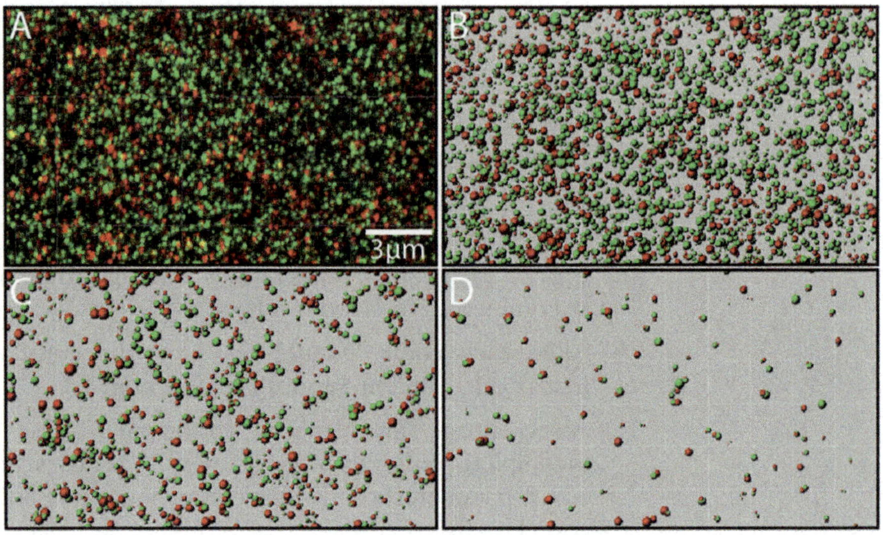

Fig. 5 Data analysis. 3D-SIM in stratum radiatum of 3 months mouse hippocampus (CA1) in which synapses are immunolabeled with presynaptic marker synaptophysin (*green*) and postsynaptic marker PSD-95 (*red*). (**a**) Maximum intensity projection (MIP) of synaptophysin and PSD-95 immunostaining in 3 months hippocampus. (**b**) MIP in which spots were created for synaptophysin (*green*) and PSD-95 (*red*), respectively. (**c**, **d**) Apposed synaptophysin and PSD-95 immunoreactive spot centers (created in **b**) that are within 0.4 μm (**c**) or 0.2 μm (**d**) distance of one another

(a) In the top tool bar, make sure the Surpass mode is selected.

(b) In the Display Adjustment window, select the channel for which to create spots.

(c) Click the Spots icon.

(d) A create window will be displayed in the bottom left. In Algorithm Settings, make sure "Segment ROI" is unchecked and the "Select Different Spot Sizes (Region Growing)" is checked. Click the blue forward button at the bottom.

(e) In the next window, choose one source channel from the pull-down menu. For Spot Detection, select the "Detect Ellipsoids" option and enter expected *xy* and *z* diameters. (These values should be determined empirically depending on the particular antibodies used and tissue regions imaged. The average values for the diameters can be determined by measuring the visualized fluorescent puncta in the Slice mode. *See* **Note 3** for some experimentally determined values.) Uncheck background subtraction.

(f) Spot detection: choose "Quality Filter." Click on histogram and drag until spots created appropriately reflect center (local maxima) of all fluorescent puncta displayed in the source channel. Note the threshold value used. This value can vary between the two channels, but within one specific channel the value should be kept consistent between sets of images that are to be compared.

(g) Choose to threshold using "Absolute Intensity" for region growing.

(h) Define spot growth boundary: confirm whether the threshold displayed in white appropriately reflects the fluorescent image of the source channel. This is determined empirically by rotating the image and moving the view up and down the *z*-plane to ensure that the white signal is appropriately overlaid on the fluorescent signal. To rotate, select "Navigate" in the Pointer menu on the upper right side of the screen, click and hold image to rotate. To move the viewing plane along the *z*-axis, click "Select" in the Pointer menu then move the highlighted plane along the *z*-axis. To return image to original orientation, select "Origin Bottom Left" from the View menu. Click the green "Finish" button.

(i) Confirm that the spots created reflect the fluorescent image in the source channel. One can change color of the spots to a different one from the original fluorescent image to better differentiate the two images.

(j) Click on the "Funnel" (i.e., Filter tab) and select volume from the pull-down menu. Turn on upper threshold and filter out any spots that are deemed too big to be considered synaptic.

(k) Go to the Graph tab and click on "Details" and select any parameters to display. For displaying multiple parameters (volume, diameters, etc.), go to the Wrench icon at the bottom left and select the parameters to record.

(l) Record the number of spots and save the file before continuing.

(m) Create spots for the second channel by following the steps above. Figure 5b displays a maximum intensity projection of spot selection in a two-channel image.

5. Quantify number of colocalized spots:

(a) Go to Image processing, then select Spot Functions, then Colocalize Spots. This will open MatLab.

(b) Select the two spots to colocalize.

(c) Enter the threshold value of the distance between spot centers. This will be dependent on the particular pre- and postsynaptic antibodies used.

(d) Apposed spots will now appear (Fig. 5c, d). Go to the Graphs tab and record number of collocated spots for each spot.

4 Notes

1. Standard immersion oil (corrected to 23 °C) can be blended with high temperature (corrected to 37 °C) oil at various ratios to fine-tune the final solution for use with your particular microscope.
2. Standard #1.5 coverslips can deviate from 170 μm; therefore, coverslip thickness should be confirmed with micrometer calipers prior to use, or high precision coverslips should be purchased. Coverslips should also be cleaned and sonicated in a basic solution.
3. For synaptic puncta marked by synaptotagmin in stratum radiatum of adult mouse hippocampus (CA1), diameters as small as 0.1–0.2 μm for *xy* and 0.2–0.5 μm for *z* diameters have been measured. These are below the resolution limit of a standard confocal microscope (~200 nm *xy* and ~600 nm *z*), emphasizing the need for SIM.

Acknowledgments

We thank Tiao Xie and Hunter Elliot at the Harvard Image and Data Analysis Core for guidance on synapse quantification. We also thank the Harvard Center for Biological Imaging for infrastructure and imaging support. These experiments were supported by funding from the Coins for Alzheimer's Research Trust (BS), Edward R. and Anne G. Lefler Fellowship (SH), and the National Institutes of Health (RO1NS071008, BS; AG000222, SH; 1S10RR029237-01).

References

1. Mucke L, Selkoe DJ (2012) Neurotoxicity of amyloid beta-protein: synaptic and network dysfunction. Cold Spring Harbor Perspect Med 2(7):a006338
2. Ehrnhoefer DE, Wong BK, Hayden MR (2011) Convergent pathogenic pathways in Alzheimer's and Huntington's diseases: shared targets for drug development. Nat Rev Drug Discov 10(11):853–867
3. Faludi G, Mirnics K (2011) Synaptic changes in the brain of subjects with schizophrenia. Int J Dev Neurosci 29(3):305–309
4. Penzes P, Vanleeuwen JE (2011) Impaired regulation of synaptic actin cytoskeleton in Alzheimer's disease. Brain Res Rev 67(1-2):184–192
5. Terry RD, Masliah E, Salmon DP et al (1991) Physical basis of cognitive alterations in Alzheimer's disease: synapse loss is the major correlate of cognitive impairment. Ann Neurol 30(4):572–580
6. DeKosky ST, Scheff SW, Styren SD (1996) Structural correlates of cognition in dementia: quantification and assessment of synapse change. Neurodegeneration 5(4):417–421
7. Selkoe DJ (2002) Alzheimer's disease is a synaptic failure. Science 298(5594):789–791
8. Koffie RM, Hashimoto T, Tai HC et al (2012) Apolipoprotein E4 effects in Alzheimer's disease are mediated by synaptotoxic oligomeric amyloid-beta. Brain 135(Pt 7):2155–2168
9. Harris KM, Sultan P (1995) Variation in the number, location and size of synaptic vesicles provides an anatomical basis for the nonuniform probability of release at hippocampal CA1 synapses. Neuropharmacology 34(11):1387–1395

10. Dani A, Huang B, Bergan J et al (2010) Superresolution imaging of chemical synapses in the brain. Neuron 68(5):843–856
11. Urban NT, Willig KI, Hell SW et al (2011) STED nanoscopy of actin dynamics in synapses deep inside living brain slices. Biophys J 101(5):1277–1284
12. Hell SW (2007) Far-field optical nanoscopy. Science 316(5828):1153–1158
13. Gustafsson MG, Shao L, Carlton PM et al (2008) Three-dimensional resolution doubling in wide-field fluorescence microscopy by structured illumination. Biophys J 94(12):4957–4970
14. Gustafsson MG (2000) Surpassing the lateral resolution limit by a factor of two using structured illumination microscopy. J Microsc 198(Pt 2):82–87
15. Schermelleh L, Carlton PM, Haase S et al (2008) Subdiffraction multicolor imaging of the nuclear periphery with 3D structured illumination microscopy. Science 320(5881):1332–1336
16. Staudt T, Lang MC, Medda R et al (2007) 2,2′-thiodiethanol: a new water soluble mounting medium for high resolution optical microscopy. Microsc Res Tech 70(1):1–9
17. Renier N, Wu Z, Simon DJ et al (2014) iDISCO: a simple, rapid method to immunolabel large tissue samples for volume imaging. Cell 159(4):896–910
18. Ke MT, Nakai Y, Fujimoto S et al (2016) Super-resolution mapping of neuronal circuitry with an index-optimized clearing agent. Cell Rep 14(11):2718–2732

Chapter 13

3D *d*STORM Imaging of Fixed Brain Tissue

Frank Herrmannsdörfer, Benjamin Flottmann, Siddarth Nanguneri, Varun Venkataramani, Heinz Horstmann, Thomas Kuner*, and Mike Heilemann*

Abstract

Central nervous system tissue contains a high density of synapses each composed of an intricate molecular machinery mediating precise transmission of information. Deciphering the molecular nanostructure of pre- and postsynaptic specializations within such a complex tissue architecture poses a particular challenge for light microscopy. Here, we describe two approaches suitable to examine the molecular nanostructure of synapses at 20–30 nm lateral and 50–70 nm axial resolution within an area of 500 μm × 500 μm and a depth of 0.6 μm to several micrometers. We employ single-molecule localization microscopy (SMLM) on immunolabeled fixed brain tissue slices. *tomo*STORM utilizes array tomography to achieve SMLM in 40 nm thick resin-embedded sections. *d*STORM of cryo-sectioned slices uses optical sectioning in 0.1–4 μm thick hydrated sections. Both approaches deliver 3D nanolocalization of two or more labeled proteins within a defined tissue volume. We review sample preparation, data acquisition, analysis, and interpretation.

Key words Super-resolution microscopy, *d*STORM, *tomo*STORM, Brain tissue fixation, Thin sectioning, STORM-data postprocessing

1 Introduction

Brain parenchyma consists of billions of densely packed neuronal and glial processes each containing a multitude of nanostructural specializations such as synapses. Synapses can be considered highly ordered molecular machines that mediate synaptic communication and information processing. Understanding the molecular nanostructural organization of synapses represents a major challenge to understand the mechanisms of synaptic transmission at all levels of complexity. Single-molecule super-resolution fluorescence microscopy [1] is a powerful technique that has the capability to unravel nanoscopic protein assemblies [2].

*Correspondence may be addressed to either author.

Alexandros Poulopoulos (ed.), *Synapse Development: Methods and Protocols*, Methods in Molecular Biology, vol. 1538,
DOI 10.1007/978-1-4939-6688-2_13,

Here, we describe two approaches suitable for 3D super-resolution imaging of immunolabeled fixed brain tissue. The first method, *tomo*STORM [3], achieves high axial resolution using the principle of array tomography [4]. A resin-embedded tissue block is serially sectioned into a ribbon of sections at a thickness of approximately 40 nm. Hence, the resolution is defined by the thickness of the section. The sections are continuously collected on a glass support, thereby forming a long ribbon. Depending on the experimental requirements and extent of automation of the imaging setup used, hundreds to several thousands of sections can be generated. This allows for the examination of large tissue volumes. The sections can be labeled with two antibodies for dual-color imaging. Stripping or bleaching of the antibodies is possible allowing for multiple rounds of dual-color labeling. Sequentially recorded image frames need to be aligned to generate a 3D image stack. With this strategy, the achievable imaging volume is only restricted by the number of sections that can be processed. Antibody labeling efficiency will be limited because the resin restricts epitope accessibility.

The second method is an adaptation of the long-established Tokuyasu approach [5] (Fig. 1). Fixed brain sections are frozen and cryo-sectioned at −80 °C into sections ranging from 100 nm to a few micrometers thick. The rehydrated sections are then immunostained and imaged with *d*STORM (direct stochastic optical reconstruction microscopy) [6] using the astigmatism approach

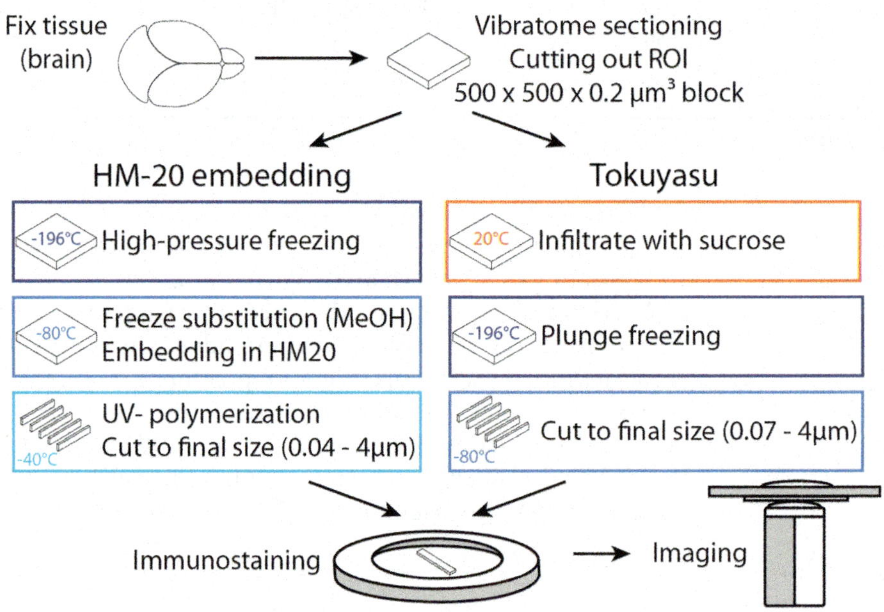

Fig. 1 Schematic overview of tissue preparation for super-resolution imaging following the *tomo*STORM (*left*) and *hcsd*STORM (*right*) approach

for 3D imaging [7]. Compared with the first method, this approach provides superior imaging and immunostaining properties while the 3D imaging volume is restricted.

In this overview and protocol, we describe both techniques and discuss their pros and cons. We also describe data acquisition and image reconstruction from localization data. Furthermore, we discuss post-processing steps and interpretation of the data.

2 Materials

2.1 Sample Preparation

1. Perfusion solution: 4% PFA in PBS.
2. Winged infusion set with two-channel tubing.
3. Standard dissection tools.
4. Cryo-ultramicrotome (e.g., Leica Ultracut S with cryo chamber).

2.2 tomoSTORM

1. Embedding resin: Lowicryl HM20 resin (LR White resin is not recommended because of poor preservation of tissue ultrastructure [3]).
2. Freeze substitution solution: 3% uranyl acetate in methanol.
3. Methanol (100% and chilled to −85 °C).
4. Ultramicrotome (e.g., Ultracut E, Reichert Jung).
5. Coverslip solution: 50% sulfuric acid, 50% hydrogen peroxide.
6. Blocking solution: 5% Fetal calf serum in PBS.
7. Primary and fluorescently conjugated secondary antibodies as desired.

2.3 Hydrated Cryo-sectioned Slices dSTORM (hcsdSTORM)

1. Cryo-protection solution: 2.1 M sucrose.
2. Liquid nitrogen.
3. Platinum loop (e.g., Perfect Loop from Diatome).

2.4 Microscope Setup

The base of a *d*STORM microscope is a conventional widefield microscope equipped with high-power laser sources, an objective with a high numerical aperture for total internal reflection (TIR) illumination and a single-photon sensitive camera.

In the case presented here, the microscope body is based on an inverted microscope body (Olympus IX71, Olympus, Japan), which is placed on an air damped optical table. The objective is an immersion oil objective suitable for TIR illumination (Plan ApoN 60×/1.45 oil, Olympus, Japan). Excitation light is provided by a multiline Ar/Kr-ion laser (InnovaI70C, Coherent, USA) and the dedicated wavelength is selected by an acousto-optical tunable filter (AOTF, Pegasus, Germany). A second laser generates light of 405 nm (Cube, Coherent, USA) and both lasers are combined by an appropriate dichroic mirror (BSR405, AHF, Germany) and focused onto the back-focal plane of the objective by a lens telescope. A manually movable mirror can be used to direct the beam into TIR illumination. A pair of long path and band path filters for the respective color channel is used to filter out any background light, while a dichroic mirror separates excitation and emission light. The emitted light is then focused onto the chip of an electron multiplying charged-coupled device (EMCCD) camera (iXon DV897DC, Andor, Ireland) by a tube lens. A cylindrical lens (f= 500 mm) can be placed in between the tube lens and the camera for 3D acquisition via astigmatism.

2.5 Imaging Protocol

1. Imaging buffer: 100 mM mercapto ethylamine or beta-mercapto ethanol, 60 mM glucose, 100 nM glucose oxidase, and 1.5 μM catalase in PBS (*see* **Note 1**).
2. STORM microscope.
3. Computer for reconstruction.

3 Methods

3.1 Principles of Single-Molecule Super-Resolution Microscopy

Due to diffraction and a limited opening angle of microscope objectives, the image of a punctual light source appears as a blurred spot. The intensity distribution in this image follows the so-called point-spread-function (PSF). The profile of a PSF in an image can be approximated with a two-dimensional Gaussian function with a precision down to a few nanometers [8, 9]. The full width at half maximum (FWHM) of a PSF depends on the emission wavelength and the integrated intensity—equaling the number of detected photons. Using visible light, the lateral FWHM of a PSF is typically at around 200–300 nm. As biomolecules and their fluorescent labels are one or two orders of magnitude smaller than their PSFs and can be highly abundant in cells, the individual PSFs start to overlap in fluorescence images. If the overlap becomes too high, individual point emitters cannot be distinguished, which results in the resolution limit of conventional light microscopes.

The idea of single-molecule localization microscopy (SMLM) is to separate the emission of fluorophores in time and to detect an image series of subsets of single fluorophores by employing photoactivation or photoswitching (e.g., STORM [10], PALM [11], FPALM [12], *d*STORM [6]). This way, a possible overlap of PSFs can be significantly reduced. The center of mass of single PSFs is approximated with nanometer precision and the examined coordinates are listed. In the end, a super-resolved image can be reconstructed from this list (Fig. 2a).

Several ways exist to provide three-dimensional information during an SMLM experiment. In 1994, Kao and Verkman showed a new approach for single-particle tracking in 3D by using a cylindrical lens with a large focal length that was placed in the detection

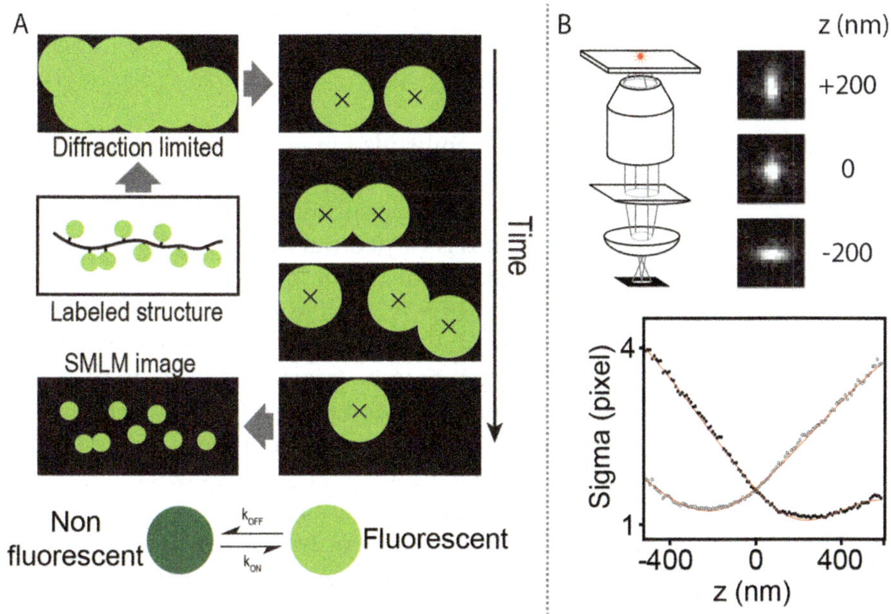

Fig. 2 Principle of SMLM and 3D image acquisition and reconstruction using astigmatism. (**a**) The emission light of individual fluorophores is temporally separated and recorded consecutively. In a final step, the center of each individual emission is located with a precision of a few nanometers. (**b**) A cylindrical lens additionally focuses light only in one spatial dimension, resulting in an asymmetric PSF that reports on the axial position of the fluorophore

pathway of an epi-fluorescence microscope [13]. The cylindrical lens introduces asymmetry into the PSF along the *z*-axis. A *z*-stack of fluorescent beads hereby acts as a reference. The width in *x* and *y* is recorded and the ratio of both determines the *z*-position. In the raw-data of a 3D experiment, this ratio is determined for every detected PSF and a *z*-value is assigned based on the reference data (Fig. 2b). Based on this work, 3D STORM imaging was demonstrated on the microtubule network and clathrin-coated pits in a cell [7]. Due to the cylindrical lens, the focus along the *x*-axis (or *y*-axis, depending on the direction of the cylindrical lens) is distorted, but remains the same in the other direction. As a consequence, the PSF of a point emitter is distorted depending on the distance to the focal plane. By extracting the width in *x*- and *y*-direction, *z*-values can be assigned with the use of adequate calibration curves (Fig. 2b). Other examples using optical effects to provide 3D information exploit effects of interference [14], generate double-helically arranged PSFs [15], or use a biplane detection scheme [16].

3.2 Perfusion

Irrespective of the strategy used, fixation of the tissue is achieved by transcardial perfusion of the animal [3, 17] (*see* **Note 2**). After transcardial perfusion the brain is removed from the skull and post fixed overnight. A block of tissue containing the target structures is then dissected out and further sectioned into 100–200 μm thick brain slices using a vibratome. From these slices, the target region can be excised (e.g., 500 μm × 500 μm).

3.3 tomoSTORM

The tissue block is embedded in embedding resin. To achieve this, the tissue blocks are frozen using a high-pressure quick-freeze apparatus. Subsequently, freeze substitution in a mixture of 3 % uranyl acetate in methanol is carried out at −85 °C for 70 h, followed by three washes in methanol at −85 °C. Then, sections are warmed up to −40 °C at a rate of 5 °C/h. Tissue blocks are then embedded in Lowicryl HM20 at −40 °C and UV-polymerized for 36 h at −40 °C. Samples can then be transferred to room temperature and UV polymerization continued for another 24 h. After achieving full polymerization, ultrathin serial sections are prepared using an ultramicrotome (Ultracut E, Reichert Jung) as described previously [3, 18]. The ribbons of sections are mounted on coverslips previously made hydrophilic by treatment with a mixture of 1:1 sulfuric acid and hydrogen peroxide. After washing the mounted ribbons in PBS for 5 min and blocking in 5 % FCS in PBS, an appropriate dilution of primary antibodies is applied in the same solution for 2 h. Samples are washed three times in PBS solution and the secondary antibody is applied for 30 min. After washing three times in PBS, the sample can be used for *d*STORM imaging.

3.4 Tokuyasu Sample Preparation for dSTORM

This procedure largely follows the established Tokuyasu method [5]. For cryo-protection, the tissue block is immersed overnight in 2.1 M sucrose, mounted on the sample holder, and plunge-frozen in liquid nitrogen. The sample is then cut on an ultramicrotome equipped with a cryo-stage at −70 °C. Depending on the aims of the experiment, sections with a thickness ranging from 100 nm to 4 μm are prepared (*see* **Note 3**). Slices are picked up with a platinum loop, placed on a glass slide, and washed in PBS until the sucrose is cleared from the sample. If sections are not evenly placed on the glass surface, they need to be unmounted and subsequently remounted on the glass. The sections are then blocked with 5 % FCS in PBS. Appropriately diluted antibodies are then added in blocking solution for two hours and subsequently washed three times with PBS. Thereafter, the secondary antibody is applied at a suitable concentration for 30 min and washed three times with PBS. Subsequently, the sample can be used for imaging.

3.5 Labeling Techniques and Imaging Conditions in Fixed Brain Tissue

3.5.1 Labels

Next to fluorescent proteins, organic fluorophores are frequently used for labeling a biomolecule of interest (*see* **Note 4**). Prominent classes of organic dyes that can also be used for SMLM are rhodamines, oxazines, and carbocyanines. They show a high relative brightness, a high photostability, and they cover the whole visible spectrum in their fluorescence emission. They can be tagged to target molecules by a variety of coupling groups. To use these dyes for SMLM, they must exhibit photoswitching [1], which is typically induced by adding an imaging buffer. This buffer often consists of millimolar concentrations of thiols like mercapto ethylamine (MEA) or beta-mercapto ethanol (β-ME) in a phosphate buffer [19]. In case of carbocyanines (such as Alexa Fluor 647 and Cy5), oxygen is removed enzymatically from the imaging buffer, which largely facilitates photoswitching [10, 20] (*see* **Note 1**).

3.5.2 Dual-Color Imaging

The most intuitive way to perform multilabeling experiments is to use spectrally distinguishable dyes. Arguably, the most frequently used organic dye is Alexa Fluor 647 or its derivative Cy5 [20]. It exhibits excellent photoswitching characteristics and emits in the far-red, where tissue autofluorescence is comparatively low. In order to avoid low signal-to-noise ratios caused by autofluorescence of the tissue, the second fluorophore should ideally emit in the yellow to red region, not overlapping with the far-red signal. The yellow-emitting dye Alexa Fluor 532 works well under the same chemical conditions as Alexa Fluor 647/Cy5 and can therefore be used in two-color experiments [21]. However, one disadvantage of this approach is that chromatic aberrations need to be corrected, for instance by using multispectral markers (for more details, see below). This requirement can be avoided by multicolor imaging using spectral demixing [22]. Here, two fluorophores with overlapping emission spectra are used (e.g., Alexa Fluor 647/Cy5 in combination with Alexa Fluor 700, CF680, or Dyomics678). A dichroic mirror with a selection wavelength between the two emission maxima of the fluorophores is placed in the imaging pathway, which splits the fluorescence emission onto two channels that are recorded simultaneously. Each fluorophore is detected in both channels; depending on the emission spectrum of the fluorophore, the intensity ratio between the two channels varies and allows assigning single fluorescence spots to one of the two fluorophores. Spectral demixing offers the advantage of negligible chromatic aberration, and facilitates imaging in the far-red where the autofluorescence of the tissue is minimal.

3.6 Imaging Protocol

Just before the start of the image recording, the buffer in which the sample is stored is replaced with imaging buffer (*see* **Note 5**). Changing the illumination between epi-fluorescence and TIR (*see* **Note 6**) helps to reduce background excitation. A region of interest can be selected by screening the sample in standard fluorescence microscopy mode. After finding the region of interest, the excitation power is strongly increased to levels of 3–10 kW/cm^2 to induce photoswitching. The em-gain of the camera needs to be set to high amplification to optimize signal-to-noise. An image series of several thousand frames is recorded until most of the fluorophores have been read out at least once. Thereby, the integration time should on average match the time a fluorophore emits photons for maximum signal-to-noise ratio (otherwise the intensity is split over multiple frames). Usually, this time averages between 10 and 200 ms (*see* **Notes 7** and **8**). If multiple channels are to be recorded, imaging will start in the red range of the spectrum and end in the blue range. This is important, because imaging in the reverse order will result in a progressive decrease of the fluorescence signal, thereby losing single-molecule information.

3.7 Image Reconstruction and Processing

3.7.1 Reconstruction

After acquisition of the raw data containing multiple frames and up to millions of PSFs (Fig. 3a), the coordinates of the emitters need to be determined as accurately as possible. This can be achieved by using one of the many freely available software packages (e.g., rapidSTORM [23] and ThunderSTORM [24]). Briefly, the most common approach to identify the coordinates is to assume a two-dimensional Gaussian model. Fitting the model to the data is either done by least squares or the maximum likelihood method. From the fitted parameters of the Gaussian model, the coordinates of each blinking event can be estimated. This estimation is error-prone mostly due to unavoidable noise present in the images. It is important to keep in mind that the emitter itself has a certain distance to the structure of interest labeled with the fluorophore. The output of this processing step is a list with an entry for each detected fluorophore which contains the coordinates, frame number, photon number, and may also contain additional information such as the fitting error. Beyond this necessary and basic processing step, further processing such as drift correction, merging, multichannel registration, and spectral demixing needs to be applied.

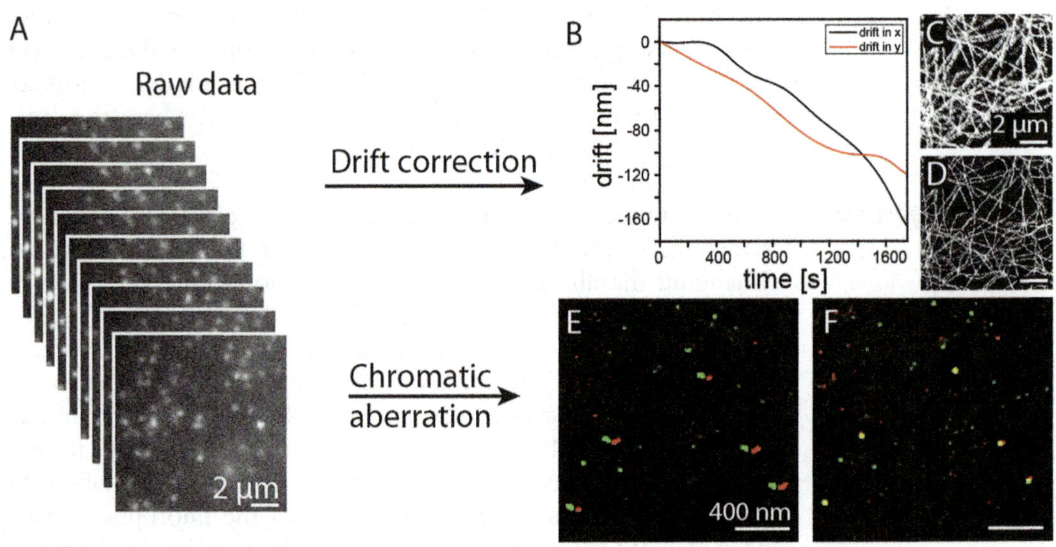

Fig. 3 Selected examples of postprocessing of SMLM data. (**a**) Single-molecule positions are extracted from the raw data image stack, and further corrected for (**b**–**d**) drift and (**e**–**f**) chromatic aberration

3.7.2 Postprocessing

1. Drift

   Mechanical drift poses a common problem in SMLM because of the long acquisition times (minutes) required to collect a sufficient number of single-molecule events. Mechanical drift is caused by thermal instabilities and other factors and results in blurred and distorted super-resolution images (Fig. 3b, c). Drift can be corrected in several ways. One option is to add multispectral beads to the sample which emit fluorescence throughout the experiment and can therefore be easily detected and tracked. Assuming that the beads are not moving relative to the bottom of the sample dish, drift can be estimated and corrected for. Spatial cross-correlation of subsets of the data represents intrinsic information that can be harnessed to estimate drift [25]. The estimated drift is then applied to correct the coordinates of all localizations (Fig. 3d).

2. Merging point spread functions

   Merging represents a postprocessing step that merges all localizations of a single blinking event into a single localization. Single fluorophore blinking events might be split over more than one frame. Hence, localizations that appear close to each other in consecutive frames originate with a high probability from the same fluorophore. Averaging of the coordinates of these events reduces the localization error.

3. Chromatic aberration

   Chromatic aberration is a key problem of multicolor SMLM. Even modern microscope lenses do not offer sufficient chromatic aberration correction to work at nanoscopic resolutions; hence, small shifts of the localizations for two spectrally separated fluorophores will be evident despite the fact that within the sample they reside at identical positions in space (Fig. 3e). Two solutions can be applied to solve this problem. One strategy is to use fluorescent markers (e.g., beads) filled with several fluorophores visible in all channels required for imaging. From the offsets of the different fluorophores, a transformation can be found that maps all bead positions of one channel to the bead positions found in the other channels (Fig. 3f). Alternatively, spectrally highly overlapping fluorophores can be used, thereby avoiding strong chromatic aberrations. Separation of the fluorophores can be reliably achieved using spectral demixing (*see* above).

3.8 Image Analysis and Visualization

The spatial resolution in an SMLM experiment is mainly determined by localization precision and labeling density. The localization precision itself depends on the number of photons detected for a single fluorophore, the pixel size, the background signal, and the emission wavelength. Despite theoretical models that provide ideal localization precisions [8, 9], an estimate of the experimental localization precision might sometimes be more useful, as this reports on effects such as inhomogeneous background that are otherwise not taken into account. One approach is to use to estimate the average localization precision from nearest-neighbor distances in adjacent imaging frames [26]. For this approach, the nearest neighbor of each data point in the consecutive frame is found, and the pair-wise distance is calculated and histogrammed. A model predicting the shape of the histogram is then fitted, and the localization precision extracted. Additional correction terms are introduced to correct for the influence of nearby localizations but originating from different fluorophores. Other approaches to determine the experimental localization precision use Fourier ring correlation between two subsets of the data [27, 28].

After completion of these image processing steps, the data is typically rendered to an intensity image. First, an empty image with the desired pixel-to-nm ratio is created. For the rendering of each localization, a two-dimensional Gaussian is centered at the fluorophore's position and sampled for the surrounding pixels. The localization error can be used as the standard deviation of the Gaussian function. This spreads a single localization over multiple pixels but enables displaying it with subpixel accuracy. For three-dimensional data a color code can be applied to a *z*-projected 2D image, ranging for example from blue for the smallest *z*-values up to red for the largest. Another option is to create an empty image stack using the same rendering procedure as in the case of 2D data, but with a three-dimensional Gaussian. This image stack can be displayed by 3D visualization software.

3.9 Examples and Biological Inferences

In the following, we present examples for *d*STORM imaging in hydrated cryo-sectioned slices, images for *tomo*STORM [3]. Figure 4a shows an overview image of the medial nucleus of the trapezoid body containing a calyx of Held presynaptic terminal and its postsynaptic cell, stained with antibodies against tubulin (red) and the synaptic vesicle protein VGluT1 (green). Tubulin is located away from the release face of the calyx while synaptic vesicles form clusters mostly between the release face and tubulin. When zooming into the image, synaptic vesicles colocalized with single microtubuli can be readily identified (Fig. 4b). These 2D images have been corrected as described above. Another example shows the distribution of synaptic vesicle clusters within the calyx using 3D STORM (Fig. 4c). An arrangement known from electron microscopy is evident, the cluster is subdivided into smaller clusters (Fig. 4d) and contains signals that may represent single synaptic vesicles. Assuming a labeling efficiency of 20 % and a copy number of 9 VGluT1 proteins per synaptic vesicle [29], about one to three labels are expected for most vesicles. While it is difficult to unequivocally identify the number of synaptic vesicles within dense vesicle clusters, the technique does provide a much more detailed

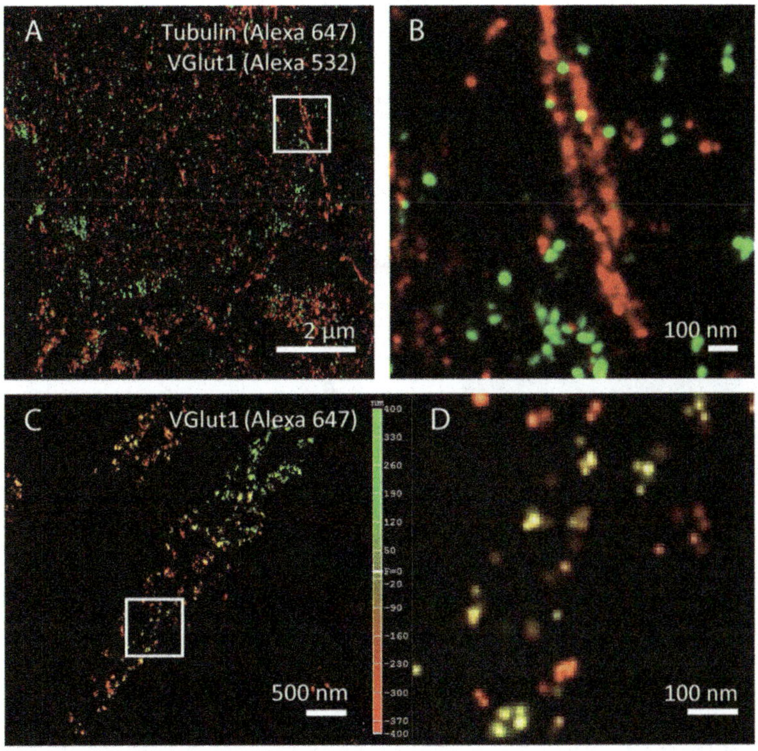

Fig. 4 (**a**) 2D super-resolution image and (**b**) zoom-in of tubulin stained with Alexa Fluor 647 (*red*) and synaptic vesicles stained with Alexa Fluor 532 (*green*) within the presynaptic terminal of a calyx of Held and its postsynaptic cell (*hcs-d*STORM). (**c**) 3D super-resolution image and (**d**) zoom-in of synaptic vesicle clusters labeled with Alexa Fluor 647 within the calyx of Held (*hcsd*STORM)

view of VGluT1 distribution than feasible with high resolution confocal microscopy [36].

Such data can be further analyzed either on the coordinates or on the rendered image (Fig. 5a). For example, based on the rendered image a connected components analysis can be used to find regions containing clusters of fluorophores (Fig. 5b, c). The size of the clusters might correspond to the number of proteins, channels, vesicles, or any other labeled target structure. Using these clusters further information like the minimal distance or the distribution of distances between these clusters can be extracted (Fig. 5d, e).

3.10 Applications and Limitations

Using either *tomo*STORM or *d*STORM on hydrated sections, it is possible to visualize the distribution of diverse proteins in tissues with nanoscopic resolution. Individual clusters of proteins can be identified and analyzed, which could not be observed with standard fluorescence techniques. However, the estimation on the number of single vesicles or exact numbers of proteins within a

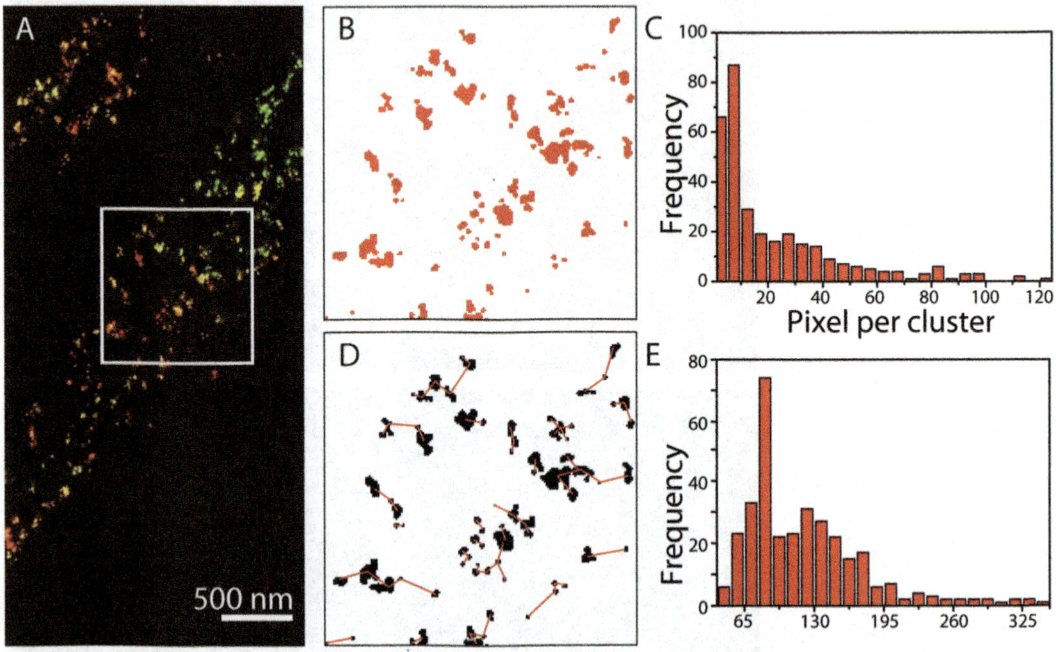

Fig. 5 Exemplary presentation of cluster analysis on SMLM data. (**a**) The super-resolution image of synaptic vesicles in the calyx of Held is analyzed using an (**b**) intensity-based cluster algorithm (connected component analysis). (**c**) The histogram shows the distribution of cluster sizes (one pixel equals 10 nm). (**d**) The inter-cluster distance was determined using a nearest-neighbor approach and (**e**) histogram

cluster remains challenging. Small structures of highly crowded areas such as the presynaptic active zone are difficult to resolve even when using 3D STORM. The error in the localization of single fluorophores, as well as sparse labeling, makes it difficult to distinguish protein clusters and assign them to distinct vesicles. Therefore, small structures such as vesicles within presynaptic terminals cannot unequivocally be resolved by 3D STORM and require examination by electron microscopy.

The main advantage of *tomo*STORM is the large volume accessible for imaging and the possibility of applying repetitive cycles of staining and imaging. The latter feature allows for the visualization of multiple target proteins within the same sample. The resolution of *tomo*STORM is slightly less than *hcsd*STORM, although on first sight one would expect that the axial resolution is better due to the small section thickness. However, it is unknown where a fluorophore is located within the 40 nm thick sections, so that a fluorophore in two adjacent sections of 40 nm thickness could be as little as 1–2 nm apart or as much as 80 nm apart, depending on the exact position of the fluorophores within the slice. On average, the axial resolution will correspond to the section thickness. In both methods, the thin sections show no or little out-of-focus background compared to thicker tissues. The main disadvantage of *tomo*STORM pertains to the high degree of unspecific labeling (*see* **Note 9**), while *hcsd*STORM exhibits excellent labeling properties and better control of dye switching. Both factors yield better localization precisions of the labeled proteins. The volume covered by *hcsd*STORM suffices to accommodate entire active zones or postsynaptic densities, so that complete entities of these molecular machines can be studied.

Main challenges ahead include improved labeling strategies that would allow the placing of the fluorophore much closer to the epitope of interest than possible with antibody-mediated labeling strategies (*see* **Note 4**). Small protein tags, such as hexa-histidine tags [30], or click chemistry approaches [31] are promising options. Furthermore, the correlation of 3D STORM labeling distributions with ultrastructure obtained by electron microscopy remains a big challenge. Preliminary experiments suggest that ultrastructure can be obtained subsequent to *d*STORM imaging in both methods described here. In particular, focused-ion-beam scanning electron microscopy can be used to generate an ultrastructural volume stack corresponding to the volume represented by the 3D STORM reconstructions.

4 Notes

1. As the chemical structures of fluorophores are different, they possess different redox potentials and thus might require a slightly different composition of the imaging buffer (concentration of thiol, pH) [19]. In addition to the buffer, the local environment of the fluorophore also has an influence on the photoswitching properties. Guanine or tryptophan residues, for instance, were shown to act as potent quenchers [32]. On the other hand, some fluorophores were demonstrated to photoswitch efficiently in the close neighborhood of a tryptophan residue [33]. This illustrates that sample preparation itself does influence the fluorescence properties of individual dyes, as the local environment of a fluorophore differs when the embedding media of tissues change. In order to ensure optimal photoswitching of organic fluorophores the composition of the imaging buffer needs to be adapted (thiol concentration, pH). In particular, multicolor experiments can be challenging, as pairs of fluorophores with matching photoswitching conditions need to be identified [21].
2. The quality of the perfusion is instrumental to obtain good tissue preservation. Structures of interest can be labeled either by using an appropriate reporter mouse line or by stereotaxic viral gene transfer [17, 34]. For example, the plasma membrane of a defined cell type can be labeled with membrane-bound GFP or mEOS2. This provides the possibility of placing the super-resolution images within the context of known structures.

3. Sections thinner than 100 nm or sections cut in series can typically not be achieved reliably with this technique. Thin sections of 100–200 nm are useful when 3D information is not required but maximal lateral resolution is needed. Sections ranging from 300 nm to 4 μm are suitable for 3D *d*STORM imaging and are sufficiently large to accommodate subcellular components such as presynaptic active zones or postsynaptic densities.
4. Important to keep in mind is that all detected signals indicate the position of the fluorophore, not the position of the labeled protein. If primary and secondary antibodies are used, the distance between the label and the protein of interest can be more than 15 nm. This distance between label and target can be larger than (or at least in the range of) the achievable resolutions of *d*STORM acquisitions and therefore cannot be neglected.
5. The imaging buffer opens disulfide bonds, which leads to detachment of antibodies from the targeted structure. Additionally, the probe decomposes faster under imaging buffer composition, compared to a probe stored in PBS. Therefore, the imaging buffer should be applied as shortly as possible before and removed directly after finishing the measurement.
6. For higher contrast, especially in thicker samples, highly inclined and laminated optical sheet (HILO [35]) illumination can be used. Therefore, the laser is not coupled into the objective right in the center but shifted parallel to the outer rim, close to total internal reflection (TIR). This causes the laser beam to leave the objective under a certain angle illuminating only a small sheet of the sample and reducing background signal from layers above or below this sheet.
7. Single PSFs have to be fitted to receive the super-resolved coordinates of the fluorophores. This fitting works best if the PSFs are not overlapping. To receive good fitting results, either use a multiemitter fitting method that can deal with overlapping PSFs to a certain amount or start image acquisition when there are only few overlapping PSFs present per frame.

   Overlapping PSFs can lead to incorrect fitting results. The overlap cannot be changed afterward, but asymmetric PSFs can be neglected for analysis in 2D experiments. However, when performing 3D experiments, the asymmetry in the PSF is used for assigning the z-value. Therefore, overlapping PSFs have to be avoided in 3D experiments as much as possible.
8. Choosing the right integration time is crucial. Prolonged integration times during image acquisition lead to a high density of PSFs per frame, while very low integration times distribute one blinking event over several frames. A good rule of thumb is that the integration time should match the mean duration of a single blinking event (10–200 ms).

9. Background labeling in conventional microscopy can appear as a lower signal-to-noise ratio and therefore the contrast of the image is reduced. In single-molecule microscopy anyPoint spread function (PSF) PSF of a fluorophore of an unspecifically bound label is identified with the same accuracy and intensity as those of correctly labeled fluorophores. As these PSFs cannot be distinguished, specific labeling is crucial for a successful experiment (in many cases where an experiment fails, it is due to poor labeling). For large structures like filaments or membranes it is easy to discriminate unspecifically bound labels from the structure, whereas it is much harder in case of proteins with a low abundance, since the specifically bound label also appears in small clusters.

References

1. Furstenberg A, Heilemann M (2013) Single-molecule localization microscopy-near-molecular spatial resolution in light microscopy with photoswitchable fluorophores. Phys Chem Chem Phys 15(36):14919–14930
2. Dani A et al (2010) Superresolution imaging of chemical synapses in the brain. Neuron 68(5):843–856
3. Nanguneri S et al (2012) Three-dimensional, tomographic super-resolution fluorescence imaging of serially sectioned thick samples. PLoS One 7(5), e38098
4. Micheva KD, Smith SJ (2007) Array tomography: a new tool for imaging the molecular architecture and ultrastructure of neural circuits. Neuron 55(1):25–36
5. Tokuyasu KT, Dutton AH, Singer SJ (1983) Immunoelectron microscopic studies of desmin (skeletin) localization and intermediate filament organization in chicken cardiac muscle. J Cell Biol 96(6):1736–1742
6. Heilemann M et al (2008) Subdiffraction-resolution fluorescence imaging with conventional fluorescent probes. Angew Chem Int Ed Engl 47(33):6172–6176
7. Huang B et al (2008) Three-dimensional super-resolution imaging by stochastic optical reconstruction microscopy. Science 319(5864): 810–813
8. Thompson RE, Larson DR, Webb WW (2002) Precise nanometer localization analysis for individual fluorescent probes. Biophys J 82(5): 2775–2783
9. Smith CS et al (2010) Fast, single-molecule localization that achieves theoretically minimum uncertainty. Nat Methods 7(5):373–375
10. Rust MJ, Bates M, Zhuang X (2006) Sub-diffraction-limit imaging by stochastic optical reconstruction microscopy (STORM). Nat Methods 3(10):793–795
11. Betzig E et al (2006) Imaging intracellular fluorescent proteins at nanometer resolution. Science 313(5793):1642–1645
12. Hess ST, Girirajan TP, Mason MD (2006) Ultra-high resolution imaging by fluorescence photoactivation localization microscopy. Biophys J 91(11):4258–4272
13. Kao HP, Verkman AS (1994) Tracking of single fluorescent particles in three dimensions: use of cylindrical optics to encode particle position. Biophys J 67(3):1291–1300
14. Shtengel G et al (2009) Interferometric fluorescent super-resolution microscopy resolves 3D cellular ultrastructure. Proc Natl Acad Sci U S A 106(9):3125–3130
15. Pavani SR et al (2009) Three-dimensional, single-molecule fluorescence imaging beyond the diffraction limit by using a double-helix point spread function. Proc Natl Acad Sci U S A 106(9):2995–2999
16. Juette MF et al (2008) Three-dimensional sub-100 nm resolution fluorescence microscopy of thick samples. Nat Methods 5(6):527–529
17. Wimmer VC, Nevian T, Kuner T (2004) Targeted in vivo expression of proteins in the calyx of Held. Pflugers Arch 449(3): 319–333
18. Horstmann H et al (2012) Serial section scanning electron microscopy (S3EM) on silicon wafers for ultra-structural volume imaging of cells and tissues. PLoS One 7(4), e35172
19. Dempsey GT et al (2011) Evaluation of fluorophores for optimal performance in localization-based super-resolution imaging. Nat Methods 8(12):1027–1036

20. Heilemann M et al (2005) Carbocyanine dyes as efficient reversible single-molecule optical switch. J Am Chem Soc 127(11):3801–3806
21. Flottmann B et al (2013) Correlative light microscopy for high-content screening. Biotechniques 55(5):243–252
22. Lampe A et al (2012) Multi-colour direct STORM with red emitting carbocyanines. Biol Cell 104(4):229–237
23. Wolter S et al (2010) Real-time computation of subdiffraction-resolution fluorescence images. J Microsc 237(1):12–22
24. Ovesny M et al (2014) ThunderSTORM: a comprehensive ImageJ plug-in for PALM and STORM data analysis and super-resolution imaging. Bioinformatics 30(16):2389–2390
25. Mlodzianoski MJ et al (2011) Sample drift correction in 3D fluorescence photoactivation localization microscopy. Opt Express 19(16):15009–15019
26. Endesfelder U et al (2014) A simple method to estimate the average localization precision of a single-molecule localization microscopy experiment. Histochem Cell Biol 141(6):629–638
27. Banterle N et al (2013) Fourier ring correlation as a resolution criterion for super-resolution microscopy. J Struct Biol 183(3):363–367
28. Nieuwenhuizen RP et al (2013) Measuring image resolution in optical nanoscopy. Nat Methods 10(6):557–562
29. Takamori S et al (2006) Molecular anatomy of a trafficking organelle. Cell 127(4):831–846
30. Wieneke R et al (2015) SLAP: small-molecule labelling of proteins for super-resolution imaging. Angew Chem Int Ed Engl 54(35):10216–10219
31. Raulf A et al (2014) Click chemistry facilitates direct labelling and super-resolution imaging of nucleic acids and proteins. RSC Adv 4(57):30462–30466
32. Doose S, Neuweiler H, Sauer M (2009) Fluorescence quenching by photoinduced electron transfer: a reporter for conformational dynamics of macromolecules. Chemphyschem 10(9–10):1389–1398
33. Nanguneri S et al (2014) Single-molecule super-resolution imaging by tryptophan-quenching-induced photoswitching of phalloidin-fluorophore conjugates. Microsc Res Tech 77(7):510–516
34. Dondzillo A et al (2010) Targeted three-dimensional immunohistochemistry reveals localization of presynaptic proteins Bassoon and Piccolo in the rat calyx of Held before and after the onset of hearing. J Comp Neurol 518(7):1008–1029
35. Tokunaga M, Imamoto N, Sakata-Sogawa K (2008) Highly inclined thin illumination enables clear single-molecule imaging in cells. Nat Methods 5(2):159–161
36. Venkataramani V, Herrmannsdörfer F, Heilemann M, Kuner T (2016) SuReSim: simulating localization microscopy experiments from ground truth models. Nat Methods 13(4): 319-321(1):25–36

Chapter 14

Photomarking Relocalization Technique for Correlated Two-Photon and Electron Microcopy Imaging of Single Stimulated Synapses

Miquel Bosch*, Jorge Castro*, Mriganka Sur, and Yasunori Hayashi

Abstract

Synapses learn and remember by persistent modifications of their internal structures and composition but, due to their small size, it is difficult to observe these changes at the ultrastructural level in real time. Two-photon fluorescence microscopy (2PM) allows time-course live imaging of individual synapses but lacks ultrastructural resolution. Electron microscopy (EM) allows the ultrastructural imaging of subcellular components but cannot detect fluorescence and lacks temporal resolution. Here, we describe a combination of procedures designed to achieve the correlated imaging of the same individual synapse under both 2PM and EM. This technique permits the selective stimulation and live imaging of a single dendritic spine and the subsequent localization of the same spine in EM ultrathin serial sections. Landmarks created through a photomarking method based on the 2-photon-induced precipitation of an electrodense compound are used to unequivocally localize the stimulated synapse. This technique was developed to image, for the first time, the ultrastructure of the postsynaptic density in which long-term potentiation was selectively induced just seconds or minutes before, but it can be applied for the study of any biological process that requires the precise relocalization of micron-wide structures for their correlated imaging with 2PM and EM.

Key words Correlated imaging, Time-lapse live two-photon fluorescence microscopy, Serial-section transmission electron microscopy, Dendritic spine, Synapse, Postsynaptic density, Photomarking, Photoetching, Photobranding, DAB

1 Introduction

Synaptic plasticity has been extensively studied because it is considered the molecular basis for learning and memory. Synapses change over time in an experience-dependent manner, both functionally and structurally [1]. These changes can be studied in vivo or ex vivo by introducing fluorescent markers (such as GFP) and longitudinally imaging synapses and dendritic spines in real time with

*These two authors contributed equally to this chapter.

Alexandros Poulopoulos (ed.), *Synapse Development: Methods and Protocols*, Methods in Molecular Biology, vol. 1538, DOI 10.1007/978-1-4939-6688-2_14,

confocal or 2-photon fluorescence microscopy (2PM). In addition to imaging, 2-photon lasers can be used to selectively stimulate individual synapses via photochemical reactions such as glutamate uncaging [2, 3]. However, light microcopy has an inherent limitation that prevents observing structural features below the light diffraction limit. Superresolution techniques can nowadays overcome this limitation and are able to reach nanoscale resolutions [4–7]. Electron microscopy (EM) is more powerful in this sense as it can reach ultrastructural imaging with higher spatial resolution, down to the order of few nanometers. EM is ideal to detect and analyze synapses, including postsynaptic, presynaptic, and extrasynaptic substructures in unaltered tissue [8]. For this reason, it is currently used for brain connectivity mapping projects [9]. However, EM has some critical limitations: it requires sample fixation, thus precluding live time-course imaging of plastic changes, and it cannot detect fluorescence, thus precluding an easy relocalization of the selected neuron or synapses under study.

We developed a powerful methodological approach that bypasses these limitations to allow the correlated imaging of the same specific synapse with both 2PM and transmission EM (TEM). We combined several procedures in a way that permits: (1) the selective stimulation of a single dendritic spine (via glutamate uncaging), (2) the longitudinal observation of the plastic changes occurring in that spine (via 2 PM), (3) the unequivocal localization of the same spine within ultrathin sections (via 2-photon-induced photomarking), and (4) the 3D reconstruction and analysis of the ultrastructural changes that selectively occurred in that spine (via serial section TEM–ssTEM-). The key point is the photomarking step, which is based on the 2-photon-induced precipitation of an electrodense compound, diaminobenzidine (DAB), that generates precise fiducial landmarks visible under both 2PM and EM imaging [10–12]. These landmarks are fine lines, one or a few microns wide, which can be drawn in any region of the tissue following any useful spatial pattern. In this case, they are drawn to flank the selected dendrite and to point to the stimulated spines. After localizing these lines on ssTEM images, it is possible to unambiguously identify the same synapses that were previously stimulated and imaged with 2PM. Two-photon precipitation of DAB was previously used to relocalize single neurons [10] and we adapted it to relocalize single synapses after long-term potentiation (LTP) was selectively induced in that spine and structural changes visualized with 2PM [12–14]. An almost identical method was developed in parallel to us, named near-infrared branding (NIRB) technique, which proved to similarly localize dendritic spines and axons in both optical and EM imaging [11]. Other methods have been used to correlate optical imaging of individually labeled neurons, dendrites, or synapses with EM imaging. However, some of these methods, such as biocytin detection [15], GFP immunostaining [16–18], GFP photo-oxidation [19, 20], or photoconversion of organic fluorescent dyes [21] suffer from the problem that the spine gets obscured by the precipitate, precluding a clear visualization of the internal subcellular structures such as the PSD. Other approaches can solve this problem but are not suitable for single synapse relocalization in thick tissue. For example, the use of endogenous landmarks, such as blood vessels patterns, is useful to reidentify neurons or groups of neurons [22, 23] but such landmarks are hard to find at microscale resolution. On the other hand, introduction of exogenous fiducial marks, such as fluorescent beads [24] or gold nanorods [25], is difficult to carry out in thick tissues like organotypically cultured slices or in vivo. We had to overcome these problems, since our goal was to image changes in PSD ultrastructure just seconds or minutes after LTP is induced in a single synapse in a hippocampal slice. Plastic changes in PSD morphology had been previously observed with EM after the classic electrical induction of LTP in multiple synapses [26, 27] but not after single-synapse LTP. Using our method we discovered that the structural plasticity of the spine and the PSD are not synchronized during LTP: PSD enlargement is 1 h delayed compared to spine enlargement [12, 28]. The same photomarking technique has recently been employed to correlate in vivo 2PM imaging of cortical neurons with focus ion beam/scanning EM [29, 30]. This method can easily be extended beyond synapse biology and used for the study of any biological process that requires precise relocalization of micron-wide structures for their correlated imaging with live 2PM and EM.

2 Materials

2.1 Hippocampal Slices

Organotypic hippocampal slice culture is a particularly useful experimental approach for studying synaptic plasticity. In this preparation, neurons mature in a relatively similar way as they do in vivo [31–35], and the local hippocampal synaptic circuitry is largely preserved [36, 37]. Acute slices from adult or adolescent transgenic mice expressing fluorescent proteins (e.g., Thy1-YFP mice) are more similar to the intact brain, but the organotypic preparation allows for a more versatile genetic manipulation, since it is easier and faster to transfect any kind of gene or combination of genes using biolistic, viral, or electroporation methods [38, 39]. Hippocampal slice cultures are typically prepared from rat (e.g., SD) or mouse (e.g., C57/BL6) pups of both sexes, at postnatal days 6–8 (P6–8; *see* **Note 1**). Slice cultures can also be prepared from other brain regions, such as cortex, striatum, cerebellum, thalamus, brainstem, etc. [40–43].

2.2 Plasmids Expressing Fluorescent Proteins

In order to longitudinally visualize dendritic spines or synaptic components in a living neuron using 2PM, it is necessary to introduce a source of fluorescence into the cell, for example, a plasmid expressing a fluorescent protein, either alone or as a fusion protein.

For time-course imaging of spine morphological changes, expression of a stable, small, and soluble marker such as EGFP is sufficient. DsRed2 is a good choice if a red-emitting marker is needed (*see* **Note 2**). Fluorescent proteins fused to naturally localized endogenous molecules can be used to target specific cell locations (*see* **Note 2**). To target the synapse or the PSD, for example, we express fluorescently labeled Homer1b, since exogenous overexpression of this protein does not alter normal synaptic physiology and Homer 1b exhibits relatively rapid synaptic turnover, which prevents loss of signal by photobleaching [44, 45]. We routinely use CMV promoter for expression of most genes, or CAG if we need stronger expression.

2.3 Two-Photon Microscope Setup

Time-lapse fluorescence imaging can be carried out with a 1-photon confocal scanning microscope or a 2-photon microscope. Two-photon is recommended for in vivo imaging or in the case of thick tissue such as brain slices, and especially in the case of photochemical reactions such as glutamate uncaging. The best way to perform experiments involving simultaneous imaging and stimulation of single spines by glutamate uncaging is to combine two mode-locked femtosecond-pulse Ti:sapphire lasers, one for imaging and one for uncaging. This expensive equipment can be purchased from several vendors, or can be custom-made provided that the adequate knowledge and equipment are possessed [46–49]. Two-photon imaging in slices is typically performed with the laser tuned between 860 nm and 980 nm, being 910–930 nm the optimum wavelength to excite both green and red fluorophores (such as EGFP and DsRed2). Glutamate uncaging is usually triggered with the same or a second laser tuned at 710–730 nm. Two-photon and 1-photon lasers can be further combined to perform more complex experiments (*see* **Note 3**).

Use an upright microscope equipped with 10× (0.3 NA) and 60× (0.9–1.1 NA) water immersion objective lenses. The microscope chamber must have a set of inlets installed to constantly perfuse artificial cerebrospinal fluid (ASCF) aerated with 95 % O_2/5 % CO_2 (carbogen). As the caged compounds are generally expensive, we recommend recirculating the same solution and using the minimum volume of liquid (e.g., 5–15 ml). This recirculation can be done using a peristaltic pump located close to the stage. It is important to carefully set up the perfusion system to ensure that the level of liquid in the chamber is always constant (*see* **Note 4**).

For such a small volume of recirculating solution, evaporation can become a problem in experiments of long duration. It is convenient to check how severe this problem is in each particular setup. A good solution to prevent evaporation is to cover the chamber with some sort of plastic shield or wet hard paper during the experiment. It is also useful to introduce humidity into the carbogen by bubbling it first through water before bubbling into the ACSF. For experiments lasting several hours it might be still necessary to add double-distilled (dd) water several times to compensate for the increase in osmolality. Pilot experiments must be carried out to measure the rate of evaporation and to determine how much water to add, and how often (*see* **Note 5**).

2.4 Solutions for Slice Preparation, 2-Photon Imaging, and Photomarking

1. Culture Medium: Modified Eagle's Medium (MEM) with 20% horse serum, 27 mM d-glucose, 6 mM $NaHCO_3$, 2 mM $CaCl_2$, 2 mM $MgSO_4$, 30 mM HEPES, 0.01% ascorbic acid, and 1 μg/ml insulin. Adjust pH to 7.3 and osmolality to 300–320 mOsm/L (*see* **Note 6**). We usually do not add antibiotics. Sterilize by filtering through a 0.22 μm filter and keep at 4 °C no longer than 4 weeks.
2. Dissection Solution: 2.5 mM KCl, 26 mM $NaHCO_3$, 1 mM NaH_2PO_4, 11 mM glucose, 238 mM sucrose, 1 mM $CaCl_2$, and 5 mM $MgCl_2$ in dd water. Prepare fresh or from partial 10× stock solution (*see* **Note 7**). Bubble with carbogen for 20 min. Use it ice-cold or, optionally, place it at −80 °C for ~10 min or at −20 °C for ~30 min until it acquires a slush-like consistency.
3. Magnesium-free ACSF: 119 mM NaCl, 2.5 mM KCl, 26.2 mM $NaHCO_3$, 1 mM NaH_2PO_4 11 mM glucose, 3 mM $CaCl_2$, 1 μM tetrodotoxin (TTX), and 50 μM picrotoxin in dd water. Prepare fresh or from a partial 10× stock solution (*see* **Note 8**). Bubble with carbogen for 20 min. Check osmolality is in the range of 320 ± 20 mOsm/L.
4. Uncaging ACSF: 2.5 mM 4-methoxy-7-nitroindolinyl (MNI)-l-glutamate (caged glutamate) in Mg-free ACSF. The concentration of caged glutamate can vary depending on the efficiency of the system and the budget of the researcher (typically 2–6 mM). This compound is expensive, so we recommend testing the intensity and duration of the uncaging laser that triggers the desired stimulation using the minimum concentration of this compound (*see* **Note 27**). Keep this compound under darkness and prepare this solution with the minimum ambient light and maximum protection from UV light (*see* **Note 26**).
5. Fixative Solution: 2% (w/v) paraformaldehyde (PFA), 2% (w/v) glutaraldehyde (GA) in 0.1 M Phosphate Buffer (PB) pH 7.4 (*see* **Note 9**). Make it fresh every day before fixation, using a dedicated glass-ware and plastic-ware. Keep at room temperature (r.t). This solution, as well as PFA and GA, is highly toxic and so the appropriate protection must be used.
6. Photomarking Solution: 7 mM 3,3′-diaminobenzidine (DAB) in ice-cold 0.1 M Tris–HCl buffer. Adjust pH to 7.4 and cool on ice. Filter with a syringe filter (0.2 μm pore size) to remove debris. This solution must be made fresh and pH should be rechecked just before the experiment. DAB is a carcinogen and has to be handled and disposed of appropriately.

2.5 Solutions for Electron Microscopy

1. Osmium Tetroxide Solution: 1% osmium tetroxide (OsO_4) in 0.1 M PB (prepared from a pre-made 2% stock solution). Osmium tetroxide is volatile and extremely toxic in contact with skin and mucosae. Prepare this solution carefully using glassware, under the fume hood to avoid inhalation, and with two layers of latex/nitrile gloves to avoid skin exposure.
2. Osmium Tetroxide/Potassium Ferrocyanide Solution: 1.5% (w/v) potassium ferrocyanide ($K_4Fe(CN)_6 \cdot 3H_2O$) in Osmium Tetroxide Solution.
3. Toluidine Blue Solution: 1% (w/v) toluidine blue in dd water. Filter with a syringe filter.
4. Durcupan Epoxy Resin: 10 parts of component A, 10 parts of component B, 0.3 parts of component C, and 0.2 parts of component D (by volume; ratios may be optimized). Durcupan must be prepared quite in advance before use. It is better to prepare it in a previously dehumidified hood (use Drierite® or a dehumidifier). In order to obtain a hard resin that will not brittle during sectioning, we recommend the proportions above, added in the same order (*see* **Note 10**). Durcupan is toxic and so the appropriate protection measures must be taken.
5. Uranyl Acetate Solution: 1% uranyl acetate in dd water. Filter it through a 0.22 μm disc filter. Uranyl acetate is light-sensitive; keep in darkness at 4 °C. All glass equipment that has been in contact with uranyl acetate should be separated from other labware.
6. Lead Citrate Solution: 3 mM lead citrate, 0.1 N NaOH in degasified dd water. Because CO_2 can precipitate lead in the form of $PbCO_3$, it is necessary to prepare degasified water (CO_2-free) by boiling dd water for 30 min and storing it in an airtight container. All material in contact with lead citrate must be made of plastic.

2.6 Pioloform Slotted Grids

To maximize the viewable area in the serial sections, use grids with oval slots of 1 mm wide by 2 mm long. To support the sections, coat the grids with pioloform (stronger than formvar) or purchase factory-coated grids, which give a highly homogenous and reproducible film thickness.

3 Methods

The complete protocol is a combination of different techniques with multiple steps each: (1) Preparation of hippocampal slice cultures; (2) Transfection of slices with fluorescent markers; (3) Two-photon imaging and stimulation of single dendritic spines; (4) Fixation of the sample; (5) Creation of landmarks by the photomarking procedure; (6) Processing of the sample for EM; (7) Serial section cutting; (8) EM imaging of the sections; (9) Relocalization and 3D reconstruction of the stimulated spines.

In this section, we will only describe in detail the steps that contain crucial or novel methodologies (steps 3, 4, 5, 9). We will provide some useful basic information but will not describe the other steps in detail (steps 1, 2, 6, 7, 8), as methods are reported in detail in this volume and elsewhere (references below).

3.1 Preparation of Hippocampal Slice Cultures

Slices of 350–400 μm thickness are prepared from P6–8 rat (or other species such as mouse, ferret, etc.) hippocampi using a McIlwain tissue chopper, and cultured at 35–37 °C on interface membranes for 1–4 weeks (*see* **Notes 11** and **12**). Follow the preparation method used in Bosch and colleagues [12, 50], originally developed by Stoppini and colleagues in 1991 [51] based on previous works [34, 52], and technically described in detail elsewhere [53–55]. The experiment is typically carried out at 7–15 days in vitro (DIV), usually 2–4 DIV after gene transfection (to allow for fluorescent protein expression).

3.2 Gene Transfer into Slices

Our preferred method for transfection of plasmids expressing fluorescent proteins is biological ballistics (biolistics, gene gun) [56]. It is easy, rapid, and highly reproducible, and generates well-isolated transfected cells, with minimum background noise. Compared with other methods, it facilitates the expression of more than one protein by combining several plasmids in the same bullet. It also allows the simultaneous downregulation of endogenous proteins (by including a shRNA expression vector), thereby facilitating replacement approaches. Follow the instructions provided by the manufacturer and/or those described in Chapter 19, by Woods and Zito [57] and Soares and colleagues [55]. Assess the quality of slices after ~5 DIV (*see* **Note 13**). Plasmids are typically transfected no earlier than 5 DIV and no later than 15 DIV (*see* **Note 14**). Allow a minimum of 2 days for the expression of exogenous proteins. Some proteins may need >4 days to reach the necessary degree of expression for comfortable imaging. On the other hand, if higher number of transfected cells is needed, or higher spatial accuracy is required (i.e., if a specific region of hippocampus has to be selectively targeted, e.g., only CA3), viral infection is a better choice [38, 58, 59]. This method however may result in too many fluorescent cells, thus making it hard to find an isolated dendrite with dark background. Single-cell electroporation and phototransfection are excellent alternative methods in which one cell can be specifically selected and transfected [60–63], but are much more time-consuming. Additionally, instead of a genetic marker, it is possible to stain the cell by particle-mediated introduction of a lipophillic dye (e.g., DiOlistics) or to fill the cell with a fluorescent organic dye through a pipette, if one is patching the neuron for electrophysiological recording (*see* **Note 15**).

3.3 Two-Photon Imaging and Single-Synapse Stimulation

1. Start the experiment by turning on the lasers to warm the whole system up (*see* **Note 16**). Because stimulation of single synapses requires a high degree of precision, the 910 nm imaging laser and the 720 nm uncaging laser must be aligned daily (with an error of <50 nm). This alignment can be performed using 0.5 μm diameter (Ø) fluorescent beads. Beads are simultaneously or sequentially imaged with both lasers and the light path of each laser must be adjusted so that both images overlap as much as possible (*see* Fig. 1a–f). It is useful to further check the alignment by selecting a group of several beads clustered together and use the uncaging beam to bleach one of them. If lasers are well aligned and reach the sample with the correct precision, only the selected bead should be bleached out but not the neighboring ones (Fig. 1g–i) (*see* **Note 17**).
2. Prepare the Mg-free ACSF (as described in Subheading 2.4; *see* **Note 8**) and the Fixative Solution. Turn on the peristaltic pump and start the Mg-free ACSF perfusion, usually at 2–4 ml/min, while constantly aerating with carbogen. Make sure the level of liquid of the chamber is constant (*see* **Note 4**). It is not necessary to recirculate the same solution at this point, as the caged glutamate is still not present: outflow solution can be discarded and inflow solution can be fresh. Turn on temperature controller and set it at the desired value (we usually set it at 30 °C, although efficient LTP and spine enlargement can be achieved at 23–35 °C).
3. When the microscope setup is ready and all conditions are stable (temperature, perfusion flow, laser power, alignment, etc.), select the appropriate sample to be placed in the chamber (*see* **Note 18**). Pick up one insert from the well and put it on the cover of a petri dish (*see* **Notes 19** and **20**). Quickly, cut a piece of the membrane containing the slices with a surgical scalpel (*see* **Notes 11** and **12**). Cut it in an irregular shape (rectangular or trapezoidal), to be able to later identify the position and the orientation of the selected neuron within the slice and within the piece of membrane. Additionally, cut a small dent in one corner of the membrane to further break symmetry and to be able to identify which side is up and which down (*see* **Note 21**).
4. Put the membrane in the chamber and use a holder to immobilize the membrane (*see* **Note 22**). Use epifluorescence illumination through the 10× objective lens to search for a good transfected pyramidal neuron to image. A good neuron is one with a moderate level of fluorescence expression and typical healthy morphology: long straight dendrites, normal spine density, and normal basal and apical dendritic tree with multiple branches (*see* the example in Fig. 2a). Avoid cells with signs of unhealthy conditions, such as dendritic blebbing, wavy dendrites, small or few spines, or fluorescent aggregates.

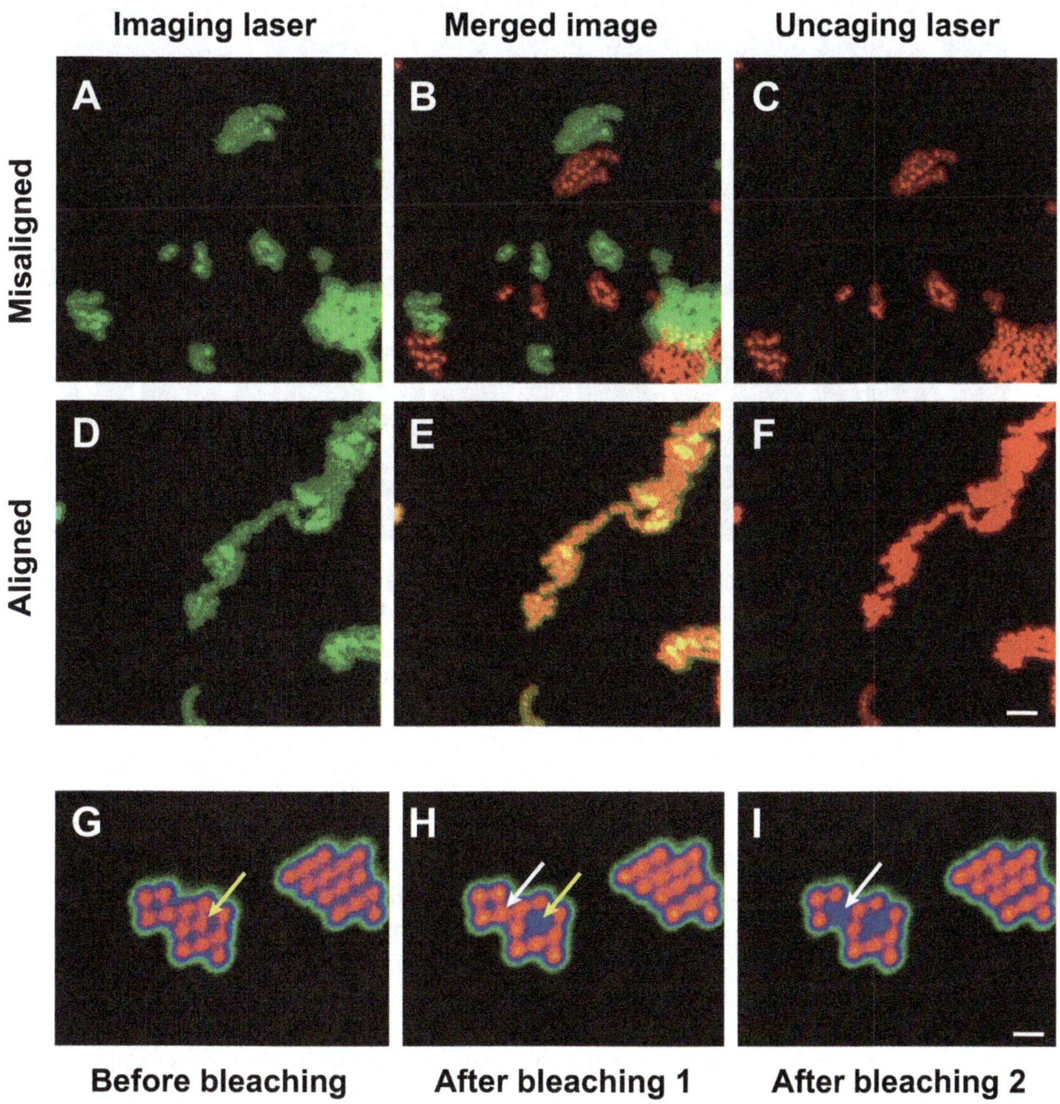

Fig. 1 Laser alignment with fluorescent beads. (**a**–**f**) Fluorescent beads of 0.5 μm Ø are used to align the imaging laser with the uncaging laser. Beads are imaged with the imaging laser (**a**, **d**) and, simultaneously or sequentially, are imaged with the uncaging laser (**c**, **f**). Both images are merged (**b**, **e**) and, if they do not completely overlap (**b**, example of large misalignment), the mirrors that bring the lasers into the microscope must be adjusted until both images match as perfectly as possible (**e**, good alignment). Scale bar in **a**–**f**, 2 μm. (**g**–**i**) Alignment can be further confirmed by imaging a cluster of fluorescent beads (**g**) and using the uncaging laser to bleach one single bead (**h**; *yellow arrow*) and then, for instance, another one (**i**; *white arrow*). Scale bar in **g**–**i**, 1 μm. Images are pseudocolored for presentation purpose

5. Incubate the slices for some time (5–60 min, typically 20 min, but this must be tested empirically) to stabilize some parameters (temperature, metabolism, mechanical drifts due to overall changes of slice shape, etc.). These parameters might change during first minutes of imaging or recording but usually stabilize soon thereafter (*see* **Note 23**). Change the objective lens to the 60× to carry out the subsequent imaging and stimulation.

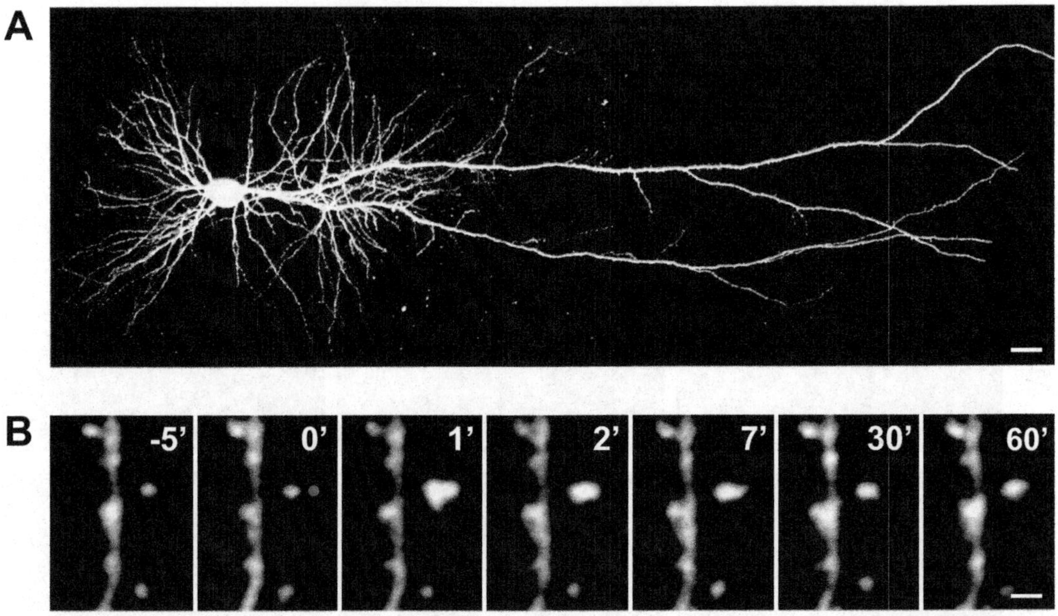

Fig. 2 Neuronal morphology and spine structural plasticity. (**a**) Example of a hippocampal CA1 pyramidal neuron from an organotypic slice culture, transfected with DsRed2. Scale bar, 10 μm. (**b**) Time-lapse 2-photon imaging of the persistent enlargement of a dendritic spine after induction of long-term potentiation by 2-photon glutamate uncaging. *Red dot*: position of the uncaging laser. Time stamp in minutes after stimulation. Scale bar, 1 μm

6. Choose a segment of a dendrite running parallel to the imaging plane, which in turn will be parallel to the bottom surface of the slice, since the slice is sitting flat on the chamber surface. This will increase the chance of the ultrathin sections being cut as parallel as possible to the dendrite and, therefore, will minimize the number of sections later required to reconstruct the whole dendrite with all the selected spines. It is also advisable to select a dendrite close to the upper slice surface to accelerate the cutting process. We choose segments of primary or secondary dendrites from the distal part of the main apical dendrite of CA1 pyramidal neurons, which we have found to have long (20–40 μm) and straight segments that run parallel to the imaging plane (*see* Figs. 2a and 3a).

Fig. 3 (continued) head enlargement of the three stimulated spines at 1 min (**c**), 1–7 min (**d**), and 1–30 min (**e**) after induction of LTP. *Red dot*: position of the uncaging laser. Scale bar, 1 μm. (**f**) Two-photon image of the same dendritic segment after fixation. Scale bar, 5 μm. (**g**) Landmarks created by 2-photon precipitation of DAB in straight fine lines. *Two lines* flanking the dendrite (labeled as "1") are initially drawn and *three semi-perpendicular lines* pointing to the stimulated spines (labeled as "2") are subsequently drawn. (**h**) Additional lines are created several microns above (labeled as "3"). (**i**) Piece of a fixed hippocampal slice visualized under bright field. *Arrowhead* points to the DAB landmarks. *SP* stratum pyramidale, *SR* stratum radiatum. *Asterisk* marks a wound made by the holder. Scale bar, 0.2 mm. (**j**) *Arrowhead* points to the same DAB landmarks (as in **h**) visualized under bright field at higher magnification. Scale bar, 25 μm. (**k–m**) Another example of the photomarking procedure on a different sample. (**k**) Two-photon live image of a dendritic segment. Three spines are also stimulated (*red dots*). (**l**) Two-photon image of the same dendrite after fixation and photomarking. Lines flanking the dendrite (labeled as "1") and semi-perpendicular lines (labeled as "2") pointing to the stimulated spines (*red dots*) are visible. (**m**) Same region under bright field where the flanking ("1") and the semi-perpendicular ("2") DAB landmarks are visible. Scale bar in **k**, 5 μm; in **l**, **m**, 10 μm

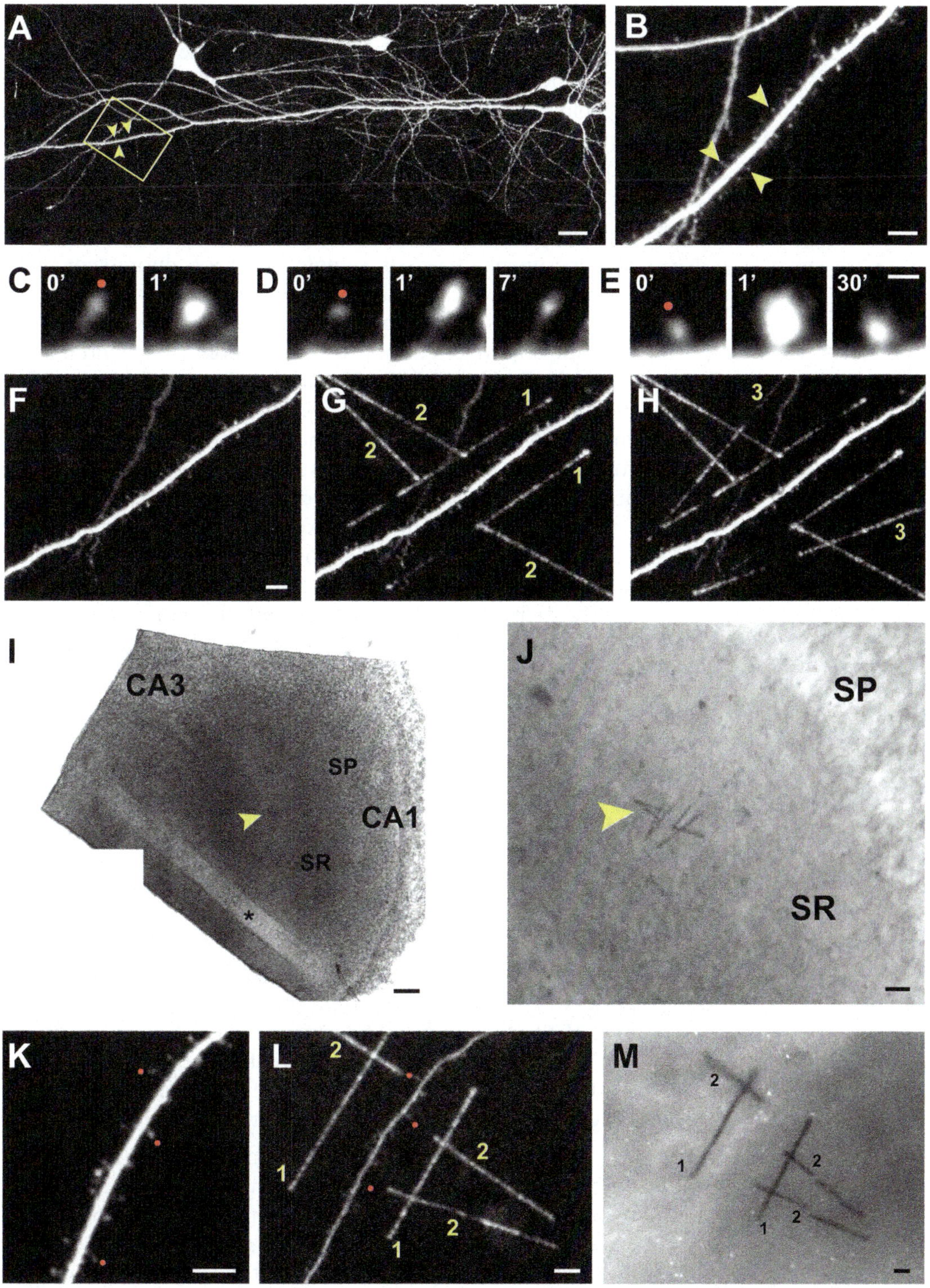

Fig. 3 Generation of landmarks by 2-photon photomarking. (**a**) Two-photon collage image of live hippocampal neurons in an organotypic slice culture. A straight and superficial segment of the dendrite of a pyramidal neuron is selected for the experiment. The *yellow square* highlights the region shown in (**b**). Scale bar, 25 μm. (**b**) Three spines (labeled with *arrowheads* in **a** and **b**) from the selected dendritic segment (*yellow square* in **a**) are chosen to be stimulated at different time points. Scale bar, 5 μm. (**c**–**e**) Two-photon time course live imaging of the spine

Choose one or several dendritic spines of medium-size diameter to stimulate (*see* **Note 24**). We typically choose small mushroom-type spines, with clear head and neck (no stubby, no filopodia, no big spines; *see* Figs. 2b and 3c–e). We target three spines in each experiment, separated at least by >10 μm, and stimulate them sequentially, in intervals not shorter than 10 min between them (*see* Fig. 3b–e; *see* **Note 25**).

7. Imaging parameters must be optimized depending on the type of experiment, type of sample, type of fluorescent gene expressed, etc. We found the following to be good parameters for time-course imaging of EGFP-transfected neurons: Use the 60× objective lens and a digital zoom of around 10× to acquire a *Z*-series of 512 × 512 pixel *XY*-scanned images taken every 0.5–1 μm in depth (*Z*). Usually, three to four images are enough to reconstruct one spine in the *Z*-dimension but it is advisable to take 7–15 images if you also want to acquire neighboring spines. Also, slow drift in the *X*, *Y*, or *Z* dimension is common during long experiments and so it might be necessary to adjust the *XYZ* positions regularly. Time-course imaging requires good signal stability. Photobleaching and phototoxicity are common problems, especially with fusion proteins of slower dynamics than free EGFP. Therefore, it is important to reduce laser intensity, exposure, dwell time, number of images taken, etc., and to enhance sensitivity by increasing PMT voltage as much as possible to minimize photodamage, even though image quality is reduced.
8. Stimulation of single spines has been performed by several methods, of which 2-photon-induced uncaging of caged glutamate is the most selective and efficient [2, 3]. Several laboratories have reported induction of LTP in single spines [12, 64–68] and a few laboratories have reported induction of long-term depression (LTD) in single spines by glutamate uncaging [69–71]. We usually induce LTP using uncaging pulses of 1 ms of duration, repeated at 1 Hz for 1 min, in Mg-free ACSF containing MNI-glutamate, TTX, and picrotoxin (*see* Subheading 2.4). We first recirculate the Mg-free solution (without caged glutamate) while we place the slice into the imaging chamber, search for the neuron, and select the spines to be stimulated.
9. Once this is done, we prepare the Uncaging ACSF and switch the tubes to the reservoir containing this Uncaging ACSF, avoiding any dilution of the solution or calculating and taking into account the expected dilution of the caged glutamate. At this point, recirculate the same solution constantly by returning the outflow liquid to the tube containing the bubbling Uncaging solution. However, you can search for the neuron and spines using epifluorescence with the caged glutamate in

the solution with no evident unwanted uncaging, provided the use of the correct fluorescence filter (typical GFP filters block the UV range of absorption of caged glutamate), avoiding UV light, keeping ambient room light to a minimum, and/or using a red or UV-filtered ambient light (*see* **Note 26**). It is possible, therefore, to replace the samples and perform several sequential experiments using the same Uncaging ACSF.

10. Laser intensity and pulse duration have to be adjusted for each microscope and for any given concentration of caged glutamate (*see* **Note 27**). Aim the laser close to the tip of the spine but not on top. As 2-photon diffraction limit (NA 1.0 lens at 720 nm) is 0.62 μm Ø, it is best to shoot ~0.3 μm away from the spine tip (*see* red dot in Figs. 2b and 3c–e) to avoid direct artefactual excitation of the spine by the light. For 3–5 mM MNI-Glutamate we typically use 3–6 mW of laser power, measured after the objective lens (30–60 mW measured at the back aperture, before the lens). At these intensities, we obtain averaged excitatory postsynaptic currents (EPSC) of ~5 pA with 1 ms pulses (*see* **Note 28**). Take baseline images for the first spine (usually every 1–5 min; *see* **Note 29**), confirm that the size of the spine is stable (*see* **Note 30**), induce LTP as described, and take subsequent time-course images, confirming that the spine has been persistently enlarged.
11. If stimulating several spines, move to the second spine, repeat the process, then to the third spine, etc. Record the position of each spine within the dendrite so that you can come back and image the second or third spine in an interleaved way. It is desirable to draw a schematic sketch of the neuron, the dendrite, and the position of each spine for later identification. If different spines have been stimulated sequentially at different time points, at the moment of fixation you will obtain spines potentiated for different periods (*see* Fig. 3c–e).

3.4 Fixation

1. The Fixative Solution must be prepared fresh each day and kept at r.t. After acquisition of the last image of the last stimulated spine, quickly turn on the red light (*see* **Note 26**), raise the objective lens, remove the plastic cover that prevents dehydration, pick up the membrane with tweezers, and dip it immediately in Fixative Solution (in a beaker or a 6-multiwell plate).
2. Shake the beaker manually or with a mechanical shaker for 5 min at r.t. to accelerate the penetration of the fixative into the tissue. Put it in a shaker at r.t. for 90 min and then at 4 °C for a minimum of 12 h, always shielded from light.

3.5 Two-Photon-Mediated DAB Photo-Marking

1. Rinse slices thoroughly in 0.1 M PB 4–5 times and transfer them back to the 2-photon microscope chamber. Since the photomarking step requires the use of toxic dirty compounds

(PFA/GA-fixed tissue and DAB), we recommend using a different dedicated set of objective lenses, chambers, perfusion tubes, and tweezers exclusively for these steps. Use the asymmetric shape or the dented mark to place the membrane in the appropriate upright position.

2. Prepare the ice-cold DAB solution just before its use (*see* Subheading 2.4). Perfuse continuously with the solution, aerated with pure oxygen (*see* **Note 31**). Use the epifluorescence light to localize the stimulated spines (*see* **Note 32**; *see* Fig. 3f).
3. Photoprecipitate the DAB creating fiducial patterns (*see* below) by line-scanning the tissue with the 720 nm 2-photon laser using line-scan mode. Other wavelengths have been also used (e.g., 910 nm; [11]). We could not achieve photoprecipitation using a 1-photon 488 nm confocal laser (*see* **Note 33**). As the DAB precipitate is also fluorescent, you should be able to see both the dendrite and the lines in the same image at 910 nm excitation (*see* Fig. 3g, h, l). The laser intensity and dwell time have to be adjusted for each sample, as photomarking efficacy can vary between experiments. Typically, laser power can be 15–25 mW as measured after the objective lens (150 and 250 mW measured at the back aperture, before the objective lens), with laser irradiation times around 15–60 s. It is recommended to perform a quick test in the same sample, selecting a similar region at the same depth but several μm away from the selected dendrite, and to draw several lines at increasing laser intensities and/or dwell times to determine which intensity and time creates the best marks, with good autofluorescence signal (not too faint, not too thick). When carefully drawing the lines at the selected region using the chosen laser intensity, it might be necessary to overwrite the same line two to three times, if the first-pass result is too faint. Too much laser intensity or too many rounds of photoprecipitation might result in too much electrodense signal and/or burning and destroying the tissue.
4. Different linear patterns can be drawn for use as landmarks for later identification of the dendrite and spines. We usually draw two lines (e.g., of 40–80 μm long), parallel to the dendrite and as equidistant to the dendrite as possible (e.g., 15 μm away; *see* Fig. 3g, l). Measure (on the computer screen, approximately) the length of the line marks, the width, the distance of each one to the dendrite, etc. Save an image of these lines and the dendrite (*see* **Note 34**).
5. Next, draw several lines, perpendicular or semi-perpendicular to the dendrite, starting ~40–60 μm away from the dendrite, and pointing to the selected spine but without reaching it, i.e., finishing 8–15 μm away from the spine, to avoid photoprecipitating

on the top of the spine. If you have stimulated three spines, draw three lines in an asymmetric pattern (e.g., each one with a different angle with respect to the dendrite; *see* Fig. 3g, l), which will greatly facilitate their unequivocal identification. Save another image and/or measure the length, angle, etc., and write down how precise the line points to the spine by extrapolation (*see* **Note 35**). These pictures and values are important to later compare the pattern of electrodense marks in the EM images to find the dendrite and spines in the absence of fluorescent signal.

6. Finally, move the focal point several microns above the dendrite (e.g., 3–4 μm) and add two more lines, drawn again in any asymmetric pattern that will help its identification (*see* Fig. 3h). These extra upper lines will be very useful in the process of sample sectioning. All these crossing lines are just examples of landmark patterns that we successfully used in our experiments, but any other geometric patterns that the experimenter finds useful can be applied (e.g., squares or rectangles [11, 30]).

3.6 Sample Processing for EM

1. Rinse the membranes in 0.1 M PB several times at r.t.
2. Detach the selected slice from the membrane with a scalpel blade slowly and carefully, avoiding any damage to the sample. Importantly, mark the slice with an asymmetric cut in one or two corners to effectively differentiate top from bottom side of the section at later steps.
3. At this point it is convenient to process several slices at the same time in 4 ml amber vials. Use plastic pipettes to remove liquid and use glass hooks to transfer slices between vials. Transfer slices to Osmium Tetroxide/Potassium Ferrocyanide solution (see Subheading 2.5) and incubate for 30 min at 4 °C in a rotator (*see* **Note 36**).
4. Rinse the slices four times for 5 min in PB at 4 °C and transfer them to the Osmium Tetroxide Solution for an additional 30 min at 4 °C.
5. Transfer slices to new vials and start the dehydration process by rinsing them in alcohol solutions at incremental concentrations: rinse twice for 5 min in 30% ethanol. Repeat with 50% ethanol, then 70% ethanol, then 90% ethanol. Rinse once in 95% ethanol for 10 min and twice in 100% ethanol for 10 min.
6. Rinse twice in propylene oxide for 10 min.
7. Transfer the slices to a 1:1 mix of fresh Durcupan Epoxy Resin and propylene oxide, and incubate them in the rotator overnight at r.t.
8. Transfer the slices into fresh Durcupan and incubate for 4 more hours in the rotator at r.t.

9. Start the process of flat-embedding the samples between two sheets of poly-chloro-tri-fluoro-ethylene (PCTFE) film (e.g., ACLAR®) with new fresh Durcupan resin. Put the slices between the plastic films and use scale calibrating weights (20–50 g) to deliver pressure. Carefully eliminate all bubbles.
10. Heat the samples in an oven at 57 °C and cure them for 2 days.
11. Remove the PCTFE film attached to the bottom side of the slice (the one furthest from the photomarked region) and glue that side with epoxy glue onto cylinders made of Durcupan. These cylinders can be made using standard BEEM cylinder molds. The top side of the section (the one closer to the photomarked region) should be facing up. This orientation is important to minimize the amount of resin we will have to remove during coarse sectioning with the glass knife. More detailed protocols on sample processing for TEM imaging can be found in Bozzola and Rusell [72].

3.7 Cutting Ultrathin Serial Sections

1. Start by manually trimming the block to create a trapezoid with parallel borders using an oil free razor.
2. In order to have perfect parallel top and bottom sides with smooth slopping surfaces in the trapezoid, continue trimming in an ultramicrotome using the angled sides of a 45° Cryotrim diamond knife at 2–3° attack angle [73]. Our block faces are typically ~ 400–500 μm wide and ~ 500–700 μm long to cover as much of the targeted area as possible.
3. Cut semi-thin sections (500 nm) with a glass knife and stain them with Toluidine Blue Solution until the two upper lines of DAB marks appear (*see* Fig. 4a).
4. At this point, initiate the ultrathin serial sectioning of the sample at 60–70 nm thickness. To minimize compression use a 35° diamond knife and try to section using attack angles of ~4°. Sections should come in a straight ribbon but, in our experience, anything longer than 15–20 sections will be hard to handle in the diamond knife boat. The best sectioning can be obtained by programming the optimal cutting speed at 0.8–1 mm/s (*see* **Note 37**). Use an eye-lash to split the ribbon into groups of five to seven sections.
5. Mount sections onto Pioloform-coated slotted grids (*see* Subheading 2.6) using a high-angle insertion of the grid in the boat and with the grid-slot parallel to the ribbon. Gently pull the grid up to prevent excessive shearing force on the ribbon.
6. Dry the grids. We recommend using absorbent paper points instead of letting the grids air dry, as the sections could get dispersed. After the sections are completely dried, proceed with staining.

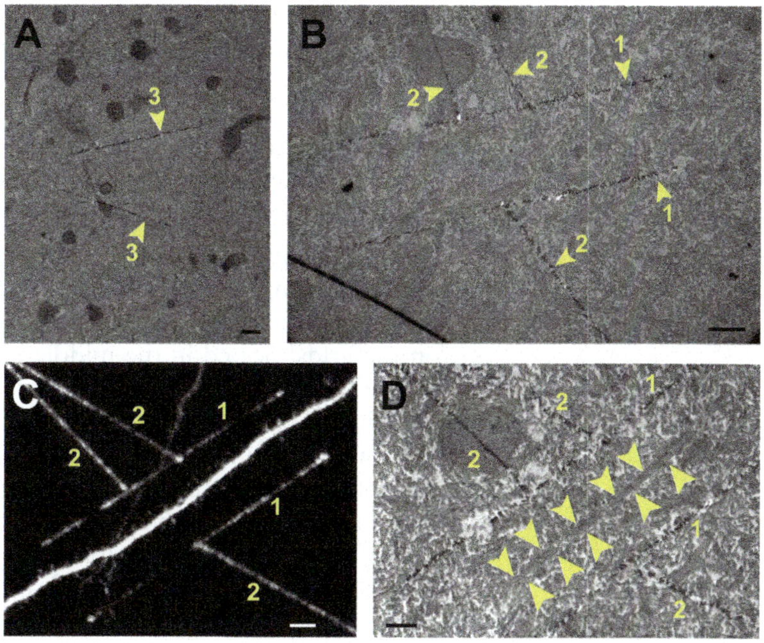

Fig. 4 Relocalization of landmarks in serial section TEM images. (**a**) Electron micrograph of the region shown in Fig. 3a–j where the two upper landmarks are visible (*arrowheads* labeled as "3"; same as those visible in Fig. 3h labeled as "3"). The appearance of these landmarks during cutting semi-thin sections announces the moment to start cutting ultrathin sections. Scale bar, 10 μm. (**b**) Electron micrograph of the same region, 3–4 μm deeper, where the next landmarks are visible, the ones parallel to the selected dendrite (*arrowheads* labeled as "1") and the ones semi-perpendicular to the dendrite pointing to the stimulated spines (*arrowheads* labeled as "2"). Scale bar, 10 μm. (**c**) Two-photon image of the selected region (also shown in Fig. 3g) where the fluorescence of the dendritic DsRed2 and the DAB landmarks is visible, to be compared to **d**. Scale bar, 10 μm. (**d**) Electron micrograph of the selected region, where the electrodense signal of the DAB landmarks is visible. The selected dendrite is delimited with *arrowheads*. Scale bar, 10 μm

7. For staining, put several drops of Uranyl Acetate Solution (*see* Subheading 2.5; *see* **Note 38**), one per grid, on top of a Petri dish covered with parafilm. Place the grid face down on the top of the drop so the sections are in direct contact with the solution. Stain for 1 min and then rinse by letting six to eight drops of dd water run over the tweezers (we found this method less perturbing for the sections than immersing the grids into beakers).
8. After completely drying the grids, repeat the same procedure with the Lead Citrate Solution, always minimizing the contact of this solution with CO_2 (*see* **Note 39**). Rinse the grids with CO_2-free dd water using the same method used with the uranyl

acetate solution. A more detailed protocol on serial sectioning for ssTEM can be found in Harris and colleagues [73] and Hayat [74].

3.8 Relocalization of the Dendrite in Serial Section Transmission EM

1. Load the grids on the sample holder and image them with the transmission electron microscope with a filament tension of 100 kV and at low magnification (3000×) [75].
2. Search for the electrodense DAB landmarks in all serial sections following an established order. Each grid and section must be sequentially numbered. Try to search the landmarks on the grids located in the middle of the sectioning sequence first. If the cutting and sectioning process has been successfully done in a parallel fashion with respect to the dendrite, very few sections should include most of the DAB lines (*see* Fig. 4b). If not totally parallel, the DAB lines will appear as small traces over many sections, which will make the relocalization more difficult, though not impossible. Once the landmarks are localized, start collecting mid-magnification images (5000–8000×) and draw an approximate map of these lines.
3. Localize the two lines that flank the dendrite in parallel and search for the dendrite that best matches the expected position of the selected dendrite, which should be situated at equal distance from these two flanking lines and at the same depth (*see* Fig. 4c, d). At this point, it is not necessary to completely identify the stimulated spines but should have a good guess as to which one is the selected dendrite by thoroughly comparing these mid-magnification images with the previous 2-photon images of the same region (*see* **Note 40**).
4. Once confident of the identification of the dendrite, search for the semi-perpendicular lines pointing to the spines and use them to find the regions where the stimulated spines are expected to be situated. Start collecting images of those regions at high magnification (23,000×) in all the sections. Collect them in a way that builds a virtual cube containing the stimulated spines and the dendrite (i.e., a 3D stack volume comprising several sections above and below the expected position of the spines). This cube will also include several unstimulated neighboring spines. The best method to prevent missing spots is a meander sampling, i.e., taking pictures uniformly spaced in a grid pattern. For that purpose, it is helpful to use the EM stage position display and to set a consecutive numbering scheme (number of section, row, and image within the row) to help during the photo-stitching step.

3.9 Morphometric 3D Reconstruction and Analysis of Spines

1. Use the automated stitching function of an image processing program (e.g., Photoshop) to stitch together all the regions imaged from the same section to create a collage image in 2D.

2. Combine these collage images from different sections to create the 3D stack volume of the whole selected dendrite containing all the stimulated spines and several neighboring spines. If stimulated spines are too far away from each other, build separated 3D stack volumes for each of the selected regions.
3. To perform image alignment, reconstruction, and analysis, we recommend the Reconstruct software package (http://synapses.clm.utexas.edu/synapseweb/software-0) [76]. Perform an initial rough reconstruction to see the overall shape of the dendrite and compare it to the 2-photon fluorescent images. If the dendrite has any feature that permits its identification, such as a turn, a point of branching, a group of spines, a non-spiny region, etc, try to match this feature between the fluorescence and the reconstructed EM images. At this point you must be quite confident that you have correctly identified the dendrite of interest.
4. Proceed to reconstruct all the spines that reside in the regions that have been imaged and stitched, in a blind fashion, i.e., before determining whether they are stimulated or unstimulated (*see* Fig. 5b–d). We focused and reconstructed the same type of stimulated or control spines, i.e., thin or mushroom spines, but not stubby or filopodia (*see* Fig. 5d–g). Reconstructions should include the dendritic shaft, as well as any other features of interest, such as the PSD, endoplasmic reticulum, presynaptic boutons, etc. (*see* Fig. 5a, e–g).
5. Localize the DAB landmarks that point to the stimulated spines and identify those spines from the set of reconstructed spines (*see* Fig. 5c, d). It is convenient to measure the relative distances of these spines to the neighboring spines or other features of the dendritic shaft (turns, branching points, etc.) and compare them to the fluorescence images. It may turn out to be impossible to unequivocally identify the stimulated spines, because they have disappeared, or because it is hard to decide between two spines. In this case, discard this spine from the study because the power of this experimental technique relies on the possibility to unequivocally identify the spines under both 2-photon and electron microscopy, and compare stimulated versus unstimulated synapses with certainty, thus avoiding any reliance on mixed population statistics.
6. Use the Reconstruct software package [76] to measure all desired parameters in each section (length and area of spine head, spine neck, PSD, endoplasmic reticulum, mitochondria, presynaptic bouton, vesicles, etc.) and use them to calculate the desired final 3D volumes and areas (*see* Fig. 5e–g).

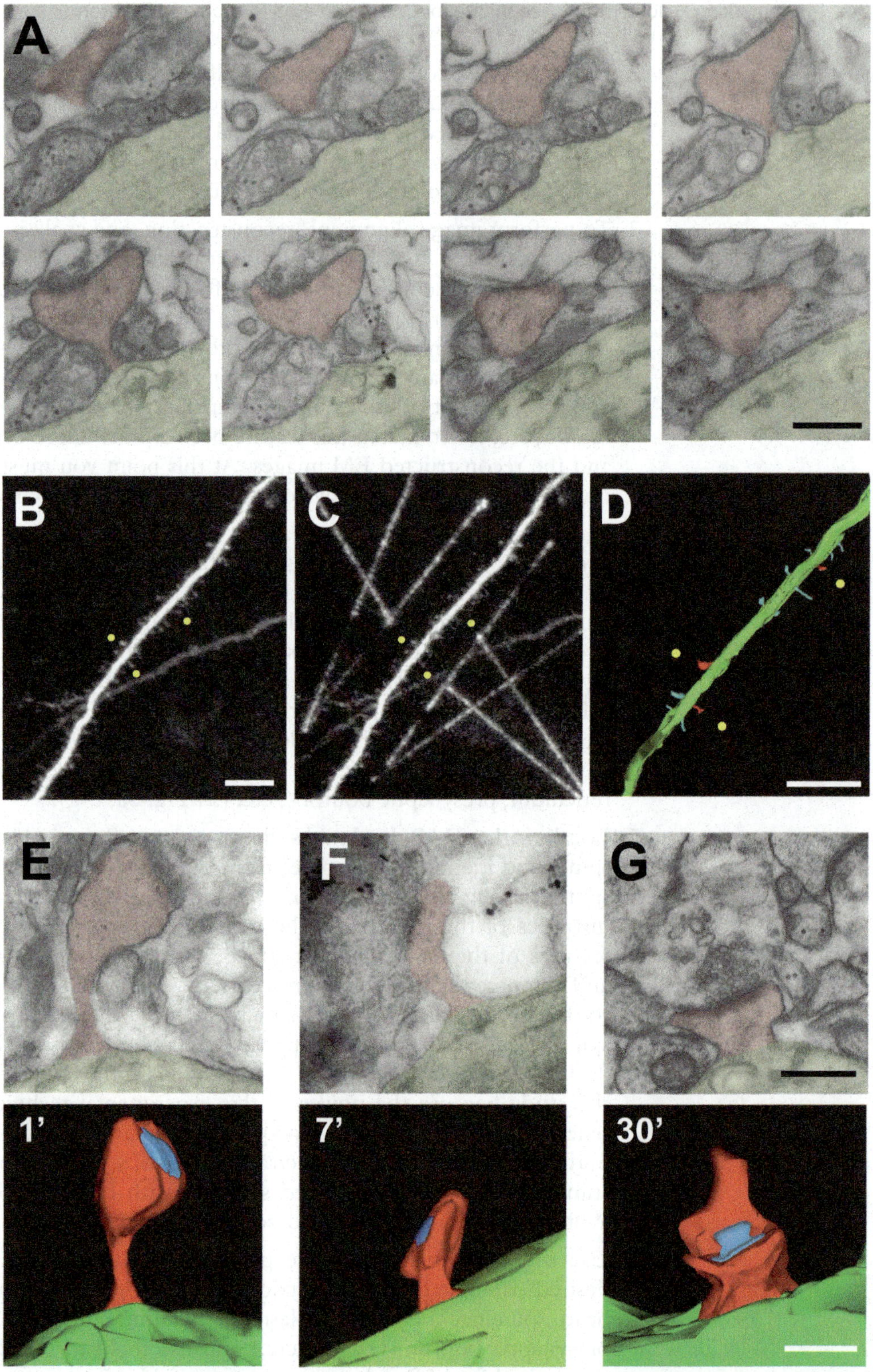

Fig. 5 Reconstruction of the dendrite and stimulated spines in 3D. (**a**) Example of a dendritic spine reconstructed from serial section EM images. *Red*, spine head and neck area; *green*, dendrite area. Scale bar, 0.5 μm. (**b**–**d**) Reconstruction of the whole selected dendrite. (**b**) Two-photon image of the dendrite after fixation (same as Fig. 3f, for comparison with **d**). Stimulated spines are highlighted with *yellow dots*.

4 Notes

1. Viable slices can be prepared from pups as young as P1, but they are less mature in terms of morphological, functional, and circuitry development. Pups older than P8 yield more mature slices but are more fragile in terms of survival and less easy to transfect.
2. Red fluorescent proteins are better excited with longer 2-photon excitation wavelengths (900–1100 nm), which offers lower resolution but deeper tissue penetration. DsRed2 is bright, stable, and well-expressed, but its tetrameric structure makes it inappropriate for fusion proteins, as there is the risk of forming aggregates. GFP can also form weak dimers. For these reasons, mCherry and monomeric-GFP are better options in the case of fusion proteins.
3. It is possible to use a single 2-photon laser for sequential, not simultaneous, imaging and uncaging, by changing the wavelength back and forth between 910 nm (imaging) and 720 nm (uncaging). Another option is to use one 720 nm 2-photon laser for uncaging and a 488 nm 1-photon laser for simultaneous imaging. Many other experimental approaches are possible with two combined 2-photon lasers (or changing wavelengths of one 2-photon laser) and one (or more) 1-photon lasers. It is possible then to perform different compatible photochemical reactions while imaging at the same time. For example, you can use: (1) a 820 nm laser to photoactivate photoswitchable proteins in one spine (e.g., to measure turnover of PAGFP-labeled proteins), (2) a 720 nm laser to photostimulate the same spine by glutamate uncaging (to trigger plastic changes) while (3) using a 488 nm laser to simultaneously image the same spine (e.g., *see* Fig. 4 in Bosch et al. [12]) In the case of 1-photon confocal imaging, always choose a dendrite close to the slice surface.
4. Avoid mechanical vibrations (and electrical noise if combining with electrophysiological recordings). Avoid fluctuations of the level of liquid in the chamber, especially the extreme cases of drying out the chamber or spilling liquid out of the chamber (which can damage the lower optical components of the microscope). A constant level in the chamber can be achieved by

Fig. 5 (continued) Scale bar, 10 μm (**c**) Two-photon image of the dendrite after photomarking (same as Fig. 3h, for comparison with **d**). (**d**) Reconstruction in 3D of the dendrite (*green*), stimulated spines (*red*), and unstimulated spines (*blue*). Scale bar, 5 μm. (**e**–**g**) Reconstruction of the stimulated spines shown in Fig. 3**c**–**e**. (**e**) EM image and 3D reconstruction of a spine stimulated 1 min before fixation. *Green*, dendritic shaft; *Red*, spine; *Blue*, PSD. (**f**) Spine stimulated 7 min before fixation. (**g**) Spine stimulated 30 min before fixation. Scale bar, 0.5 μm. Adapted from [12] forcing the outflow rate to be faster than the inflow rate. For example, use the pump to remove the liquid from the chamber and elevate it to the reservoir and let gravity to return it to the chamber; or use a higher flow speed for outflow than inflow; or if using the same pump and thus the same speed, use a softer plastic tube for the outflow and a harder tube for the inflow, thus facilitating the pump to more easily displace the liquid through the outflow tube. It is important to replace the tubes often (monthly) with new ones. Check that the pump and tubes are correctly installed and in good condition every day, as this is a common source of problems. Avoid precipitation of salt from ACSF, especially on objective lenses. Recirculate dd water at the end of the day to thoroughly wash all tubes, chambers, and objectives lenses that have been in contact with ACSF or any solution.

5. Measure osmolality every 30 min and determine when it increases significantly. In our setup, we could perform 1.5–2 h long experiments without any major increase in osmolality. For longer experiments, we usually kept a constant osmolality by adding ~100–200 μl of dd water every 1 h, approximately.

6. We recommend testing several batches of horse serum, selecting the one that gives better survival, slice quality, synaptic plasticity, etc., and purchasing large quantities of that batch. Adjust medium pH to 7.3 by adding consecutive small amounts of 10 M NaOH. Typically, up to ~100–200 μl of 10 M NaOH is needed to increase pH from around ~6.9 up to 7.3. Adjust osmolality by adding dd water, even though concentration of all components might be reduced below the suggested values. Osmolality is more important than final concentration of these components. Typically, consecutive small amounts of dd water, up to a final volume of ~90–120 ml, are added to reach 300–320 mOsm/L.

7. Prepare a 10× partial stock solution: 25 mM KCl, 260 mM $NaHCO_3$, 10 mM NaH_2PO_4, and 110 mM glucose. Filter and keep at 4 °C. Before dissection, dilute it 10 times with dd water, and add 238 mM sucrose, 1 mM $CaCl_2$, and 5 mM $MgCl_2$.

8. Prepare a Ca-free Mg-free 10× ACSF stock solution in dd water with 1190 mM NaCl, 25 mM KCl, 260 mM $NaHCO_3$, 10 mM NaH_2PO_4, and 110 mM glucose. Filter and keep at 4 °C. On the day of the experiment, dilute 10 times with dd water, bubble with carbogen for 20 min, and add 3 mM $CaCl_2$. Just before its use, add 1 μM TTX and 50 μM picrotoxin. TTX is added to block the spontaneous neuronal activity that would otherwise increase too much due to the lack of Mg. Picrotoxin is optional, in case you also want to block GABAergic transmission.

9. First prepare 4% EM-grade PFA solution (w/v) in 0.2 M PB. Dissolve it by heating at 55 °C and slightly raising the pH with a few drops of 1 M NaOH. Filter and let cool down to room temperature (r.t.). Mix this solution with a pre-made stock solution of 10% GA and dd water with these proportions (4% PFA:10% GA:dd water; 10:4:6). Adjust pH to 7.4.
10. Because of their viscosity, add components A and B by directly pouring them in a beaker thoroughly cleaned with acetone that will only be used for this purpose. For the less viscous components C and D, use plastic disposable pipettes. Use a scale to measure the quantities and mix thoroughly with a perfectly cleaned glass rod. Transfer to a vacuum bell jar and proceed to degasify the mix. When applying vacuum one should see major bubbles that disappear after a few minutes and a small film of foam in the interphase between Durcupan and air. Use 10 ml syringes to take the resin, avoiding the foamy interphase. Wrap the syringes in parafilm and store in the fridge in a dark compartment to prevent further polymerization. The remaining foamy resin can be used to prepare the 1:1 mix of propylene oxide and resin.
11. Slices are usually cultured on top of semi-permeable (0.4 μm pore) membrane inserts (Millipore) in 6-well plates. Some money can be saved by culturing the slices on top of already-cut pieces of Biopore Millicell hydrophilic polytetrafluoroethylene membrane placed onto these membrane inserts. Membrane inserts can then be resterilized and recycled [55]. Another economical alternative is to use Omnipore (Millipore) membranes [54].
12. It is convenient to culture four slices in each well, placed close to each other in the center of the well (to facilitate the transfection by gene gun), but without touching each other. Culturing more than four slices per well would consume medium nutrients too fast. Less than four slices is possible but then more wells and medium will be required, making the experiment more expensive.
13. Assess the quality of slices by DIV 5 by confirming that: (1) microglia has cleared and the slice becomes transparent, with no dark spots; (2) hippocampal subregions (CA1, CA3, Dentate Gyrus) are clearly visible, and (3) CA3 thickness is comparable to CA1 thickness (CA3 large neurons are more vulnerable to damage).
14. Dead surface neurons will disappear over the first days of culturing and glial response will decrease after that. Transfection of slices too early may result in many unwanted transfected cells: dying neurons or glial cells. On the other hand, efficiency of transfection decreases with time: one will need to prepare and transfect many more slices to have good chances of finding the appropriate fluorescent neuron.

15. If the neuron is whole-cell patched, long-term experiments might be precluded because of the dilution of cellular components into the pipette solution. Some extra experimental conditions might be necessary. For instance, maintenance of spine enlargement after LTP induction during whole-cell patch clamp requires adding 5 μM actin monomers to the pipette [77].
16. Empirically determine how long it takes for the system to reach stable conditions (temperature, etc.) by measuring laser power at final point (chamber) and by checking laser alignment. It can vary from a few minutes to a few hours.
17. This checking point is also useful to confirm the lasers are well aligned in the *Z* dimension. Try to change the position of the objective lens by focusing on several points along the *Z* dimension and confirm that maximum bleaching is achieved when the bead is well focused. If maximum bleaching is shifted and occurs at another *Z* position, you can fix that by a major realignment all optical components, or you can introduce the same *Z* shift when uncaging the dendritic spine.
18. It is possible to prepare the Uncaging ACSF at this point and recirculate it in the system before placing the sample (*see* also **Note 26**). We recommend, however, keeping the working Mg-free ACSF, place the sample, and replace the solution later with Uncaging ACSF.
19. It can sometimes be useful to check the location of fluorescent cells with an epifluorescence microscope before cutting out the membrane from the culture dish, in order to know which well and which slice might contain well-transfected neurons. Alternatively, one can prepare a high number of wells with enough slices. Just cut out a piece of membrane containing the four slices, put it in the chamber and look for the appropriate neuron. By chance, after checking several slices, it is usually possible to find one slice with an appropriate neuron in the appropriate region expressing the appropriate level of fluorescence signals.
20. Minimize the time that the slices are out of the incubator or out of the imaging chamber. Thus, it is worthwhile having a small incubator next to the microscope in which to store the plates on the imaging day, also to avoid disturbing the remaining cultures in the main incubator.
21. Take notes of the shape of the membrane piece, the position of the slice and the position and orientation of the neuron. It will be necessary to know later where to start looking for the landmarks.
22. Slice holders can be purchased or custom-made. Bend a 1 mm Ø platinum wire and make a ~10–14 mm Ø semicircle or U-shape. Cut the vertical strings of a nylon mesh (e.g., stockings) and keep

the horizontal ones. Glue the nylon strings on the two sides of the metal semicircle so that ~5 strings cross it, and cut the ends.

23. Some metabolic parameters, such as neuronal protein synthesis rate, might need >3 h of preincubation at 32 °C [78, 79] to reach a true stable state.
24. If you choose a secondary dendrite or the distal part of the main dendrite, the diameter of the dendrite will be smaller and the spine will be easier to image.
25. This way, you can have three stimulated spines in the same sample and thus reduce the number of samples to process by the factor of 3. However, if spines are too close in space or stimulated too close in time, you may induce metaplastic interactions [65].
26. Keep the caged glutamate under dark conditions. Prepare the Uncaging ACSF under the minimum illumination possible. It is useful to use a red light bulb or LED to illuminate the room or inside the microscope dark box, when needed, and to keep the chamber and the circulating uncaging solution covered all the times (with black curtains or yellow cellophane or UV filters) to protect it from ambient light, computer monitors, etc. If neurons start deteriorating just after the addition of caged glutamate, it could mean the batch is contaminated with free glutamate. This rarely happens and it can be detected by standard amino acid analysis.
27. We recommend performing pilot experiments where simultaneous intracellular recordings are made together with 2-photon imaging. In these pilot experiments one wants to adjust the caged glutamate concentration, laser intensity, and pulse duration to obtain an averaged EPSC response within a range similar to mini EPSCs [12, 36, 45, 80]. Once this is firmly established, subsequent imaging experiments can be performed without simultaneous recordings, which are highly time-consuming and preclude long-term experiments.
28. The efficiency of uncaging depends on the depth in the tissue and on the type of neuron. Therefore, this laser intensity value can be further adjusted for each neuron and for each depth by selecting a test spine, located in the same neuron and at a similar depth but in another dendrite or away from the selected region where the experimental spines will be stimulated, and inducing LTP to obtain a peak increase in spine volume of around 200–300% in the first 1–2 min and a subsequent sustained increase in spine volume of 50–100% for at least 10 min (*see* Figs. 2b and 3c–e) [12].
29. Neurons transfected with EGFP, DsRed2, or any free soluble fluorescent protein are less susceptible to bleaching, and so one can take images more often than when using slow-dynamics

synaptic fusion proteins as markers (e.g., EGFP-PSD-95). Typically, one can take an image every 1–5 min if the total experiment lasts for 30–60 min, or every 10–30 min if the experiment lasts for hours.

30. A variation of 10–20 % of spine head volume between consecutive images is typical in these experiments using Mg-free solutions, possibly due to the endogenous plasticity of spines, and to a lesser extent, also due to technical variations of the microscopic imaging.
31. Do not bubble with carbogen, as the CO_2 will decrease pH and will prevent DAB photoprecipitation. Do not expose fixed slices to hostile conditions that could erase the fluorescence signal, such as drastic changes in pH or strong photobleaching illumination.
32. A decrease in fluorescence signal is expected after the fixation step. A small degree of morphological distortion is also expected. Dendrites might look a little bit wavy and some blebbing may appear. Spine morphology should look as it was in the last live image. This distortion, if it happens, should not preclude a valid comparison between stimulated and control spines, since it should homogeneously affect all of them. However, if the neuron looks unhealthy (too much blebbling, dendrites that are too curvy) and/or the spines look too distorted or disappear, it means the fixation step has not been quick or strong enough. We recommend discarding such samples.
33. This photoprecipitation can be done in any region of the tissue and is probably achieved by the light-dependent generation of free radicals from endogenous intracellular oxidizable molecules. Photoprecipitation seems to occur more efficiently in regions that resemble the cell nuclei (*see* Fig. 3l).
34. Do not take many images and do not use high intensity of imaging laser, as bleaching is now a more significant problem because the sample and the fluorescent proteins are now fixed.
35. It is rather difficult to draw the lines pointing exactly to the spine. And it is actually not totally necessary. So, if the line points not exactly to the spine but X μm away, we will just have to search for the spine $-X$ μm away from the line on the EM images.
36. Osmium tetroxide solutions are highly susceptible to contaminants and easily reduced by light, so even if using amber vials, cover them with aluminum foil. Any solution that develops black clouding or deposits should be discarded and a fresh solution should be made.
37. It is important to prevent vibrations and air drafts that may interfere with the sectioning process by having the ultramicrotome setup in an enclosed area and minimizing movement around it. It is also important to minimize the levels of static by keeping a humid atmosphere with a humidifier.

38. Use a syringe covered with aluminum foil attached to a 0.22 µm disc filter to drop the Uranyl Acetate Solution. Always discard the first drop.
39. To obtain a CO_2-free environment, wrap the rim of the petri dish with kimwipes soaked in 1 N NaOH solution and add some NaOH pellets. It is also critical to reduce breathing when handling the grids and wear a surgical mask while staining to prevent exposure of the samples to exhaled CO_2.
40. While taking the EM images, it is convenient to simultaneously look at printed copies of the 2-photon fluorescence images of the selected region (or display them on a monitor) and constantly compare the pattern of landmarks to confirm you are in the correct region and correct depth.

Acknowledgments

This work was supported by a "Beatriu de Pinós" fellowship (AGAUR, "Generalitat de Catalunya"), the FRAXA Foundation, a Marie Curie Reintegration Grant (H2020-MSCA-IF) (to M.B.), the "Fundación Caja Madrid" (to J.C.), the Anne Punzak Marcus Fund (to M.S.), RIKEN, a NIH grant (R01DA17310), a Grant-in-Aid for Scientific Research (A), and a Grant-in-Aid for Scientific Research on Innovative Area "Foundation of Synapse and Neurocircuit Pathology" from the Ministry of Education, Culture, Sports, Science and Technology of Japan, the Human Frontier Science Program, and The Key Recruitment Program of High-end Foreign Experts of the Administration of Foreign Experts Affairs of Guangdong Province (to Y.H.). Conflict of interest statement: Y.H. is partly supported by Takeda Pharmaceutical Co. Ltd. and Fujitsu Laboratories.

References

1. Bosch M, Hayashi Y (2011) Structural plasticity of dendritic spines. Curr Opin Neurobiol 22:1–6
2. Matsuzaki M, Ellis-Davies GC, Nemoto T et al (2001) Dendritic spine geometry is critical for AMPA receptor expression in hippocampal CA1 pyramidal neurons. Nat Neurosci 4:1086–1092
3. Amatrudo JM, Olson JP, Agarwal HK et al (2014) Caged compounds for multichromic optical interrogation of neural systems. Eur J Neurosci 41:5–16
4. Nägerl UV, Willig KI, Hein B et al (2008) Live-cell imaging of dendritic spines by STED microscopy. Proc Natl Acad Sci U S A 105:18982–18987
5. Tønnesen J, Nägerl UV (2013) Superresolution imaging for neuroscience. Exp Neurol 242:33–40
6. Chéreau R, Tønnesen J, Nägerl UV (2015) STED microscopy for nanoscale imaging in living brain slices. Methods 88:57–66
7. Sigrist SJ, Sabatini BL (2012) Optical super-resolution microscopy in neurobiology. Curr Opin Neurobiol 22:86–93
8. Harris KM, Weinberg RJ (2012) Ultrastructure of synapses in the mammalian brain. Cold Spring Harb Perspect Biol 4:7
9. Marx V (2013) Brain Mapping in high resolution. Nature 503:147–152

10. Tanaka J-i (2005) Number and density of AMPA receptors in single synapses in immature cerebellum. J Neurosci 25:799–807
11. Bishop D, Nikić I, Brinkoetter M et al (2011) Near-infrared branding efficiently correlates light and electron microscopy. Nat Methods 8:568–570
12. Bosch M, Castro J, Saneyoshi T et al (2014) Structural and molecular remodeling of dendritic spine substructures during long-term potentiation. Neuron 82:444–459
13. Castro J, Bosch M, Hayashi Y et al (2009) Ultrastructural reorganization after long-term potentiation at a single dendritic spine. Poster communication at Neuroscience Meeting. Society for Neuroscience, Chicago, IL
14. Bosch M, Castro J, Narayanan R et al (2009) Structural and molecular reorganization of a single dendritic spine during long-term potentiation. Poster communication at Neuroscience Meeting. Society for Neuroscience, Chicago, IL
15. Nägerl UV, Köstinger G, Anderson JC et al (2007) Protracted synaptogenesis after activity-dependent spinogenesis in hippocampal neurons. J Neurosci 27:8149–8156
16. Zito K, Knott G, Shepherd GMG et al (2004) Induction of spine growth and synapse formation by regulation of the spine actin cytoskeleton. Neuron 44:321–334
17. Knott GW, Holtmaat A, Trachtenberg JT et al (2009) A protocol for preparing GFP-labeled neurons previously imaged in vivo and in slice preparations for light and electron microscopic analysis. Nat Protoc 4:1145–1156
18. Chen JL, Lin WC, Cha JW et al (2011) Structural basis for the role of inhibition in facilitating adult brain plasticity. Nat Neurosci 14:587–594
19. Monosov EZ, Wenzel TJ, Lüers GH et al (1996) Labeling of peroxisomes with green fluorescent protein in living P. pastoris cells. J Histochem Cytochem 44:189–581
20. Grabenbauer M, Geerts WJC, Fernadez-Rodriguez J et al (2005) Correlative microscopy and electron tomography of GFP through photooxidation. Nat Methods 2:857–862
21. Nikonenko I, Boda B, Alberi S et al (2005) Application of photoconversion technique for correlated confocal and ultrastructural studies in organotypic slice cultures. Microsc Res Tech 68:90–96
22. Bock DD, Lee W-CA, Kerlin AM et al (2011) Network anatomy and in vivo physiology of visual cortical neurons. Nature 471:177–182
23. Briggman KL, Helmstaedter M, Denk W (2011) Wiring specificity in the direction-selectivity circuit of the retina. Nature 471:183–188
24. Watanabe S, Punge A, Hollopeter G et al (2011) Protein localization in electron micrographs using fluorescence nanoscopy. Nat Methods 8:80–84
25. Sochacki KA, Shtengel G, van Engelenburg SB et al (2014) Correlative super-resolution fluorescence and metal-replica transmission electron microscopy. Nat Methods 11:305–308
26. Desmond NL, Levy WB (1986) Changes in the postsynaptic density with long-term potentiation in the dentate gyrus. J Comp Neurol 253:476–482
27. Ostroff LE, Fiala JC, Allwardt B et al (2002) Polyribosomes redistribute from dendritic shafts into spines with enlarged synapses during LTP in developing rat hippocampal slices. Neuron 35:535–545
28. Meyer D, Bonhoeffer T, Scheuss V (2014) Balance and stability of synaptic structures during synaptic plasticity. Neuron 82:430–443
29. Blazquez-Llorca L, Hummel E, Zimmerman H et al (2015) Correlation of two-photon in vivo imaging and FIB/SEM microscopy. J Microsc 259:129–136
30. Maco B, Cantoni M, Holtmaat A et al (2014) Semiautomated correlative 3D electron microscopy of in vivo-imaged axons and dendrites. Nat Protoc 9:1354–1366
31. Buchs PA, Stoppini L, Muller D (1993) Structural modifications associated with synaptic development in area CA1 of rat hippocampal organotypic cultures. Dev Brain Res 71:81–91
32. Collin C, Miyaguchi K, Segal M (1997) Dendritic spine density and LTP induction in cultured hippocampal slices. J Neurophysiol 77:1614–1623
33. Zafirov S, Heimrich B, Frotscher M (1994) Dendritic development of dentate granule cells in the absence of their specific extrinsic afferents. J Comp Neurol 345:472–480
34. Gähwiler B (1997) Organotypic slice cultures: a technique has come of age. Trends Neurosci 20:471–477
35. Muller D, Buchs PA, Stoppini L (1993) Time course of synaptic development in hippocampal organotypic cultures. Dev Brain Res 71:93–100
36. De Simoni A, Griesinger CB, Edwards FA (2003) Development of rat CA1 neurones in acute versus organotypic slices: role of experience in synaptic morphology and activity. J Physiol 550:135–147
37. Debanne D, Guérineau NC, Gähwiler BH et al (1995) Physiology and pharmacology of unitary synaptic connections between pairs of cells in areas CA3 and CA1 of rat hippocampal slice cultures. J Neurophysiol 73:1282–1294

38. Washbourne P, McAllister AK (2002) Techniques for gene transfer into neurons. Curr Opin Neurobiol 12:566–573
39. Karra D, Dahm R (2010) Transfection techniques for neuronal cells. J Neurosci 30:6171–6177
40. Bolz J, Novak N, Götz M et al (1990) Formation of target-specific neuronal projections in organotypic slice cultures from rat visual cortex. Nature 346:359–362
41. Molnár Z, Blakemore C (1991) Lack of regional specificity for connections formed between thalamus and cortex in coculture. Nature 351:475–477
42. Seidl AH, Rubel EW (2010) A simple method for multiday imaging of slice cultures. Microsc Res Tech 73:37–44
43. Umeshima H, Hirano T, Kengaku M (2007) Microtubule-based nuclear movement occurs independently of centrosome positioning in migrating neurons. Proc Natl Acad Sci U S A 104:16182–16187
44. Hayashi MK, Tang C, Verpelli C et al (2009) The postsynaptic density proteins homer and shank form a polymeric network structure. Cell 137:159–171
45. Sala C, Futai K, Yamamoto K et al (2003) Inhibition of dendritic spine morphogenesis and synaptic transmission by activity-inducible protein Homer1a. J Neurosci 23:6327–6337
46. Nikolenko V, Nemet B, Yuste R (2003) A two-photon and second-harmonic microscope. Methods 30:3–15
47. Majewska A, Yiu G, Yuste R (2000) A custom-made two-photon microscope and deconvolution system. Pflugers Arch 441:398–408
48. Tsai PS, Nishimura N, Yoder EJ et al (2002) Principles, design and construction of a scanning microscope for in vitro and in vivo brain imaging. In: Frostig RD (ed) In vivo optical imaging of brain function. CRC Press, Boca Raton
49. Tsai PS, Kleinfeld D (2009) In vivo two-photon laser scanning microscopy with concurrent plasma-mediated ablation principles and hardware realization. In: Frostig RD (ed) In vivo optical imaging of brain function. CRC Press, Boca Raton, pp 59–115
50. Okamoto K-I, Nagai T, Miyawaki A et al (2004) Rapid and persistent modulation of actin dynamics regulates postsynaptic reorganization underlying bidirectional plasticity. Nat Neurosci 7:1104–1112
51. Stoppini L, Buchs P-A, Muller D (1991) A simple method for organotypic cultures of nervous tissue. J Neurosci Methods 37:173–182
52. Yamamoto N, Kurotani T, Toyama K (1989) Neural connections between the lateral geniculate nucleus and visual cortex in vitro. Science 245:192–194
53. Gogolla N, Galimberti I, DePaola V et al (2006) Preparation of organotypic hippocampal slice cultures for long-term live imaging. Nat Protoc 1:1165–1171
54. Koyama R, Muramatsu R, Sasaki T et al (2007) A low-cost method for brain slice cultures. J Pharmacol Sci 104:191–194
55. Soares C, Lee KFH, Cook D et al (2014) Patch-clamp methods and protocols. 1183:205–219
56. O'Brien JA, Lummis SCR (2006) Biolistic transfection of neuronal cultures using a hand-held gene gun. Nat Protoc 1:977–981
57. Woods G, Zito K (2008) Preparation of gene gun bullets and biolistic transfection of neurons in slice culture. J Vis Exp e675:3–6
58. Ehrengruber MU, Hennou S, Bueler H et al (2001) Gene transfer into neurons from hippocampal slices: comparison of recombinant Semliki Forest Virus, adenovirus, adeno-associated virus, lentivirus, and measles virus. Mol Cell Neurosci 17:855–871
59. Malinow R, Hayashi Y, Maletic-Savatic M et al (2010) Introduction of green fluorescent protein (GFP) into hippocampal neurons through viral infection. Cold Spring Harb Protoc 2010:pdb.prot5406
60. Haas K, Sin W, Javaherian A et al (2001) Single-cell electroporation for gene transfer in vivo. Neuron 29:583–591
61. Pagès S, Cane M, Randall J et al (2015) Single cell electroporation for longitudinal imaging of synaptic structure and function in the adult mouse neocortex in vivo. Front Neuroanat 9:36
62. Antkowiak M, Torres-Mapa ML, Witts EC et al (2013) Fast targeted gene transfection and optogenetic modification of single neurons using femtosecond laser irradiation. Sci Rep 3:3281
63. Barrett LE, Sul JY, Takano H et al (2006) Region-directed phototransfection reveals the functional significance of a dendritically synthesized transcription factor. Nat Methods 3:455–460
64. Matsuzaki M, Honkura N, Ellis-Davies GCR et al (2004) Structural basis of long-term potentiation in single dendritic spines. Nature 429:761–766
65. Harvey CD, Svoboda K (2007) Locally dynamic synaptic learning rules in pyramidal neuron dendrites. Nature 450:1195–1200
66. Steiner P, Higley MJ, Xu W et al (2008) Destabilization of the postsynaptic density by PSD-95 serine 73 phosphorylation inhibits spine growth and synaptic plasticity. Neuron 60:788–802

67. Govindarajan A, Israely I, Huang SY et al (2011) The dendritic branch is the preferred integrative unit for protein synthesis-dependent LTP. Neuron 69:132–146
68. Lee S-JR, Escobedo-Lozoya Y, Szatmari EM et al (2009) Activation of CaMKII in single dendritic spines during long-term potentiation. Nature 458:299–304
69. Oh WC, Hill TC, Zito K (2013) Synapse-specific and size-dependent mechanisms of spine structural plasticity accompanying synaptic weakening. Proc Natl Acad Sci 110:E305–E312
70. Holbro N, Grunditz A, Oertner TG (2009) Differential distribution of endoplasmic reticulum controls metabotropic signaling and plasticity at hippocampal synapses. Proc Natl Acad Sci 106:15055–15060
71. Hayama T, Noguchi J, Watanabe S et al (2013) GABA promotes the competitive selection of dendritic spines by controlling local Ca2+ signaling. Nat Neurosci 16:1409–1416
72. Bozzola J, Russell L (1999) Specimen preparation for transmission electron microscopy. In: Bozzola J, Russell L (eds) Electron microscopy, 2nd edn. Jones & Bartlett Publishers, Sudbury, MA, pp 16–47
73. Harris KM, Perry E, Bourne J et al (2006) Uniform serial sectioning for transmission electron microscopy. J Neurosci 26:12101–12103
74. Hayat M (1985) Basic techniques for transmission electron microscopy. Cambridge University Press, Cambridge
75. Williams D, Carter B (1996) Transmission electron microscopy: a textbook for materials science. Plenum Press, New York
76. Fiala JC (2005) Reconstruct: a free editor for serial section microscopy. J Microsc 218:52–61
77. Tanaka J-i, Horiike Y, Matsuzaki M et al (2008) Protein synthesis and neurotrophin-dependent structural plasticity of single dendritic spines. Science 319:1683–1687
78. Sajikumar S, Navakkode S, Frey JU (2005) Protein synthesis-dependent long-term functional plasticity: methods and techniques. Curr Opin Neurobiol 15:607–613
79. Osterweil EK, Krueger DD, Reinhold K et al (2010) Hypersensitivity to mGluR5 and ERK1/2 leads to excessive protein synthesis in the hippocampus of a mouse model of fragile X syndrome. J Neurosci 30:15616–15627
80. Hirokawa J, Sadakane O, Sakata S et al (2011) Multisensory information facilitates reaction speed by enlarging activity difference between superior colliculus hemispheres in rats. PLoS One 6, e25283

Chapter 15

3D Analysis of Synaptic Ultrastructure in Organotypic Hippocampal Slice Culture by High-Pressure Freezing and Electron Tomography

Cordelia Imig and Benjamin H. Cooper

Abstract

Transmission electron microscopy serves as a valuable tool for synaptic structure-function analyses aimed at identifying morphological features or modifications associated with specific developmental stages or dysfunctional synaptic states. By utilizing cryo-preparation techniques to minimize the introduction of structural artifacts during sample preparation, and electron tomography to reconstruct the 3D ultrastructural architecture of a synapse, the spatial organization and morphological properties of synaptic organelles and subcompartments can be quantified with unparalleled precision. In this chapter, we present an experimental approach combining organotypic slice culture, high-pressure freezing, automated freeze-substitution, and electron tomography to investigate spatial relationships between synaptic vesicles and active zone release sites in synapses from lethal mouse mutants.

Key words Transmission electron microscopy (TEM), Electron tomography (ET), Organotypic slice culture, High-pressure freezing (HPF), Automated freeze-substitution (AFS), Synaptic ultrastructure

1 Introduction

Synaptic vesicles (SVs) fuse with the presynaptic plasma membrane at specialized sites called active zones (AZs), thereby releasing their neurotransmitter cargo into the synaptic cleft (reviewed in refs. 1, 2). In the field of molecular neurobiology, transmission electron microscopy (TEM) has served as an invaluable tool for the visualization and morphometric analysis of synaptic structures important for the maintenance and functional integrity of synaptic neurotransmission, including morphologically and functionally distinct SV pools [3–15] as well as pre- [3, 13, 16–21], trans- [18, 22–27], and postsynaptic [18, 28–30] cytoskeletal components or protein complexes.

The ability to resolve SVs and their spatial organization within the presynaptic terminal is of particular importance in studies attempting to identify morphological correlates of functional deficits resulting from the loss or perturbation of key components

Alexandros Poulopoulos (ed.), *Synapse Development: Methods and Protocols*, Methods in Molecular Biology, vol. 1538,
DOI 10.1007/978-1-4939-6688-2_15, © Springer Science+Business Media LLC 2017

of the neurotransmitter release machinery. Experimental approaches combining high-pressure freezing (HPF) and automated freeze-substitution (AFS) cryo-preparation techniques with classical TEM or 3D electron tomography (ET) have demonstrated that the morphological correlates of synaptic dysfunction can be manifest in the nanometer range, and that their detection is highly dependent on both the quality of ultrastructural preservation and the accuracy with which spatial relationships between synaptic entities can be resolved [31–36].

Conventional sample preparation techniques for TEM analysis, including aldehyde fixation and room-temperature dehydration, can introduce morphological artifacts that potentially influence the spatial organization and functional status of SVs at the presynaptic AZ and limit the interpretation of ultrastructural data (reviewed in ref. 37). Aldehyde exposure was found to induce SV exocytosis during fixation to different degrees depending on the experimental setting [10, 38–40]. A recent comparative analysis of synaptic ultrastructure in transcardially perfused vs cryo-fixed mouse somatosensory cortex revealed a specific reduction in membrane-proximal SVs at aldehyde-fixed AZs [41], thus identifying vesicular fusion as a potentially confounding consequence of chemical fixation in synaptic structure-function analyses.

Cryo-preparation methods combining HPF and AFS for the analysis of presynaptic ultrastructure offer two major advantages. Firstly, cryo-fixation occurs extremely rapidly, halting dynamic synaptic processes and immobilizing their structural arrangements within the millisecond timescale. Secondly, cellular ultrastructure is preserved in a near-native state, thereby enhancing the sensitivity and the reliability with which morphological correlates of perturbed synaptic function can be detected (reviewed in refs. 37, 42, 43). During HPF, an elevation in atmospheric pressure (~2100 bar) is synchronized with rapid liquid nitrogen cooling of the specimen to inhibit the volumetric expansion of water, thus suppressing ice crystal formation and the exothermic release of energy produced as a by-product: this effectively increases the rate of heat extraction, such that free water molecules lack the time required to assemble into large, structurally deleterious ice crystals and preferentially form dense, amorphous, or microcrystalline ice [44]. Although suboptimal freezing conditions can lead to the formation of crystalline ice within intracellular compartments, affected samples exhibit conspicuous ultrastructural artifacts that can be readily identified by visual inspection and excluded from analysis by quality control at the TEM [45].

Subsequent AFS at low temperatures substitutes vitrified water with organic solvent in the presence of fixative and contrasting agents, thereby minimizing shrinkage artifacts as the sample is gradually brought to room-temperature for plastic-embedding.

The preparation of tissue for HPF and ultrastructural analysis in neurotransmission-deficient mouse mutants is subject to certain limitations, since only small sample volumes can be effectively vitrified without incurring ice crystal damage, and mutations severely perturbing synaptic function often result in perinatal mortality. Acute dissection of neonatal brain tissue immediately prior to cryo-fixation restricts analyses to immature developmental timepoints and potentially exposes neurons to both mechanical and oxidative stresses. The organotypic hippocampal slice culture system circumvents many of these issues by generating slices of suitable dimensions for rapid cryo-fixation by HPF, providing an opportunity for tissue to recover from mechanical trauma induced by dissection during the culture period, and allowing neurons from lethal mouse mutants to establish mature synaptic connections in an in-situ-like environment.

Multiple factors limit the accuracy with which the spatial organization of SVs can be resolved in standard TEM analyses of ultra-thin plastic sections (reviewed in ref. 46). Such limitations, coupled with suboptimal tissue preservation, likely explain why subtle morphological consequences of synaptic perturbation have historically proven extremely difficult to detect by conventional TEM sample preparation and imaging approaches [47–52]. By revealing the ultrastructural architecture of the synapse in a three-dimensional context, electron tomography (ET) enables mapping of the position of SVs with respect to active zone release sites with nanometer precision [3, 33, 36]. In doing so, ET enables not only the reliable discrimination of morphologically distinct SV pools, but the ability to normalize quantified parameters to internal reference volumes or areas (e.g., numbers of docked SVs per unit area of active zone), thereby controlling for variations in section thickness or individual active zone sizes.

In the following sections, we provide methodological protocols outlining how organotypic hippocampal slice culture, HPF, AFS, and ET on plastic sections can be practically employed to reveal fine ultrastructural details and morphological correlates of synaptic dysfunction in lethal genetic mouse mutants.

2 Materials

2.1 Preparation of Cell Culture Inserts for Organotypic Slice Culture

1. Millicell Cell Culture Inserts and membrane confetti.
2. McIllwain tissue chopper (e.g., The Mickle Engineering Company Ltd, UK).
3. Culture Medium: 50% minimum essential medium (MEM), 25% basal medium Eagle (BME without Glutamine), 37 mM glucose, 2 mM glutamax, 25% heat-inactivated horse serum (example after ref. 53).
4. Six-well cell culture plates.

2.2 High-Pressure Freezing of Organotypic Slices

1. High-Pressure-Freezer (e.g., Leica EM HPM100).
2. Sample holder half cylinder (Leica 16770135) and middle plate (Leica 16770136).
3. Specimen carriers Type A (Leica 16770141) and B (Leica 16770142), 3 mm diameter.
4. Cryopreparation box containing sample carrier release device (Leica 16770111114).
5. 1-Hexadecene.
6. External cryo-protectant: 20% Bovine Serum Albumin (BSA) Culture Medium.
7. Stereomicroscope.
8. Perforated cryotubes.

2.3 Freeze Substitution and Plastic Embedding

1. Automated Freeze-Substitution (AFS) device (e.g., Leica AFS2).
2. Aluminum AFS containers with mesh-based plastic inserts (Leica).
3. Styrofoam box and liquid nitrogen-cooled aluminum stage for separation and transfer of HPF sample carriers.
4. Tannic Acid Solution: 0.1% tannic acid in acetone (glass-distilled EM-grade), prepare fresh.
5. Osmium Solution: 2% Osmium tetroxide (OsO_4) in acetone, prepare fresh.
6. EPON components: 14.4 g 2-Dodecenylsuccinic acid anhydride (DDSA), 21.4 g Glycidether 100, 11.3 g Methylnadic anhydride (MNA), 0.84 ml 2,4,6-Tris(dimethylaminomethyl) phenol (DMP-30) (*see*, for example, ref. 45), prepare when ready to use as described in Subheading 3.3.
7. Gelatin capsules, Size 00.
8. Disposable Tri-pour® polypropelene beakers.
9. Paper labels to fit into gelatin capsules.
10. 60 °C Incubator/Oven.

2.4 Sectioning

1. Leica-EM TRIM (Leica Microsystems).
2. Ultramicrotome (e.g., Leica Ultracut UCT ultramicrotome).
3. Diatome histo Jumbo 45° diamond knife (Diatome AG).
4. Diatome ultra 45° diamond knife (Diatome AG).
5. Formvar-filmed and carbon-coated copper 100 mesh EM grids.
6. Hair tool (e.g., a fine eyelash glued to a wooden stick).
7. Single-edged carbon steel razor blades with grip (for trimming off planchette).

8. Double-edged stainless razor blades (for trimming plastic block face).
9. Grid boxes.

2.5 Contrasting and Fiducial Marker Application

1. Uranyl Acetate Solution: 1 % uranyl acetate in water.
2. Reynold's Lead Citrate Solution [54]: 80 mM Lead(II) nitrate (N_2O_6Pb), 120 mM Sodium citrate dihydrate in decarbonated water. See for example [45].
3. Protein A (ProtA) coupled to 10 nm gold particles (e.g., Cell Microscopy Core Products, University Medical Center Utrecht, The Netherlands).
4. 5 ml syringes with Luer-lock.
5. Syringe-driven filter units 0.22 μm pore size.

2.6 2D-TEM Analysis

1. Electron Microscope (e.g., Zeiss LEO 912-Omega 80 kV).
2. CCD camera (e.g., Sharp:eye Troendle, TRS).
3. Image acquisition software (e.g., iTEM, Olympus Soft Imaging Solutions GmbH).
4. ImageJ/Fiji software for image processing, visualization, and quantification (refs. 55, 56; http://imagej.nih.gov/ij/).

2.7 3D-Electron Tomography Analysis

1. Electron Microscope (e.g., JEM-2100 200 kV, JEOL) equipped with a high tilt specimen holder.
2. SerialEM software for automated tilt series acquisition (ref. 57; University of Colorado, Boulder, Colorado, US; http://bio3d.colorado.edu/SerialEM/).
3. CCD camera (e.g., Orius SC1000 Gatan Inc.).
4. IMOD software package for tomogram generation (refs. 58, 59; University of Colorado, Boulder, Colorado, US; http://bio3d.colorado.edu/imod/).

3 Methods

3.1 Preparation of Organotypic Slice Cultures

1. Prepare a six-well cell culture plate by pipetting 1 ml of Culture Medium and place a Millipore Cell culture insert into each well.
2. Carefully place precut and autoclaved membrane confetti with fine forceps onto cell culture inserts (Maximum of four confetti pieces per membrane). Avoid air bubbles between the membranes (*see* **Note 1**).
3. Pre-warm medium by placing the culture plate into the cell culture incubator (here 37 °C and 5 % CO_2).
4. Prepare the organotypic slice culture. We follow a protocol for P0 mice described in Kerr et al. [53]. 300 μm thick hippocampal

slices of E18 – P0 mice are produced using a tissue chopper. Slices still attached to the entorhinal cortex are transferred onto the membrane confetti. Medium is changed every 2–3 days and slices are used for experiments between 4 and 5 weeks (*see* **Note 2**). Please follow the appropriate protocol for your brain region of interest.

3.2 High-Pressure-Freezing

1. Change slice Culture Medium 1 day before freezing.
2. HPF requires work with liquid nitrogen and has to therefore be performed in a well-ventilated room. Fill the HPF device according to manufacturer's instructions and allow ample time for the device to cool down (Leica HPM100; please refer to instructions of the manufacturer for details about the equipment and its proper usage).
3. Fill the small sample dewar with liquid nitrogen and place it into the side chamber of the HPF device.
4. Prepare external cryo-protectant and place it on ice (*see* **Note 3**). To prepare cryo-protectant, warm 5 ml Culture Medium to 37 °C and dissolve 1 g BSA.
5. Pre-warm an aliquot of slice Culture Medium in the cell culture incubator at 37 °C.
6. Take slice cultures out of the incubator (one six-well plate with only one membrane insert at a time). Transfer one of the tissue explants into a petri dish containing cell culture medium by carefully handling the membrane confetti with a fine pair of forceps.
7. Place the dish under the stereomicroscope and carefully trim off excess membrane around the tissue slice using a scalpel.
8. Immerse the tissue slice briefly in a drop of cryo-protectant on parafilm and place it into the 100 μm cavity of the specimen carrier Type A with the membrane confetti facing upwards (Fig. 1a) (*see* **Note 4**).
9. Place the specimen carrier into the plastic middle plate of the sample holder. Dip the flat side of the specimen carrier Type B briefly into 1-hexadecene and close the specimen carrier (Fig. 1a) (*see* **Note 5**). Wick off excess liquid using Whatman's paper.
10. Freeze samples according to the instructions of the manufacturer of your HPF device.
11. Remove samples from the small liquid nitrogen dewar located at the side of the instrument. Release sample carriers in liquid nitrogen using the release device provided by Leica and store samples in perforated cryotubes in liquid nitrogen (*see* **Note 6**).

3.3 Freeze Substitution and Plastic Embedding

1. Fill the automated freeze-substitution device (Leica AFS2) with liquid nitrogen and program the AFS protocol [16], as seen in Fig. 1c. Start the program and pause it while allowing the specimen chamber to cool down to −90 °C.

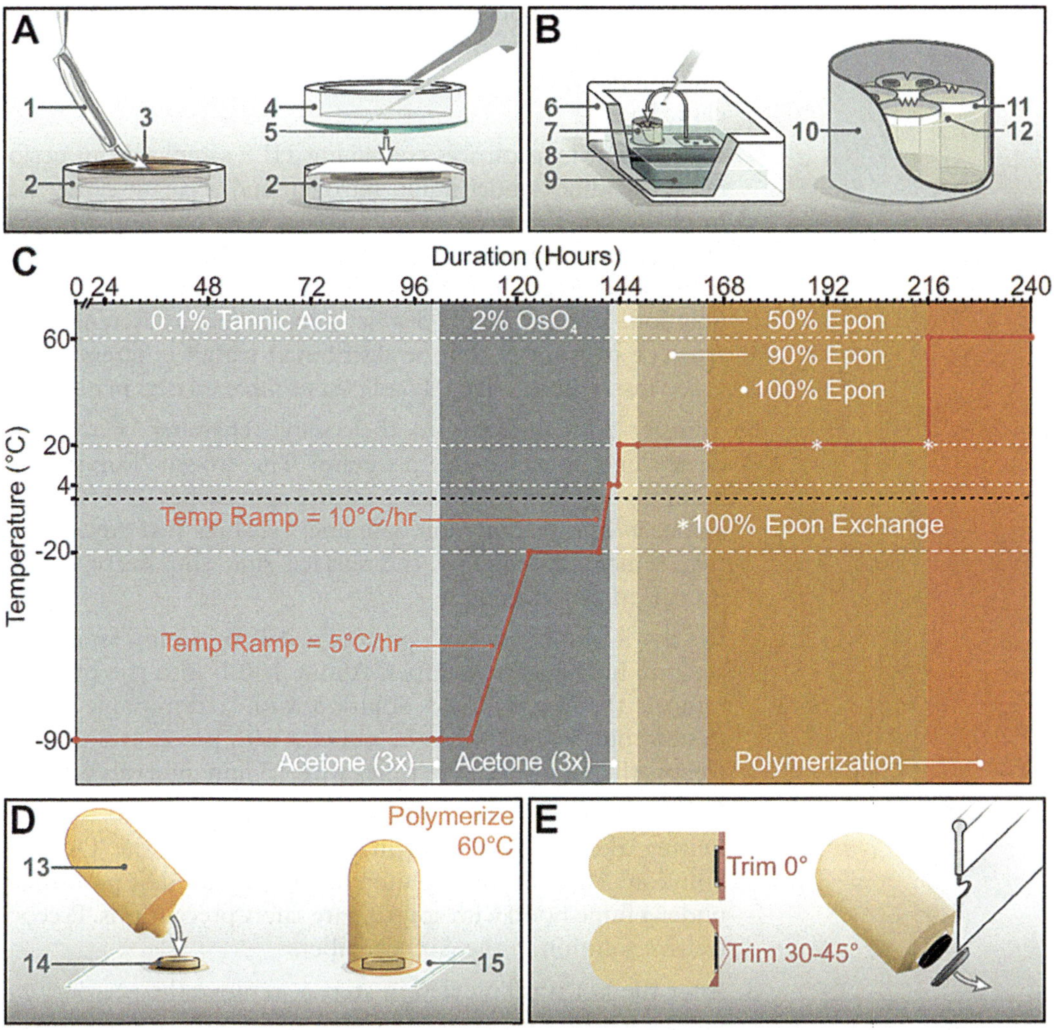

Fig. 1 Sample preparation. (**a**) Loading of hippocampal organotypic slice into specimen carriers for HPF. (**b**) Transfer of HPF samples from liquid nitrogen storage to freeze substitution media. (**c**) AFS protocol indicating the temperatures and substitution media to which samples are exposed during freeze substitution and subsequent resin infiltration. (**d**) Plastic embedding of resin-infiltrated sample with aluminum specimen carrier. (**e**) Plastic block trimming and aluminum specimen carrier removal in preparation for ultramicrotomy. *1, organotypic hippocampal slice on membrane confetti*; *2, aluminum specimen carrier* (*Leica, Type A*); *3, external cryoprotectant*; *4, aluminum specimen carrier* (*Leica, Type B*); *5, 1-hexadecene*; *6, styrofoam box* (*cut-away view*); *7, aluminum AFS container assembly* (*cut-away view*); *8, aluminum cryo-transfer block*; *9, liquid nitrogen*; *10, enlargement of AFS container assembly* (*cut-away view*); *11, mesh-based plastic capsule*; *12, frozen substitution solution* (*e.g., 0.1 % tannic acid in anhydrous acetone*); *13, resin-filled gelatin capsule*; *14, freeze substituted and plastic infiltrated sample in carrier*, *15, Parafilm-covered glass slide*

2. Prepare 20 ml of Tannic Acid Solution for the first freeze substitution in a 50 ml conical tube. Aliquot ~10 ml of the solution into each aluminum container (bucket). Four mesh-based plastic AFS sample containers (16702734, Leica) will fit into each bucket and can be individually identified by cutting notches in their rims with a razor blade (Fig. 1b).

3. Cover the AFS bucket with a plastic lid and allow the Tannic Acid Solution to freeze by placing it on a precooled aluminum stage submerged in liquid nitrogen in a shallow styrofoam box (Fig. 1b).
4. Extract specimen carriers containing HPF samples from perforated cryotubes under liquid nitrogen and separate carrier lids from bases on the precooled metal block (Fig. 1b). Rapidly transfer specimen carriers containing samples from the liquid nitrogen-cooled stage onto the surface of the frozen Tannic Acid Solution. Experimentally distinct samples (e.g., different genotypes or treatments) can be distributed within different mesh-based inserts accordingly. Load two to three carriers per insert.
5. Transfer AFS bucket into the sample chamber of the AFS device and unpause the program. The frozen Tannic Acid Solution will melt as the temperature approaches –90 °C. Brief agitation of the aluminum containers ensures that any carriers that remain floating on the surface sink and become fully immersed in solution.
6. On day 5 of the protocol, precool a vial containing anhydrous acetone in the AFS chamber. About 100 h into the protocol remove the Tannic Acid Solution with a disposable plastic Pasteur pipette and wash the samples with precooled acetone. Repeat this step two more times with 15 min intervals between washes.
7. During the washing steps in **step 6**, prepare the Osmium Solution. All work with osmium tetroxide should be performed under a fume hood with appropriate safety precautions. Precool fixative solution in the AFS chamber.
8. After the last washing step in **step 6**, remove the acetone and replace it with the precooled Osmium Solution for fixation.
9. When the temperature reaches 4 °C on day 7 of the protocol, wash the samples three times at 4 °C with anhydrous acetone as described in **step 6**.
10. Prepare EPON resin for infiltration by mixing 21.4 g DDSA, 21.4 g glycide ether and11.3 g MNA in a Tri-pour® beaker. Mix components with a magnetic stirrer at moderate velocity for 15 min. Avoid excessive stirring speeds, which can introduce unwanted air bubbles. After mixing for 15 min, add 840 μl of the accelerator DMP-30 and stir for another 15 min to allow full incorporation.
11. Prepare 50% and 90% dilutions of EPON resin in acetone. Aliquot the 50% EPON resin into labeled 2 ml microcentrifuge tubes at RT and transfer substituted samples (within aluminum carriers) from the mesh-based inserts in acetone into respective microcentrifuge tubes and allow resin to infiltrate

for 3 h. Transfer samples into microcentrifuge tubes containing 90% EPON in acetone and allow resin to infiltrate overnight.

12. Prepare fresh EPON resin (as described in **step 10**) on day 8. Incubate samples in 100% EPON. Change EPON three times during the day.
13. Prepare fresh EPON resin as described in **step 10** on day 9. Transfer specimen carriers sample-side-up in a small EPON drop onto object slides covered with parafilm (Fig. 1d). In case the sample becomes separated from its specimen carrier during the AFS, place the sample membrane-side facing up onto the parafilm. Fill fresh 100% EPON into a gelatin capsule, add a small paper label identifying the sample into the capsule and invert it on the specimen carriers. Samples are then polymerized for 24 h at 60 °C (Fig. 1d).

3.4 Sectioning

1. Trim off excess EPON around and on top of the aluminum specimen carrier using a Leica-EM TRIM (Fig. 1e). Alternatively use razor blades.
2. Dislodge and remove the sample carrier from the block using a razor blade. The surface of the slice that was facing the base of the sample carrier should now be exposed. Carefully trim off excess EPON using a razor blade.
3. Mount the sample block into an ultramicrotome device. Follow the instructions of the manufacturer for cutting sections.
4. Cut a few 500 nm thick sections on a Leica Ultracut UCT ultramicrotome (or equivalent) using, for example, a Diatome histo Jumbo 45° diamond knife until the first tissue appears in the semithin sections.
5. Collect four or five 200 nm thick sections for ET (*see* **Note 7**) onto Formvar-filmed, carbon-coated copper mesh grids (*see* **Note 8**) using, for example, a Diatome ultra 45° diamond knife.
6. Subsequently collect ultrathin sections (e.g., 60 nm) for 2D-EM analysis.
7. Store grids in a grid box until further use.

3.5 Contrasting and Fiducial Marker Application

1. Place a piece of parafilm onto a glass plate.
2. Prepare small (~30 μl) droplets of Uranyl Acetate Solution filtered through a syringe-driven filter unit.
3. Place grids with ultrathin sections facing down onto the droplets and incubate for 30 min at RT in the dark. Do not contrast tomography sections.
4. Wash grids multiple times in ddH_2O and wick off excess fluid with filter paper.

5. Dispense small droplets of Reynold's Lead Citrate Solution on parafilm filtered through a syringe-driven filter unit. (Place KOH pellets in the vicinity of the droplets to absorb CO_2 from air. Avoid exhalation onto droplets to prevent precipitation of lead carbonate). Incubate for 2 min.
6. Wash grids multiple times in ddH_2O, wick off excess fluid with filter paper, and allow them to air dry for at least 60 min.
7. To introduce fiducial markers for tomographic reconstructions, label grids containing the 200 nm thick sample sections with ProtA coupled to 10 nm gold. We incubate grids for 1–2 min on each side on small droplets of ProtA solution in water. The optimal dilution of ProtA solution is determined empirically for each batch of ProtA solution and for the intended magnification at which tomographic tilt-series are to be acquired. Approximately 20 fiducial gold particles on each surface within the observed field of view will support tomographic reconstruction with adequate correction of local distortions incurred by electron irradiation during the series acquisition.

3.6 TEM of Spine Synapses

1. Perform regular TEM imaging of spine synapses on ultrathin plastic sections. In our case, electron micrographs (2048 × 2048 pixels) of synaptic profiles were acquired at 12,500-fold magnification and with a pixel spacing of 0.95 nm. Glutamatergic, asymmetric spine synapses were identified by the presence of an accumulation of presynaptic SVs, a prominent PSD and the absence of mitochondria or microtubules in the postsynaptic compartment (Examples shown in Fig. 2a–e; For a more detailed description, please refer to the SI of ref. 36).
2. Quantify various 2D synaptic parameters as shown in (Fig. 2f) using image analysis software (e.g., iTEM software).

3.7 3D-Electron Tomography Analysis

1. Perform electron tomography on the 200 nm thick plastic sections. We recorded tilt series from −60° to + 60° with 1° tilt increments and binned by a factor of two at 30,000-fold magnifications using SerialEM for automated tilt series acquisition [57] and with an Orius SC1000 camera (Gatan). The target defocus during image acquisition is set to −0.2 to 0.4 μm. The individual images in the tilt series before tomogram reconstruction have 1336 × 1336 pixels with an individual pixel-spacing of 0.52 nm. To maximize the quality of tomographic reconstructions from single-axis tilt series, synaptic profiles were only selected for reconstruction if they fulfilled the following criteria: (1) the synaptic cleft is clearly visible at 0° tilt angle (Fig. 3b), (2) the active zone membrane is oriented more or less parallel to the tilt rotation axis (Fig. 3b), and (3) gold fiducial markers are evenly distributed with an appropriate density on both surfaces of the section (Fig. 3b).

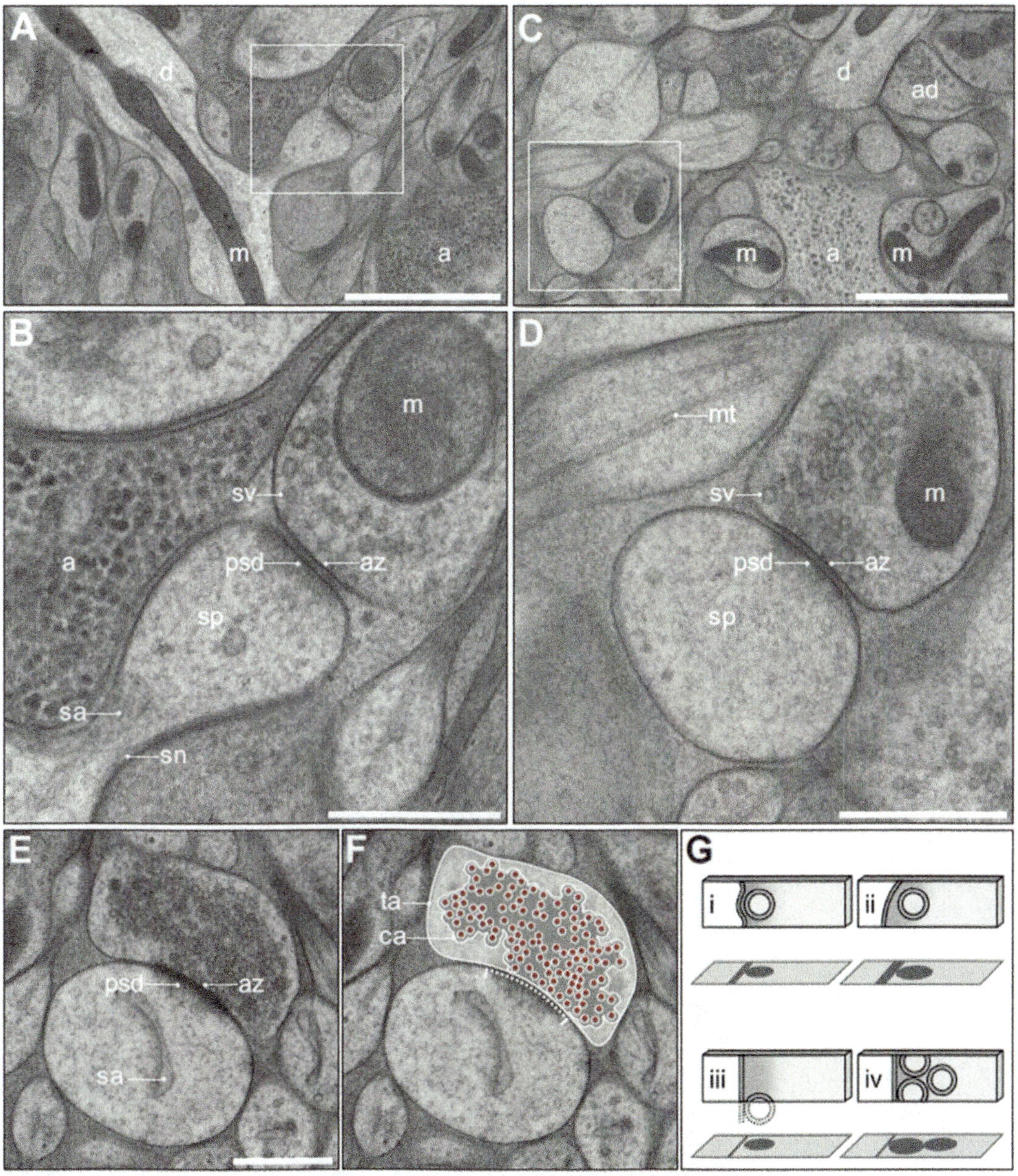

Fig. 2 Ultrastructural organization and 2D analysis of hippocampal spine synapses. (**a**–**f**) Electron micrographs of synaptic neuropil from high pressure frozen organotypic hippocampal slices. (**a**) Longitudinally sectioned hippocampal spine synapse. (**b**) Enlargement of white frame in (**a**). (**c**) Cross section of a hippocampal spine synapse. (**d**) Enlargement of white frame in (**c**). (**e**) Cross section of a hippocampal spine synapse. (**f**) Synaptic parameters quantified by 2D analysis. (**g**) Scenarios potentially confounding the 2D analysis of SV docking in transmission electron micrographs acquired from ultrathin sections. Scale bars: **a**, **c**, 1 μm; **b**, **d**, and **e**, 500 nm. *d* dendrite, *m* mitochondrion, *a* astrocyte, ad axodendritic synapse, *sv* synaptic vesicle, *az* active zone, *psd* postsynaptic density, *mt* microtubule, *sa* spine apparatus, *sn* spine neck, *ta* presynaptic terminal area, *ca* synaptic vesicle cluster area

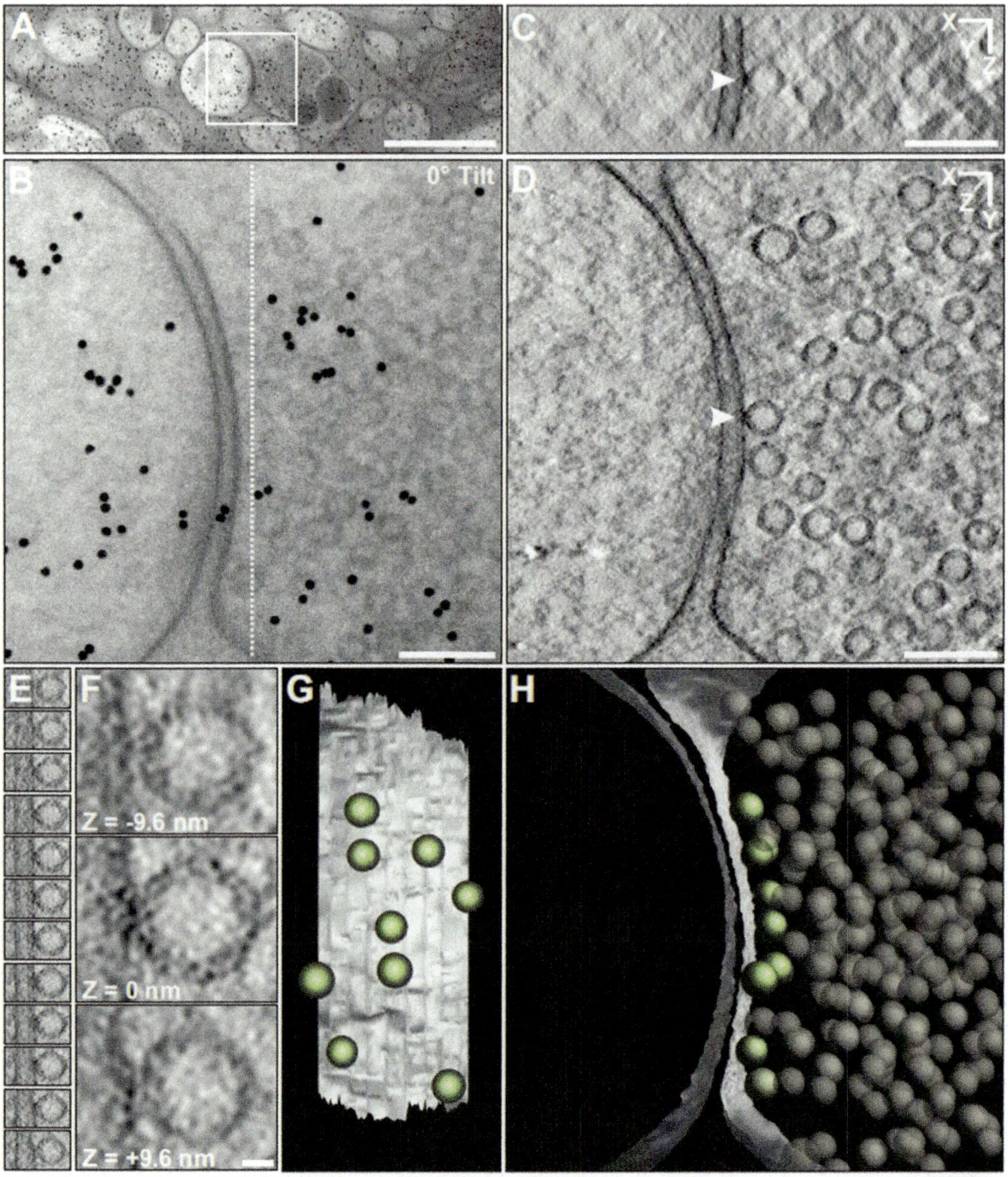

Fig. 3 3D electron tomography. (**a**) Low magnification overview of hippocampal neuropil acquired from a fiducially coated 200 nm thick plastic section. The white frame indicates a spine synapse selected for tomographic reconstruction. (**b**) High magnification micrograph of pre- and postsynaptic appositions (white frame in **a**) obtained during 2D tilt series acquisition (at 0° tilt). *White dashed line* indicates tilt axis orientation. (**c**, **d**) Tomographic slices presenting orthogonal perspectives of a membrane-attached synaptic vesicle (indicated by *white arrowheads*). (**e, f**) Changes in the apparent spatial relationship between a synaptic vesicle and the active zone membrane occur with deviation from the vesicle midline. (**g**, **h**) 3D models in which the spatial arrangement of docked synaptic vesicles (*green*) at the active zone (*white*) can be appreciated. Scale bars: **a**, 1 μm; **b–d**, 100 nm; **f**, 20 nm

2. Reconstruct tomographic volumes from the tilt series using the IMOD package [58]. Tomograms are reconstructed using a back projection algorithm. In our case, the projections are binned three times during the reconstruction of the final aligned stack, which ultimately produces reconstructed tomographic volumes results with an isotropic voxel size of 1.55 nm.
3. Export tomographic volumes into ImageJ. Please note that tomographic stacks (*.*rec*) can either be opened directly using Fiji or exported as a series of TIFF images (IMOD program *mrc2tif*) in ImageJ. Set the right scale in ImageJ.
4. Identify the midline of each SV, indicated by maximal diameter and a clearly defined lipid bilayer, and measure the smallest distance between the outer leaflet of the SV membrane and the inner leaflet of the AZ membrane in ImageJ using the straight line tool (Fig. 3e). Only consider SVs with no measurable distance between the two membranes as docked. The AZ is defined here as the stretch of presynaptic plasma membrane opposing the PSD of the postsynaptic compartment. The AZ area is quantified by measuring the AZ length on each consecutive virtual z-slice using the freehand line tool. The values are multiplied in our case by a factor of 1.55 (nm; voxel depth) and added to obtain the final AZ area per tomogram. We normalized the number of SVs to the AZ area (SVs/0.01 μm^2 AZ) for each tomogram in the analysis.

4 Notes

1. We employ the interface method for organotypic cultures of hippocampal tissue [60]. This technique is based on the idea that tissue explants are cultured on a porous membrane at the interface between medium and air [60]. However, older slices firmly grow onto the membrane making it difficult to remove explants from culture inserts. This requires for example the use of razor blades to detach slices from the membrane [33]. We therefore culture slices on small pieces ("confetti") of hydrophilic membrane (Millipore cat. #. FHLC04700, 0.45 μm pore size) [61]. These membranes are not sterile upon purchase and have to be cut and autoclaved prior to use. Carefully handle the membrane confetti with sterile forceps to make sure the underlying membrane of the cell culture insert is not contaminated or damaged.
2. We freeze slices after 4–5 weeks in culture, although freezing at earlier time points (e.g., 2–3 weeks) [62] is also possible. During the culture period, tissue explants are allowed to recover from the slicing procedure, which is crucial for subsequent ultrastructural analysis. During the culturing period,

dead cells disappear from the tissue surface, resulting in a flattening of the slice [61, 63]. Previous studies have shown that hippocampal organotypic explants from 400 μm slices thin out to ~150 μm within a week of culturing, reducing the slice thickness to only a few cells [63]. We estimated that our slices thin out within our culturing period from 300 μm to < 100 μm.

3. The choice of an appropriate external cryo-protectant, or "filler," for freezing biological samples and especially neuronal tissue has turned out to be a critical one. During pressurizing it is important that the cavity of the specimen carrier is entirely occupied—air bubbles in the external solution will result in suboptimal freezing and pressure damage. Whereas culture systems of neuronal monolayers have successfully been frozen in culture medium without any additives [8], a method that has also been employed by a few studies using organotypic slices for HPF [33, 62], HPF of tissue sections required in our hands external cryo-protection to achieve optimal freezing results. Different extracellular fillers are routinely used for biological samples (discussed, for example, in refs. 64, 65). They are commonly classified as either penetrating (e.g., sucrose, glycerol) or nonpenetrating substances (e.g., BSA, dextran, low-melt agarose, 1-hexadecene). However, the most commonly used filler for rapid cryo-fixation of hippocampal organotypic slice cultures has been a combination of 20 % dextran and 5 % sucrose dissolved in culture medium [3, 22, 62]. This cryo-protectant appears to yield extremely well-preserved neuronal ultrastructure. However, the addition of dextran and sucrose increases substantially the osmolarity of the culture medium by ~300 mOsm [3, 22] resulting in a final osmolarity of ~600 mOsm. Application of hypertonic solutions to neurons triggers SV release and a concentration of 500 mM sucrose is sufficient to deplete the RRP of SVs in neuronal cultures [66]. Even lower sucrose concentrations (i.e., 250 mM) have been shown to have an effect on SV release [67]. For this reason, the use of hypertonic cryo-protectants has been viewed skeptically with respect to the correlation of synaptic ultrastructure and function in high-pressure frozen neuronal material. To circumvent this issue, we employ a cryo-protectant composed of 20 % BSA in culture medium (~340 mOsm) as an external filler. The addition of BSA has yielded excellent ultrastructural preservation for SV analyses in CNS synapses [36] and in retinal photoreceptor synapses [9, 45]. Please note that when using BSA, both the sample and cryo-protectant will become osmified during AFS and may therefore make it more difficult to identify the position and orientation of tissue in the specimen carrier prior to ultramicrotomy. Alternatively, it has been shown that acute cortical slices can be quickly frozen directly in 1-hexadecene yielding excellent tissue preservation [41].

4. Ensure that the slice fits into the cavity of the carrier without compression. Leica specimen carriers Type A are designed with 100 μm- and 200 μm-deep cavities on opposing sides. For our hippocampal organotypic slices we use the 100 μm-deep sample carrier. Some excess membrane will exceed the cavity and touch the rim of the aluminum carrier. Please note that the majority of the tissue explants from hippocampal organotypic slice cultures prepared from E18 and P0 mouse pups fit into the cavity of a 3 mm diameter specimen carrier. However, organotypic slices from older animals or rats might require punching out regions-of-interest with a biopsy punch [62]. Alternatively, the Leica HPM100 allows freezing of tissue in larger (6 mm diameter) aluminum carriers.
5. 1-Hexadecene is immiscible with water and is therefore considered inert and as a nonpenetrating cryo-protectant. It is used here to facilitate separation of frozen specimen carriers and lids prior to freeze-substitution.
6. If the AFS will be started on the same day of freezing, samples can be immediately loaded into FS vials/buckets.
7. The maximal thickness of the section for tomography depends on the acceleration voltage of the electron microscope used for imaging (*see* for example ref. 68). We imaged tilt-series on 200 nm thick sections using a JEM-2100 (JEOL) TEM operating at 200 kV.
8. The making of Formvar-coated grids has been previously described [69].

References

1. Heuser JE (1989) Review of electron microscopic evidence favouring vesicle exocytosis as the structural basis for quantal release during synaptic transmission. Q J Exp Physiol 74:1051–1069
2. Südhof TC (2012) The presynaptic active zone. Neuron 75:11–25
3. Fernández-Busnadiego R, Zuber B, Maurer UE et al (2010) Quantitative analysis of the native presynaptic cytomatrix by cryoelectron tomography. J Cell Biol 188:145–156
4. Schikorski T, Stevens CF (1997) Quantitative ultrastructural analysis of hippocampal excitatory synapses. J Neurosci 17:5858–5867
5. Schikorski T, Stevens CF (2001) Morphological correlates of functionally defined synaptic vesicle populations. Nat Neurosci 4:391–395
6. Rizzoli SO, Betz WJ (2004) The structural organization of the readily releasable pool of synaptic vesicles. Science 303:2037–2039
7. Harata N, Ryan TA, Smith SJ et al (2001) Visualizing recycling synaptic vesicles in hippocampal neurons by FM 1–43 photoconversion. Proc Natl Acad Sci U S A 98:12748–12753
8. Watanabe S, Rost BR, Camacho-Pérez M et al (2013) Ultrafast endocytosis at mouse hippocampal synapses. Nature 504:242–247
9. Cooper B, Hemmerlein M, Ammermüller J et al (2012) Munc13-independent vesicle priming at mouse photoreceptor ribbon synapses. J Neurosci 32:8040–8052
10. Satzler K, Sohl LF, Bollmann JH et al (2002) Three-dimensional reconstruction of a calyx of Held and its postsynaptic principal neuron in the medial nucleus of the trapezoid body. J Neurosci 22:10567–10579
11. Heuser JE, Reese TS (1973) Evidence for recycling of synaptic vesicle membrane during transmitter release at the frog neuromuscular junction. J Cell Biol 57:315–344
12. de Lange RPJ, de Roos ADG, Borst JGG (2003) Two modes of vesicle recycling in the rat calyx of Held. J Neurosci 23:10164–10173

13. Graydon CW, Zhang J, Oesch NW et al (2014) Passive diffusion as a mechanism underlying ribbon synapse vesicle release and resupply. J Neurosci 34:8948–8962
14. Zampighi GA, Zampighi LM, Fain N et al (2006) Conical electron tomography of a chemical synapse: vesicles docked to the active zone are hemi-fused. Biophys J 91:2910–2918
15. Watanabe S, Liu Q, Davis MW et al (2013) Ultrafast endocytosis at Caenorhabditis elegans neuromuscular junctions. Elife 2, e00723
16. Rostaing P, Real E, Siksou L et al (2006) Analysis of synaptic ultrastructure without fixative using high-pressure freezing and tomography. Eur J Neurosci 24:3463–3474
17. Siksou L, Rostaing P, Lechaire J-P et al (2007) Three-dimensional architecture of presynaptic terminal cytomatrix. J Neurosci 27:6868–6877
18. Burette AC, Lesperance T, Crum J et al (2012) Electron tomographic analysis of synaptic ultrastructure. J Comp Neurol 520:2697–2711
19. Zampighi GA, Fain N, Zampighi LM et al (2008) Conical electron tomography of a chemical synapse: polyhedral cages dock vesicles to the active zone. J Neurosci 28:4151–4160
20. Kittel RJ, Wichmann C, Rasse TM et al (2006) Bruchpilot promotes active zone assembly, Ca2+ channel clustering, and vesicle release. Science 312:1051–1054
21. Frank T, Rutherford MA, Strenzke N et al (2010) Bassoon and the synaptic ribbon organize Ca^2+ channels and vesicles to add release sites and promote refilling. Neuron 68:724–738
22. Zuber B, Nikonenko I, Klauser P et al (2005) The mammalian central nervous synaptic cleft contains a high density of periodically organized complexes. Proc Natl Acad Sci U S A 102:19192–19197
23. High B, Cole AA, Chen X, Reese TS (2015) Electron microscopic tomography reveals discrete transcleft elements at excitatory and inhibitory synapses. Front Synaptic Neurosci 7:9
24. Hajós F (1980) The structure of cleft material in spine synapses of rat cerebral and cerebellar cortices. Cell Tissue Res 206:477–486
25. Ichimura T, Hashimoto PH (1988) Structural components in the synaptic cleft captured by freeze-substitution and deep etching of directly frozen cerebellar cortex. J Neurocytol 17:3–12
26. Lučić V, Yang T, Schweikert G et al (2005) Morphological characterization of molecular complexes present in the synaptic cleft. Structure 13:423–434
27. Perez de Arce K, Schrod N, Metzbower SWR et al (2015) Topographic mapping of the synaptic cleft into adhesive nanodomains. Neuron 88:1165–1172
28. Chen X, Winters C, Azzam R et al (2008) Organization of the core structure of the postsynaptic density. Proc Natl Acad Sci U S A 105:4453–4458
29. Linsalata AE, Chen X, Winters CA, Reese TS (2014) Electron tomography on γ-aminobutyric acid-ergic synapses reveals a discontinuous postsynaptic network of filaments. J Comp Neurol 522:921–936
30. Chen X, Nelson CD, Li X et al (2011) PSD-95 is required to sustain the molecular organization of the postsynaptic density. J Neurosci 31:6329–6338
31. Hammarlund M, Palfreyman MT, Watanabe S et al (2007) Open syntaxin docks synaptic vesicles. PLoS Biol 5, e198
32. Weimer RM, Gracheva EO, Meyrignac O et al (2006) UNC-13 and UNC-10/rim localize synaptic vesicles to specific membrane domains. J Neurosci 26:8040–8047
33. Siksou L, Varoqueaux F, Pascual O et al (2009) A common molecular basis for membrane docking and functional priming of synaptic vesicles. Eur J Neurosci 30:49–56
34. Fernández-Busnadiego R, Asano S, Oprisoreanu A-M et al (2013) Cryo-electron tomography reveals a critical role of RIM1α in synaptic vesicle tethering. J Cell Biol 201:725–740
35. Stigloher C, Zhan H, Zhen M et al (2011) The presynaptic dense projection of the Caenorhabditis elegans cholinergic neuromuscular junction localizes synaptic vesicles at the active zone through SYD-2/liprin and UNC-10/RIM-dependent interactions. J Neurosci 31:4388–4396
36. Imig C, Min S-W, Krinner S et al (2014) The morphological and molecular nature of synaptic vesicle priming at presynaptic active zones. Neuron 84:416–431
37. Siksou L, Triller A, Marty S (2009) An emerging view of presynaptic structure from electron microscopic studies. J Neurochem 108:1336–1342
38. Smith JE, Reese TS (1980) Use of aldehyde fixatives to determine the rate of synaptic transmitter release. J Exp Biol 89:19–29
39. Fahim MA, Robbins N, Price R (1987) Fixation effects on synaptic vesicle density in neuromuscular junctions of young and old mice. Neurobiol Aging 8:71–75
40. Rosenmund C, Stevens CF (1997) The rate of aldehyde fixation of the exocytotic machinery

in ultured hippocampal synapses. J Neurosci Methods 76:1–5

41. Korogod N, Petersen CC, Knott GW (2015) Ultrastructural analysis of adult mouse neocortex comparing aldehyde perfusion with cryo fixation. Elife 4, e05793
42. Fernández-Busnadiego R, Schrod N, Kochovski Z et al (2011) Insights into the molecular organization of the neuron by cryo-electron tomography. J Electron Microsc (Tokyo) 60(Suppl 1):S137–148
43. Frotscher M, Zhao S, Graber W et al (2007) New ways of looking at synapses. Histochem Cell Biol 128:91–96
44. Moor H (1987) Theory and practice of high pressure freezing. In: Steinbrecht RA, Zierold K (eds) Cryotechniques in biological electron microscopy. Springer-Verlag, Berlin, pp 175–191
45. Möbius W, Cooper B, Kaufmann WA et al (2010) Electron microscopy of the mouse central nervous system. Methods Cell Biol 96:475–512
46. Verhage M, Sørensen JB (2008) Vesicle docking in regulated exocytosis. Traffic 9:1414–1424
47. Richmond JE, Davis WS, Jorgensen EM (1999) UNC-13 is required for synaptic vesicle fusion in C. elegans. Nat Neurosci 2:959–964
48. Bronk P, Deák F, Wilson MC et al (2007) Differential effects of SNAP-25 deletion on Ca2+-dependent and Ca2+-independent neurotransmission. J Neurophysiol 98:794–806
49. de Wit H, Cornelisse LN, Toonen RFG, Verhage M (2006) Docking of secretory vesicles is syntaxin dependent. PLoS One 1, e126
50. Hunt JM, Bommert K, Charlton MP et al (1994) A post-docking role for synaptobrevin in synaptic vesicle fusion. Neuron 12:1269–1279
51. Broadie K, Prokop A, Bellen HJ et al (1995) Syntaxin and synaptobrevin function downstream of vesicle docking in Drosophila. Neuron 15:663–673
52. Deák F, Schoch S, Liu X et al (2004) Synaptobrevin is essential for fast synaptic-vesicle endocytosis. Nat Cell Biol 6:1102–1108
53. Kerr AM, Reisinger E, Jonas P (2008) Differential dependence of phasic transmitter release on synaptotagmin 1 at GABAergic and glutamatergic hippocampal synapses. Proc Natl Acad Sci U S A 105:15581–15586
54. Reynolds ES (1963) The use of lead citrate at high pH as an electron-opaque stain in electron microscopy. J Cell Biol 17:208–212
55. Schneider CA, Rasband WS, Eliceiri KW (2012) NIH image to ImageJ: 25 years of image analysis. Nat Methods 9:671–675
56. Schindelin J, Arganda-Carreras I, Frise E et al (2012) Fiji: an open-source platform for biological-image analysis. Nat Methods 9:676–682
57. Mastronarde DN (2005) Automated electron microscope tomography using robust prediction of specimen movements. J Struct Biol 152:36–51
58. Kremer JR, Mastronarde DN, McIntosh JR (1996) Computer visualization of three-dimensional image data using IMOD. J Struct Biol 116:71–76
59. Mastronarde DN (1997) Dual-axis tomography: an approach with alignment methods that preserve resolution. J Struct Biol 120:343–352
60. Stoppini L, Buchs P-A, Muller D (1991) A simple method for organotypic cultures of nervous tissue. J Neurosci Methods 37:173–182
61. De Simoni A, Yu LMY (2006) Preparation of organotypic hippocampal slice cultures: interface method. Nat Protoc 1:1439–1445
62. Studer D, Zhao S, Chai X et al (2014) Capture of activity-induced ultrastructural changes at synapses by high-pressure freezing of brain tissue. Nat Protoc 9:1480–1495
63. Muller D, Toni N, Buchs P-A et al (2001) Interface organotypic hippocampal slice cultures. In: Fedoroff S, Richardson A (eds) Protocols for neural cell culture, 3rd edn. Humana Press, Totowa, NJ, pp 13–27
64. Dahl R, Staehelin LA (1989) High-pressure freezing for the preservation of biological structure: theory and practice. J Electron Microsc Tech 13:165–174
65. McDonald KL, Morphew M, Verkade P, Müller-Reichert T (2007) Recent advances in high-pressure freezing: equipment- and specimen-loading methods. Methods Mol Biol 369:143–173
66. Rosenmund C, Stevens CF (1996) Definition of the readily releasable pool of vesicles at hippocampal synapses. Neuron 16:1197–1207
67. Basu J, Betz A, Brose N, Rosenmund C (2007) Munc13-1 C1 domain activation lowers the energy barrier for synaptic vesicle fusion. J Neurosci 27:1200–1210
68. Koster AJ, Grimm R, Typke D et al (1997) Perspectives of molecular and cellular electron tomography. J Struct Biol 120:276–308
69. Slot JW, Geuze HJ (2007) Cryosectioning and immunolabeling. Nat Protoc 2:2480–2491

Part IV

Following Synapse Dynamics

Chapter 16

Analyzing Endosomal Docking, Fusion, Sorting, and Budding Mechanisms in Isolated Organelles

Sina V. Barysch and Ioanna Bethani

Abstract

Due to their central role in the reception and sorting of newly internalized material, early endosomes undergo extensive membrane remodeling. They dock and fuse with endocytic carrier vesicles originating from the plasma membrane, sort the internalized material in internal microdomains, and allow the budding of new carrier vesicles from their membrane, destined to fuse with the plasma membrane (recycling) or other organelles. Early endosomal compartments might also be involved in the recycling of synaptic vesicles in nerve terminals. The present protocol describes a technique allowing to assess the mechanistic and molecular aspects of the membrane remodeling processes of docking, fusion, sorting, and budding in early endosomes of neuron-like (and other) cells. It involves the fluorescent labeling and isolation of endosomal organelles, the setup of assays allowing for docking/fusion or sorting/budding in vitro, and finally the assessment and quantification of the membrane remodeling events by fluorescent microscopy. The technique can be easily manipulated by the addition of inhibitors or activators, and can be combined with other techniques, such as immunostaining and high-resolution microscopy, expanding the experimental possibilities in the investigation of early endosomal characteristics.

Key words Endosomes, Fusion, Budding, In vitro, Post-nuclear supernatant (PNS), Rat brain cytosol

1 Introduction

In the endomembrane system of eukaryotic cells, early endosomes function as a sorting hub coordinating the cellular distribution of newly internalized material. Material arrives from the plasma membrane and is packed into carrier vesicles, which dock and subsequently fuse with early endosomes. Sorting events then take place within the endosomal compartment, organizing the material into distinct microdomains. Cargo destined to recycle back to the plasma membrane or be delivered to the trans-Golgi network is packed in carrier vesicles, which subsequently bud from the endosome. Cargo destined for degradation, on the other hand, remains in the endosome until the maturation of the organelle and its

Alexandros Poulopoulos (ed.), *Synapse Development: Methods and Protocols*, Methods in Molecular Biology, vol. 1538, DOI 10.1007/978-1-4939-6688-2_16, © Springer Science+Business Media LLC 2017

fusion with the lysosome. Therefore, early endosomes are involved in a constant circle of membrane remodeling events, consisting of the docking and fusion of endosomes with plasma-membrane carrier vesicles or with each other (homotypic fusion), as well as the sorting and budding of carrier vesicles from the endosomes themselves (summarized in [1, 2]).

Synaptic vesicle recycling shares many mechanistic similarities with the homeostatic recycling of protein and lipids, as described above, mainly regarding the membrane remodeling events of docking/fusion and sorting/budding. However, despite extensive research in the field of synaptic membrane dynamics and the considerable amount of knowledge regarding fusion and budding of synaptic vesicles to and from the plasma membrane, the exact role of the endosomal compartment in synaptic vesicle recycling and function remains elusive. Organelles of endosomal identity have been indeed observed in nerve terminals based on the presence of protein and lipid identifiers such as Rab5 and PI3P on the membrane of these organelles, whereas endosomal proteins have been localized on synaptic vesicles (summarized in [3]). In addition, there is evidence that newly internalized synaptic vesicles can fuse with early endosomes, although very little is known on the actual role of endosomal sorting and the mechanisms of budding involved in the redistribution of synaptic components to new synaptic vesicles (summarized in [4]). Therefore, appropriate methodology allowing the investigation of endosomal dynamics at the synapse would be of considerable help in filling the gaps in understanding synaptic vesicle recycling and function.

The mechanisms and proteins involved in organelle fusion have been extensively studied in the past using cell-free biochemical approaches (summarized in [5]) The main principle of these assays relies on the distinct labeling of two endosome populations with different markers, subsequent fusion of these populations with each other, and biochemical analysis of the content of the fused organelles containing both markers. On the other hand, the investigation of sorting and budding mechanisms has been mainly based on the physical differences between mother- and daughter-budding organelles regarding their size and buoyancy (see assays using permeabilized cells or density gradients described in [6–9]). Nevertheless, none of these techniques could address the fourth vital element of endosomal dynamics, the initial docking of organelles prior to fusion.

The technique presented here provides the significant advantage of allowing the study of all four fundamental steps of endosomal membrane dynamics: docking/fusion and sorting/budding (*see* Fig. 1 for a schematic representation). It is based on the separate labeling of endosomal populations with distinct fluorescent labels. In fusion reactions, two differentially labeled populations of organelles are allowed to fuse in vitro, the two different fluorescent

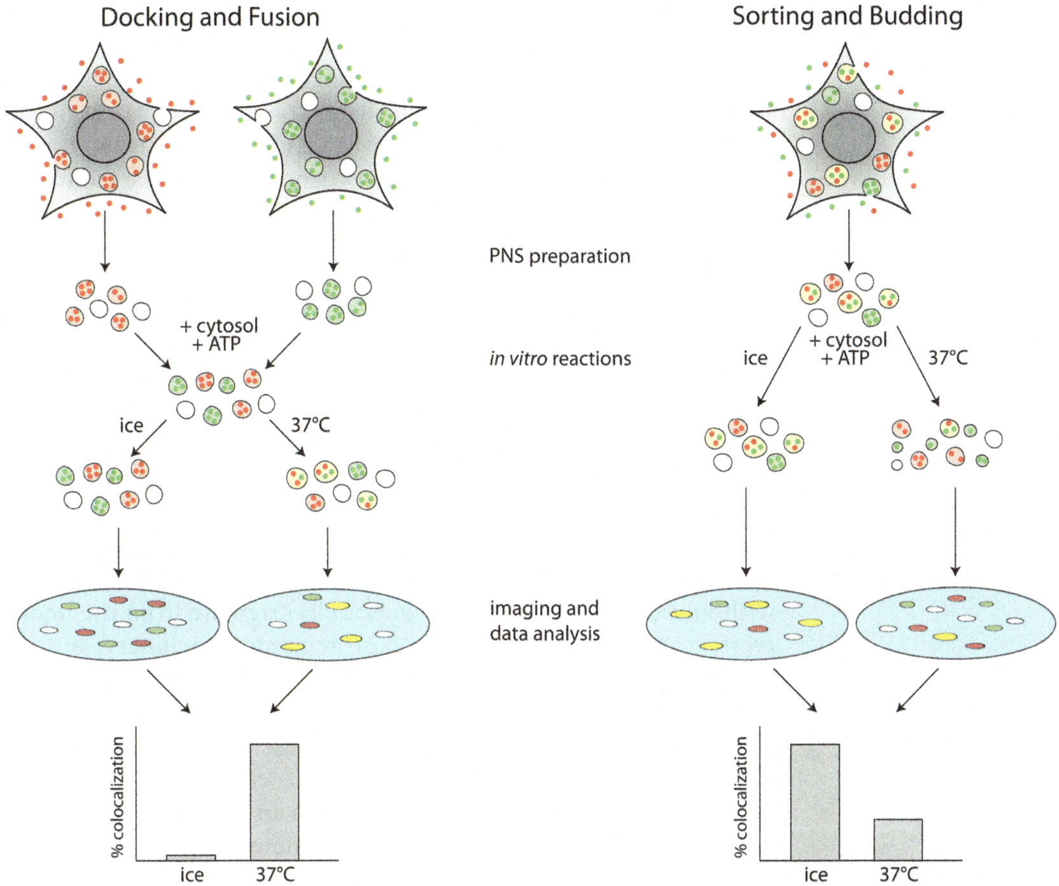

Fig. 1 Schematic representation of the fluorescence-based endosomal docking/fusion and sorting/budding assay. Neuroendocrine PC12 cells (or any other cell type) internalize defined fluorescently labeled endocytic markers, resulting in labeled endosomes that are subsequently isolated: To later study docking and fusion (left panel), two sets of cells separately take up green- and red-labeled dextran (10 kDa); for sorting and budding assays (right panel), cells are simultaneously labeled with 2–3 different fluorescent markers (transferrin, LDL and cholera toxin subunit B). Cells are then homogenized and post-nuclear supernatants (PNS) containing the fluorescent endosomes are prepared. In vitro incubation of the PNS samples with cytosol and an ATP-regenerating system either leads to docking and fusion of the green- and red-labeled endosomes (left), or to sorting/budding of the multiple labeled endosomes (right). This can be visualized by attaching them onto glass coverslips, followed by fluorescence microscopy. While an increase in colocalization is measured for docking and fusion (from single red- and green-labeled to double-labeled endosomes), a decrease is observed for sorting and budding (from multiple-labeled to single-labeled endosomes. The precise position of each endosome is determined and the distance between endosomes of different labels allows the distinction between closely apposed endosomes and truly double-labeled ones

labels end up co-residing in the same organelle and their colocalization is monitored by fluorescent microscopy. Furthermore, calculating the distance between the centers of intensity of the differentially labeled organelles allows distinction between docked and fused organelles in the same assay [10]. The reverse procedure is followed for investigating sorting and budding: organelles are simultaneously labeled with two or more different fluorescent dyes and allowed to undergo budding, which will lead to separation of the different fluorescent labels in distinct organelles [11]. It should be noted that the technique still shares some of the disadvantages of the biochemical assays. Fusion of two identically labeled organelles, or budding of double labeled organelles cannot be identified; neither can fusion of multiple organelles into one. However, the great advantage of this technique is its ability to measure quantitatively and simultaneously docking and fusion, and with a small variation sorting and budding, without depending on the physical properties of the organelles or the efficiency of the biochemical reactions. In addition, it provides “visual” access to individual organelles (in contrast to bulk biochemical reactions), and in combination with immunostaining and advanced imaging techniques, can provide information on the identity and localization of the molecular components involved in endosomal dynamics.

Interestingly, the current technique has been initially developed and extensively applied in the neuroendocrine cell line PC12; without being actual neurons, these cells originate from the neural crest, share many of the neuronal proteins and machinery, and can be easily differentiated to neuronal-like cells upon appropriate treatment with nerve growth factor.

In summary, the current protocol presents a convenient technique of assessing the mechanisms of endosomal membrane dynamics and allows for the exploration of unknown territories, especially in the endosomal dynamics of neuronal cells.

2 Materials

Prepare all solutions using ultrapure water and analytical grade reagents. All solutions must be detergent-free and should thus be prepared in plastic ware, or in glassware that has been thoroughly cleaned with ethanol. Prepare and store all reagents and solutions at room temperature, unless otherwise stated.

2.1 General Components

1. Plastic flasks (15-ml and 50-ml tubes and flasks of different sizes) and plastic pipettes (5, 10, 25, and 50 ml).
2. Protein determination assay, e.g., Bradford (Quick Start^{TM} Bradford 1× dye reagent, Bio-Rad, 500–0205).
3. PBS: 150 mM NaCl, 20 mM NaH_2PO_4/Na_2HPO_4, pH 7.4 (with HCl).
4. Homogenization buffer: 250 mM sucrose, 3 mM imidazole, pH 7.4 (with HCl); can be stored at 4 °C and used within 1 week after preparation. Immediately before use, freshly add 1 μg/ml Pepstatin A, 0.2 mM PMSF, 10 μg/ml leupeptin and 10 μg/ml aprotinin.
5. Refrigerated table top centrifuge (e.g., Eppendorf, 5415 R).

2.2 Preparing Rat Brain Cytosol

1. Dissection tools.
2. Sucrose buffer: 320 mM sucrose, 5 mM HEPES, pH 7.4 (adjusted with NaOH); can be stored at 4 °C for up to 1 week after preparation. Immediately before its use, freshly add 1 μg/ml Pepstatin A and 0.2 mM PMSF.
3. Centrifuge (e.g., Sorvall RC5B Plus) with a SS-34 rotor and polycarbonate flanged tubes (Sorvall, 03146).
4. Ultracentrifuge (e.g., Beckman TL-100) with a TLA-100.3 rotor and polycarbonate tubes (Beckman, 349622).
5. Glass/Teflon Potter Elvehjem Tissue Grinder, 55 ml (e.g., OMNI International, 07–358054) attached to an overhead rotor.

2.3 Preparing Post-Nuclear Supernatants (PNS) from PC12 Cells

1. Standard cell culture equipment (laminar flow hoods, incubators, cooling centrifuges, water baths, refrigerator and freezers, cell culture plates, and flasks).
2. PC12 cells (clone 251).
3. Trypsin-EDTA (0.25%).
4. PC12 culture medium: DMEM medium supplemented with 10% (vol/vol) horse serum, 5% (vol/vol) fetal calf serum, 4 mM glutamine, and 100 units/ml each of penicillin and streptomycin; can be stored at 4 °C and should be used within 1–2 weeks after preparation.
5. Pre-internalization medium: 1 g d-glucose in 1 bottle (500 ml) OptiMEM (Invitrogen, 31985) can be stored in aliquots at −20 °C.
6. Internalization medium: pre-internalization medium with fluorescent cargoes, i.e., 1 mg/ml dextran (10 kDa) Alexa 488, 1 mg/ml dextran (10 kDa) Alexa 594, 100 μg/ml transferrin Alexa 488, 6 μg/ml LDL DiI, and/or 6 μg/ml cholera toxin subunit B Alexa 647. Internalization medium should be prepared immediately before use and cannot be stored (*see* **Note 1**).
7. PBS-BSA: 5 mg/ml BSA in PBS; should be prepared freshly on the day of the experiment and cannot be stored.
8. Ball homogenizer with a clearance of 20 μm, e.g., from Isobiotech, Heidelberg, Germany (No. 202 09 547.9). The design and functionality has been described in the original paper from Balch and coworkers [12], as well as in a recent publication for semiautomated cell rupture [13].

2.4 Docking/Fusion or Sorting/Budding Assay

1. DHM buffer: 625 mM HEPES, 75 mM magnesium acetate, 10 mM DTT, pH 7.4 (with KOH); can be stored in aliquots at −20 °C.
2. KAc buffer: 1 M potassium acetate in H_2O; can be stored in aliquots at −20 °C.

3. ATP solution: 100 mM ATP in H_2O, pH 7.4 (with KOH); can be stored in aliquots at –20 °C.
4. CP solution: 800 mM creatine phosphate in H_2O; can be stored in aliquots at –20 °C.
5. CK solution: 4 mg/ml creatine kinase (3200 U/ml) in H_2O; can be stored in aliquots at –20 °C.
6. Glucose solution: 250 mM glucose in H_2O; can be stored in aliquots at –20 °C.
7. Hexokinase solution: Centrifuge 5 μl hexokinase (1500 U/ml, e.g., Roche, supplied as ethanol precipitate) at 16,000 × *g* for 2 min at 4 °C, discard the supernatant, and resuspend the hexokinase pellet in 5 μl glucose solution (250 mM). This buffer is prepared immediately before use and cannot be stored.
8. Bead solution: Sonicate multifluorescent TetraSpeck™ microspheres, 0.2 μm (Invitrogen, T7280) for 2–5 min in the bath sonicator (e.g., Bandelin Sonorex RK 100) and dilute them 1:1000 (vol/vol) in PBS to prepare a bead stock solution (2 μl beads in 2 ml PBS); can be kept at 4 °C for up to 2 weeks.
9. 12-well plates.
10. 18-mm glass coverslips.
11. Plate centrifuge (e.g., Heraeus® Multifuge® 4 KR) with a rotor and inlays for plates (e.g., Heraeus® HIGHplate® windshielded rotors).
12. 200 μl reaction tubes (from polycarbonate, e.g., Beckman, 343775).
13. Lid for reaction tubes: rubber sealing from 1 ml syringes (Dispomed, 22009).
14. Dark water bath with an agitation device (e.g., GFL 1086).

2.5 Imaging and Data Analysis

1. Open imaging chamber for 18-mm coverslips (e.g., Warner Instruments, QR-41LP).
2. Epi-fluorescence microscopy setup with a 100×, 1.4 numerical aperture oil objective, and the appropriate filter sets to detect dapi, Alexa 488, DiI, and Alexa 647, e.g.:

   dapi: excitation filter 350/50 D, beamsplitter 400 DCLP and emission filter 460/50 D.

   Alexa 488: excitation filter 480/40 HQ, beamsplitter 505 LP Q and emission filter 527/30 HQ.

   DiI: excitation filter 545/30 HQ, beamsplitter 570 LP Q and emission filter 610/75 HQ.

   Alexa 647: excitation filter 620/60 HQ, beamsplitter 660 LP Q and emission filter 700/75 HQ.

3. Software for data analysis, e.g., ImageJ (NIH, Bethesda, MD, USA) or Matlab (The Mathworks Inc., Natick, MA, USA).

3 Methods

3.1 Preparation of Rat Brain Cytosol

1. Sacrifice 10–40 rats, remove their brains, transfer them into a beaker with sucrose buffer, and wash them 2–3 times with ice-cold sucrose buffer until most of the blood is removed.
2. Homogenize 10 brains using a Glass/Teflon Potter Elvehjem Tissue Grinder. Brains are efficiently homogenized by 10 strokes (1 stroke is 1 up and down movement) at 900 rpm with an overhead rotor. It is important to perform the first stroke slowly without applying too much pressure; the glass homogenizer might break otherwise.
3. Centrifuge homogenate at 3000 × *g* (5000 rpm in a Sorvall SS-34 rotor) for 10 min at 4 °C. Remove the supernatant and discard the pellet.
4. Centrifuge the supernatant at 32,000 × *g* (16,500 rpm in a Sorvall SS-34 rotor) for 15 min at 4 °C. Remove the supernatant and discard the pellet.
5. To obtain pure cytosol, centrifuge the supernatant at 330,000 × *g* (90,000 rpm in a Beckman 100.3 rotor) for 30 min at 4 °C. (*see* **Note 2**).
6. Measure and record the protein concentration of the cytosol using any standard protein determination assay (the concentration usually ranges from 6 to 10 mg/ml).
7. The cytosol can be snap-frozen and stored at −80 °C in small aliquots (e.g., 0.3–1 ml) for at least 1 year.

3.2 Preparation of Post-Nuclear Supernatants (PNSs) from PC12 Cells

Handling of PC12 cells for culturing should be carried out under a sterile laminar flow hood at room temperature, all steps for PNS preparation should be performed at 4 °C.

1. Split PC12 cells 1:6 and prepare 6–30 plates (Ø 15 cm). For that, remove medium, wash cells with 10 ml PBS, and add 3 ml trypsin-EDTA per plate. When cells start detaching from the plate, add 9 ml PC12 cell culture medium. It is important to dissociate cells thoroughly, to avoid the presence of cell clumps that may interfere with the efficiency of cargo uptake; break cell clumps by passing the cells 20 times through a 200 μl pipette tip that is attached to a 10 ml plastic pipette. Seed 2 ml of the cell suspension on a fresh plate containing 20 ml PC12 cell culture medium and let cells grow at 37 °C, 10% CO_2, 90% humidity for several days until they reach 80% confluence.

2. To prepare PNS, make the following solutions available before starting the experiment: PBS (37 °C, for cell culture), trypsin-EDTA (37 °C), PC12 cell culture medium (4 °C), PBS (4 °C, for PNS preparation), pre-internalization medium (4 °C), homogenization buffer (4 °C), and PBS-BSA (4 °C).
3. Collect PC12 cells from up to six plates. For that, remove the old PC12 cell culture medium, wash each plate with 5 ml PBS, and add 2 ml trypsin-EDTA to each plate. Shortly incubate at room temperature while shaking the plates strongly (until cells start to detach) and stop the reaction by adding 5 ml cold PC12 cell culture medium to each plate. Remove cells from the plate completely by pipetting, collect in 50-ml plastic tubes, and store on ice until the end of this step. Repeat this procedure until all remaining cells are collected.
4. Centrifuge cells at 250×*g* for 5 min at 4 °C, remove the supernatant, and collect the cell pellet with 10 ml cold PBS into 15-ml plastic tubes (use as many tubes as differently labeled PNS samples are needed, e.g., two for the docking/fusion assay). Centrifuge cells at 250×*g* for 5 min at 4 °C and wash pellets with 10 ml cold pre-internalization medium. Centrifuge again at 250×*g* for 5 min at 4 °C, remove supernatants, estimate the volume of the pellets (it should be around 0.5–1 ml per six plates and is needed for **step 8**), and pre-warm them at 37 °C.
5. For internalization of fluorescent cargoes, incubate the pre-warmed PC12 cell pellets with warm internalization medium in a 1:1 ratio for 5 min at 37 °C. Note that thereby, the concentration of fluorescent cargoes in the internalization medium is reduced by 50%. For later docking/fusion reactions, one set of cells is typically labeled with 0.5 mg/ml dextran (10 kDa) Alexa 488 and another one with 0.5 mg/ml dextran (10 kDa) Alexa 594. For later sorting/budding assays, cells are typically simultaneously labeled with 50 μg/ml transferrin Alexa 488 and 3 μg/ml LDL DiI, or in addition with 3 μg/ml cholera toxin subunit B Alexa 647.
6. To stop the reaction, chill internalization reactions on ice for 5 min, centrifuge at 250×*g* for 5 min at 4 °C, and remove the internalization medium.
7. Wash the cells three times with 10 ml cold PBS-BSA and once with homogenization buffer (centrifuge each time at 250×*g* for 5 min at 4 °C and remove the supernatant).
8. Resuspend the cell pellet in 4× its volume of homogenization buffer and crack the cells with 20 strokes through a pre-cooled ball homogenizer. Note that it is important to wash the ball homogenizer thoroughly with homogenization buffer to remove air bubbles that would cause damage to the endosomes. Repeat this step as many times as is necessary to

homogenize all cells and wash the ball homogenizer thoroughly with homogenization buffer in between two differently labeled sets of cells (*see* **Notes 3** and **4**).

9. Transfer cell homogenates into 1.5-ml tubes and centrifuge them at 1200 × *g* for 15 min at 4 °C. This centrifugation step will lead to the formation of two fractions: a bottom fraction corresponding to the nuclear pellet and an upper fraction corresponding to the PNS (*see* **Note 5**).
10. Remove the PNS into a fresh tube and determine its protein concentration (it should usually range between 10 and 15 mg/ml). Snap-freeze and store small aliquots at −80 °C (*see* **Note 6**).

3.3 Docking/Fusion or Sorting/Budding Assay

All steps for the preparation of docking/fusion and sorting/budding assays should be performed on ice water to avoid any membrane remodeling events during pipetting. The actual reaction is then performed at 37 °C.

1. Thaw aliquots of cytosol, PNS, DHM buffer, KAc buffer, ATP solution, CP solution, CK solution, and homogenization buffer slowly on ice, without vortexing or extensive pipetting. Here, homogenization buffer without protease inhibitors can be used since most proteases from the cytosol and PNS should have been inactivated during their preparation. Depending on the desired assay, docking/fusion, or sorting/budding, the following PNS samples should be used: for docking/fusion assays, dextran (10 kDa) Alexa 488-labeled PNS and dextran (10 kDa) Alexa 594-labeled PNS is used. For sorting/budding assays, a double- or triple-labeled PNS (with transferrin Alexa 488, LDL DiI and optional cholera toxin subunit B Alexa 647) is used.
2. Wash glass coverslips in pure ethanol and place one into each well of a 12-well plate. Dilute bead stock solution 1:100 in PBS (for one plate, 12 ml PBS + 120 μl bead solution), add 1 ml in each well, centrifuge the plate at 5880 × *g* for 45 min at 4 °C (we use 5460 rpm in a Heraeus® Multifuge®), and do NOT remove the solution after centrifugation. Note that air bubbles between the bottom of the well and the coverslips should be removed thoroughly; otherwise, coverslips might float up and bead attachment might be inefficient.
3. Prepare cytosol mixes on ice in the order described in Table 1, depending on whether an ATP-regenerating system or an ATP-depleting system is required:
4. Prepare docking/fusion or sorting/budding reactions in 200 μl reaction tubes in the following order on ice:
   Close the reaction tubes with their lids and incubate the tubes for 45 min at 37 °C under slow agitation (speed 20 shakes per minute in the GFL 1086 water bath) in the dark (to avoid bleaching of the fluorescent dyes). Control tubes are left on ice for 45 min (in the dark to avoid bleaching of fluorescent dyes).

Table 1
Recommended concentrations and volumes for docking/fusion or sorting/budding reactions

Component	ATP-regenerating system	Amount for 1 reaction	ATP-depleting system	Amount for 1 reaction
Cytosol	100 μg	10–17 μl	100 μg	10–17 μl
DHM buffer	0.18 mM DTT 11.25 mM HEPES 1.35 mM MgAc	0.9 μl	0.18 mM DTT 11.25 mM HEPES 1.35 mM MgAc	0.9 μl
KAc buffer	45 mM KAc	2.25 μl	45 mM KAc	2.25 μl
ATP solution	3.3 mM ATP	1.67 μl	–	–
CP solution	26.7 mM CP	1.67 μl	–	–
CK solution	6.7 μg/5.3 U CK	1.67 μl	–	–
Hexokinase solution	–	–	7.5 U hexokinase 25 mM glucose	5 μl

(a) Cytosol mix for 1 reaction (18–25 μl, depending on the cytosol concentration).

(b) (optional) inhibitory components, e.g., chemicals, proteins, peptides, or antibodies.

(c) 200 μg PNS (10–16 μl, depending on the PNS concentration); for budding assays, use double- or triple-labeled PNS; for docking/fusion assays, use 100 μg of each of the two single-labeled PNS.

(d) Homogenization buffer up to 50 μl.

5. Place the reaction tubes on ice for some minutes to cool, and then add 7 μl of each reaction onto the bead-containing solution on the coverslips (**step 2**) and centrifuge the 12-well plates at 5880×*g* (5460 rpm in a Heraeus® Multifuge®) for 45 min at 4 °C. Note that beads and docking/fusion or sorting/budding reactions cannot be centrifuged simultaneously; in order to visualize beads properly, they should be attached in a first centrifugation step (**step 2**).

3.4 Imaging and Data Analysis Components

1. To visualize the fluorescently labeled organelles, place every single coverslip into the open imaging chamber, add PBS so that the coverslip is covered with liquid, focus the samples and acquire approximately 10 images in the following channels: blue (dapi) channel (to only record the fluorescent beads), as well as in the respective PNS channels (i.e., Alexa 488, DiI, Alexa 595 or Alexa 647). Examples for bead- and endosome-containing images are shown in Fig. 2 (left panel). Note that the exposure time for each channel should be a compromise of maximal signal

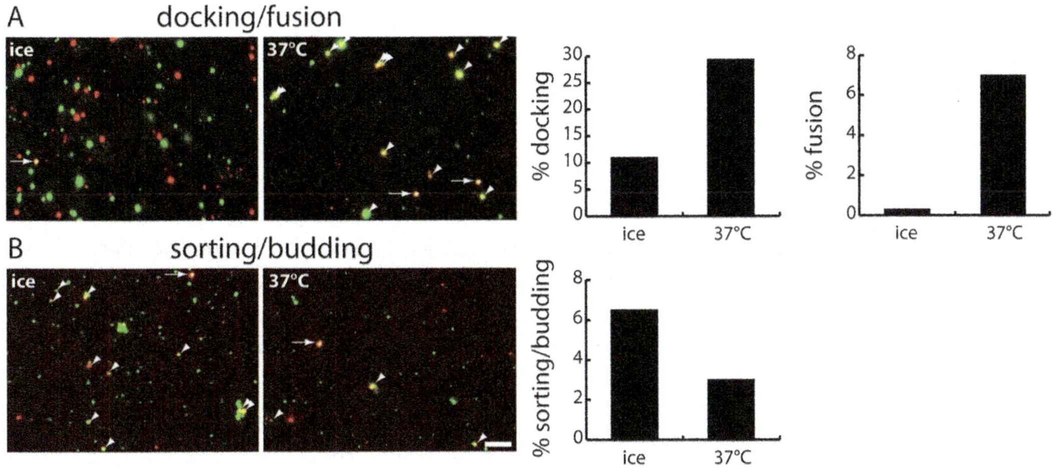

Fig. 2 Imaging and data analysis of endosomal docking/fusion and sorting/budding. Overlay of images taken in the green, red, and blue channels shows multifluorescent beads (*arrows*) and double-labeled endosomes (*arrowheads*). Scale bar is 5 μm (**a**) Docking/fusion assays are performed with endosomes separately labeled with dextran (10 kDa) Alexa 488 and dextran (10 kDa). On ice (negative control), only *green*- or *red*-labeled endosomes are visible, whereas at 37 °C (positive control), several double-labeled endosomes appear. Data analysis reveals that 7 % of the endosomes in this experiment (approximately 15 images) undergo fusion and 29 % undergo docking (compared to 11 % docking on ice). Note that the exact numbers vary between experiments, with fusion rates up to 25 % and docking rates up to 40 % being possible. (**b**) Sorting/budding assays are performed with endosomes simultaneously labeled with transferrin Alexa 488 and LDL DiI. On ice (negative control), several double-labeled endosomes are visible; at 37 °C, their number decreases significantly. Data analysis in this experiment (approximately 15 images) reveals that 6.5 % of the transferrin-containing endosomes are positive for LDL and this decreases to 3 % after treatment at 37 °C. Note that due to the lower amount of LDL-containing endosomes, this number is usually three to five times higher when looking specifically at LDL-positive endosomes

and bleed-through from other channels. For that it is essential to image endosomes (PNS) of every color individually when the experiment is performed for the first time.

2. To process images, first treat all images with an *unsharp* filter that sharpens the images by subtracting an unsharp (smoothed) version of the image from the original image (different sharpening operators can be used, e.g., the *imopen* function in Matlab). Then, threshold images so that only those spots, which correspond to fluorescent objects (i.e., beads and endosomes), are retained. It is important not to use too low thresholds (that will allow the inclusion of background objects) or too high thresholds (that will only allow the inclusion of the brightest objects).
3. To identify the beads and endosomes, calculate the intensity centers in the x- and y-dimensions for every spot in every fluorescent channel according to the following formula

$$x_i = \frac{\sum_{j=1}^{N} x_j I_j}{\sum_{j=1}^{N} I_j}, y_i = \frac{\sum_{j=1}^{N} y_j I_j}{\sum_{j=1}^{N} I_j}$$

where N is the number of pixels in the spot, x and y are the coordinates of each pixel in the spot, and I the intensity of each pixel in the spot. Note that in the blue channel, the spots represent the beads (since they are the only fluorescent objects in this channel). In the other channels, however, the spots resemble a mixture of (few) beads and (plenty of) endosomes.

4. Identify the intensity centers of the beads in the endosome color channels (green and red); these are the spots in the green and red channels that are closest to the blue bead spots. In the following text, the bead coordinates in green and red will be referred to as *GBx*, *GBy* and *RBx*, *RBy*, respectively (where *B* stands for "bead," *G* for "green" and *R* for "red").
5. Correct the shift in x- and y-direction across the endosome color channels, keeping in mind that any shift between the green and red coordinates of the beads also applies to all other spots in the image. The shift in the x-direction is *RBx–GBx*; the shift in the y-direction is *RBy–GBy*. To correct for the shift, apply the following formulas for each set of endosome coordinates in the red channel (referred to as *REx* and *REy*, where *E* stands for "endosome"): *REx=REx–(RBx–GBx)* and *REy=REy–(RBy–GBy)*.
6. To identify docked/fused or sorted/not-budded (still double-labeled) endosomes, first calculate for every endosome (*E*) in one channel the distance (*d*) to its closest neighbor in the other channel. This is performed by calculating for each endosome in one channel the distances to all endosomes in the other channel, using the formula of Pythagoras:

$$d = \sqrt{\left(GE_x - \mathrm{Re}_x\right)^2 + \left(GE_y + \mathrm{Re}_y\right)^2}$$

where *e* stands for an endosome in the second channel. The smallest calculated distance is then the distance to the closest neighbor. Note that the beads should be excluded from the calculation, as they will interfere with the readout by introducing false positive (colocalized) values. Calculate how many of those spots have their intensity centers within a distance of 100 nm; these spots correspond to fused organelles in the docking/fusion assay or organelles that had not undergone sorting or budding (still double-labeled) in the sorting/budding assay. For the assessment of docked organelles in the docking/fusion assay, calculate how many spots have their intensity centers within a distance of 150–500 nm (*see* **Note 7**).

7. To control for random colocalization, apply a random distribution of the intensity centers from one color channel (e.g., by creating a "mirror image") and repeat the previous step.
8. To generate final numbers for colocalized, double-labeled endosomes, subtract the values from the control "mirror images" (**step 7**) from the original ones (**step 6**). Examples for typical docking, fusion, and sorting/budding graphs are shown in Fig. 2 (right panel).

4 Notes

1. The efficiency of the uptake can be compromised due to LDL oxidizing easily – therefore, the exact concentration of LDL DiI may need to be adjusted up to 10 μg/ml after several weeks of storage.
2. Cytosol devoid of membranes does not require 330,000 × *g* but may be obtained with lower centrifugal force, with the minimum being 100,000 × *g* for 1 h.
3. If cell types other than PC12 cells are used, it might not be clear which clearance should be used for homogenization. In those cases, use different balls that result in different clearances (e.g., from 10 μm to 30 μm) and prepare PNS samples according to this protocol. Analyze PNSs and nuclear pellets (NPs) for typical cytosolic markers, as well as nuclear markers with SDS-PAGE/western blot. Use the ball that shows most cytosolic proteins in the PNS and most nuclear proteins in the NP.
4. When larger volumes of cells are used, it can sometimes be difficult to homogenize them in later rounds. In those cases, it is recommended to perform a round of homogenization with only homogenization buffer.
5. It is sometimes difficult to see the separation of the two layers properly. In case problems are encountered, illuminate the sample or further dilute the samples with cold homogenization buffer and repeat the step.
6. In case PNS concentrations are too low, perform the experiment suggested in **Note 3**. If the concentrations remain low, increase cell concentration (i.e., use less homogenization buffer) before homogenization.
7. While 100 nm is the distance that should be used to identify double-labeled endosomes from PC12 cells (*see* references [10, 11, 14] for detailed descriptions), in larger types of cells (e.g., fibroblast) larger distances between two centers of intensity may resemble truly double-labeled endosomes. This can be analyzed by comparing distance-histograms of negative controls (e.g., sample left on ice or without ATP) with the positive control of

the docking/fusion assay. In the negative control (when no fused organelles are available), a gap up to 100 nm in PC12 cells (or larger distances in other cell types) should be visible. If the concentration of labeled endosomes on every image is relatively high, closely apposed or docked endosomes should show a sudden increase in densities. The threshold should be set at the end of the mentioned gap. Likewise, distances for docking may need to be adjusted for different cell types.

Acknowledgments

This protocol was developed in the laboratory of R. Jahn (Göttingen). We thank R. Jahn for his contribution in the experimental design of the method, and S.O. Rizzoli (Göttingen) for his contribution in the design and execution of the experiments described in this method. We thank also D. Brandhorst, D. Zwilling, and U. Geumann (Göttingen) for their contributions to the development of the technique, and M. Lengauer and F. Kötting (Göttingen) for help with the ball homogenizer.

References

1. McDermott H, Kim K (2015) Molecular dynamics at the endocytic portal and regulations of endocytic and recycling traffics. Eur J Cell Biol 94:235–248
2. Granger E, McNee G, Allan V et al (2014) The role of the cytoskeleton and molecular motors in endosomal dynamics. Semin Cell Dev Biol 31:20–29
3. Saheki Y, De Camilli P (2012) Synaptic vesicle endocytosis. Cold Spring Harb Perspect Biol 4:a005645
4. Rizzoli SO (2014) Synaptic vesicle recycling: steps and principles. EMBO J 33:788–822
5. Gruenberg J, Howell KE (1989) Membrane traffic in endocytosis: insights from cell-free assays. Annu Rev Cell Biol 5:453–481
6. Desnos C, Clift-O'Grady L, Kelly RB (1995) Biogenesis of synaptic vesicles in vitro. J Cell Biol 130:1041–1049
7. Clift-O'Grady L, Desnos C, Lichtenstein Y et al (1998) Reconstitution of synaptic vesicle biogenesis from PC12 cell membranes. Methods 16:150–159
8. Prekeris R, Klumperman J, Chen YA et al (1998) Syntaxin 13 mediates cycling of plasma membrane proteins via tubulovesicular recycling endosomes. J Cell Biol 143:957–971
9. Pagano A, Crottet P, Prescianotto-Baschong C et al (2004) In vitro formation of recycling vesicles from endosomes requires adaptor protein-1/clathrin and is regulated by rab4 and the connector rabaptin-5. Mol Biol Cell 15:4990–5000
10. Geumann U, Barysch SV, Hoopmann P et al (2008) SNARE function is not involved in early endosome docking. Mol Biol Cell 19:5327–5337
11. Barysch SV, Aggarwal S, Jahn R et al (2009) Sorting in early endosomes reveals connections to docking- and fusion-associated factors. Proc Natl Acad Sci U S A 106:9697–9702
12. Balch WE, Rothman JE (1985) Characterization of protein transport between successive compartments of the Golgi apparatus: asymmetric properties of donor and acceptor activities in a cell-free system. Arch Biochem Biophys 240:413–425
13. Schmitt S, Saathoff F, Meissner L et al (2013) A semi-automated method for isolating functionally intact mitochondria from cultured cells and tissue biopsies. Anal Biochem 443:66–74
14. Barysch SV, Jahn R, Rizzoli SO (2010) A fluorescence-based in vitro assay for investigating early endosome dynamics. Nat Protoc 5:1127–1137

Chapter 17

Concurrent Imaging of Receptor Trafficking and Calcium Dynamics by Spinning Disk Confocal Microscopy

DeLaine D. Larsen, Regina Wai-Yan Choy, and Minjong Park

Abstract

Synaptic activity is modulated by the activation of neuromodulator receptors present in dendrites of neurons. The majority of neuromodulator receptors are G protein coupled receptors (GPCRs), in which membrane trafficking regulates their activities. Membrane trafficking of neuromodulator receptors and their signaling occurs on a rapid time scale and emerging studies indicate that neuromodulator receptors function not just from the plasma membrane but also from the endocytic compartments. Here, we describe a live cell imaging approach using spinning disk confocal microscopy to investigate the effect of neuromodulator receptor activation on synaptic activity by measuring calcium dynamics in primary rat striatal neurons. The advantages of spinning disk confocal microscopy and recent improvements in the genetically encoded calcium sensor, GCaMP6, provide an imaging approach to image both the receptor membrane trafficking to endocytic compartments, and calcium dynamics at a high spatial and temporal resolution. We believe this approach of imaging both the neuromodulator receptor membrane trafficking and synaptic activity using GCaMP6 is a powerful tool to address many questions regarding possible roles of membrane trafficking of neuromodulator receptors in synaptic activity.

Key words Calcium imaging, Neuromodulator receptor trafficking, Delta opioid receptor, Live-cell imaging, Spinning disk confocal microscopy, Endocytosis, Genetically encoded calcium sensor (GCaMP6)

1 Introduction

Information is propagated through neural circuits upon the release of neurotransmitters at synapses. The strength of synaptic activity elicited by neurotransmitters is modulated by neuromodulator receptors available at the postsynaptic membrane, with the strength of the modulation dependent on the number of active receptors present [1]. The level of active neuromodulator receptors is regulated by endocytic membrane trafficking of the receptors with the internalization of surface neuromodulator receptors to the endocytic compartments induced by agonist binding [2]. This endocytic membrane trafficking is a rapid process that occurs in seconds to minutes and is the major mechanism to regulate signaling from the plasma membrane [3].

Alexandros Poulopoulos (ed.), *Synapse Development: Methods and Protocols*, Methods in Molecular Biology, vol. 1538,
DOI 10.1007/978-1-4939-6688-2_17, © Springer Science+Business Media LLC 2017

After internalization the intracellular sorting of these signaling receptors from the endocytic compartments further modulates the strength of synaptic activity by either being recycled back to the postsynaptic membrane or entering the lysosomal degradation pathway [4]. Interestingly, recent studies indicate that some of the neuromodulator receptors continue to actively signal from the endocytic compartments [5, 6]. One way to monitor how the internalization and trafficking of neuromodulator receptors impact the activity of neurons is to utilize calcium imaging [7]. Calcium can be used as readout of synaptic activity since calcium influx in the presynaptic terminals causes rapid release of neurotransmitters from synaptic vesicles [8] and changes in calcium levels at the postsynaptic terminal cause activity-dependent synaptic plasticity [9]. In this chapter, we describe a method to simultaneously image neuromodulator receptor membrane trafficking and changes in cytosolic calcium level in primary rat striatal neurons using spinning disk confocal microscopy and the protein-based genetically encoded calcium indicator, GCaMP6.

Spinning disk confocal microscopy is a powerful technique for live cell imaging which allows for high speed imaging with lower laser excitation powers than required for laser point-scanning confocal microscopy. The fast acquisition ability of the spinning disk confocal microscope comes from the utilization of a disk that has multiple pinholes, each functioning as a confocal aperture, which spins at a rate of over 1000 times per second, illuminating the entire field view [10, 11]. A camera is used for detection instead of photomultiplier tubes (PMTs) and the rate of acquisition is then limited to the frame rate of the camera being utilized and can be as high as 30–60 frames per second when using electron multiplying charged coupled device (EMCCD) cameras [12]. Furthermore, the use of the spinning disk lowers the peak excitation light density increasing the fluorescence efficiency and decreasing the photobleaching effects compared to widefield or laser scanning confocal microscopy [13]. The high spatial and temporal resolution available with a spinning disk confocal enables better imaging of the complex morphology of neurons, with elaborate branches of dendrites and axons off of the soma, as well as capturing the rapid events involved of neuromodulator receptor trafficking and calcium dynamics. In addition to the spinning disk confocal microscope, we have utilized a protein-based genetically encoded calcium sensor, GCaMP6. GCaMP6 is based on GFP, a green fluorophore (excitation/emission : 488/509 nm) linked to the calcium-binding protein calmodulin (CaM) and the CaM-interacting M13 peptide, which increases in brightness after calcium binding. GCaMP6 is vastly improved in sensitivity and kinetics compared to the previous GCaMP progeny (details are extensively reviewed in [14–16]). Here, we demonstrate the use of GCaMP6 to image changes in calcium dynamics after agonist-induced activation of δ-opioid receptor and simultaneously follow the endocytic membrane trafficking of δ-opioid receptor.

2 Materials

2.1 Neuron Culture

1. Hank's Balanced Salt Solution (BSS).
2. Roswell Park Memorial Institute (RPMI) 1640 Medium.
3. DMEM-FBS: Dulbecco's Modified Eagle Medium, 10% Fetal Bovine Serum.
4. Neurobasal medium: Neurobasal® Medium, without phenol red (e.g., Life Technologies, Cat. no 12348–017), 1× B-27 serum-free supplement (e.g., Life Technologies, Cat. no 17504–044), 1× GlutaMAX™ (e.g., Life Technologies, Cat. no 35050–061).
5. Imaging medium: Combining 1:1 conditioned Neurobasal medium with fresh Neurobasal medium with 30 mM HEPES buffer.
6. Trypsin-EDTA solution: 0.05% trypsin, 0.02% EDTA, in Hanks' BSS, phenol red, pH 7.2–8.
7. Poly-D-lysine solution.
8. Cell strainer with 40 μm pores.
9. Hemocytometer.
10. Amaxa Nucleofector II Device (optional, Lonza, Cat. No. AAB-1001).
11. Amaxa Rat Neuron Nucleofector® Kit (optional, Lonza, Cat. No. VPG-1003).
12. Antibody against extracellular epitope of native receptor or epitope tag of exogenous receptor (here used monoclonal ANTI-FLAG M1 antibody Sigma, Cat. No. F3040).
13. Antibody labeling reagent for covalently conjugating fluorophore to primary antibody (here used NHS ester Alexa Fluor 555 Life Technologies, Cat. no A20187).
14. Extracellularly tagged receptor construct (optional, here used pCAGGs-SSF-δ-opioid receptor [17].
15. GCaMP6 expression construct (e.g., pGP-CMV-GCaMP6m Addgene, Cat. No. 40754).
16. 35 mm glass-bottom dish (e.g., MatTek, Cat. No. P35G-1.5-14-C).
17. Receptor agonist (here used DPDPE [D-Pen^2,D-Pen^5] Enkephali Tocris Cat. No. 1431).

2.2 Instrumentation

1. Inverted fluorescence microscope (optimally with Perfect Focus, e.g., Nikon Eclipse Ti).
2. Spinning Disk Confocal (e.g., Yokogawa CSU-22).
3. Plan Apo VC 60X/1.4 NA oil immersion objectives.

4. Four laser lines including 491 nm and 561 nm DPSS lasers (e.g., Laser launch with NEOS AOTF and controller).
5. Electron multiplying CCD camera (e.g., Photometrics Evolve EMCCD).
6. Emission filters; ET525/50 m for GCaMP6 and ET610/60 m for Alexa Fluor 555.
7. Immersion oil (e.g., Nikon Type A).
8. Cage incubator for microscope temperature control (e.g., OKO Lab, BoldLine H301).
9. Software for image acquisition (e.g., Micro-manager [18], http://www.micro-manager.org).
10. Software for image processing and visualization (e.g., ImageJ [19], http://imagej.nih.gov/ij).

3 Methods

3.1 Preparation of Glass-Bottom Dish

1. In tissue culture hood, open the glass-bottom dishes and coat with poly-D-lysine for 24 h.
2. Rinse coverslips three times in sterilized ddH_2O. It is important to remove unbound poly-D-lysine.
3. Air dry glass-bottom dishes in the tissue culture hood.

3.2 Rat Striatal Neuron Culture Preparation

Carry out all procedures at room temperature unless otherwise specified. All animals must be treated in accordance with institutional and national guidelines for the Care and Use of Laboratory Animals and all protocols must be institutionally approved (i.e., by the Institutional Animal Care and Use Committee (IACUC).

1. Euthanize embryonic day 18 (E18) timed pregnant rat with CO_2.
2. Sterilize rat with 75 % ethanol, and cut along the abdomen.
3. Remove the uterus, and place the pups in petri dish containing Hank's BBS buffer.
4. Decapitate the pups with scissors, and place the heads in a new petri dish containing ice-cold Hank's BBS buffer on ice.
5. From this step onward, all the procedures are done on ice and under a dissection microscope. Stabilize the head by piercing through snout with a pair of forceps, and peel off the skull using the other forceps to detach the brain. Note: Be careful not to press on the brain, which could damage it.
6. Transfer the brains to a new petri dish containing ice-cold Hank's BBS on ice.
7. Take each brain into a new petri dish containing ice-cold Hank's BBS, separate hemispheres, remove pia mater, and dissect out the striatum. Trim away any cortical tissue as much as possible to obtain a pure striatal culture.

8. Transfer striata to a 15-mL conical tube containing 10 mL ice-cold Hank's BBS.
9. After collection of all striata, aspirate the buffer without disturbing the tissue that has settled at the bottom of the tube.
10. Add 1.5 mL Trypsin-EDTA solution to the tube, and incubate at 37 °C for 20 min, mixing every 5 min.
11. During trypsination, fire-polish the tips of Pasteur pipettes by rotating over a Bunsen flame until the edges become smooth. Prepare Pasteur pipettes with three different tip opening sizes (regular size opening, 1/2 the size opening, and 1/4 the size opening).
12. Add 10 mL DMEM-FBS, centrifuge at 1000 rpm for 5 min in a clinical centrifuge.
13. Remove the supernatant with a pipette without disturbing the striatal tissue. Do not aspirate.
14. Repeat the DMEM-FBS wash (**steps 4–6**) three more times.
15. Resuspend striatal tissue in 3 mL of DMEM-FBS.
16. Triturate striatal tissue gently with polished Pasteur pipette (regular size opening) ten times, then with polished Pasteur pipette (1/2 the size opening) another ten times.
17. Allow big chunks of striatal tissue to settle at the bottom of the tube, transfer the cell suspension over a cell strainer.
18. Pass 2 mL fresh DMEM-FBS through the cell strainer to collect any residual triturated cells to the cell suspension. Triturate cell suspension with polished Pasteur pipette (1/4 the size opening) five times.
19. Count cells using a hemocytometer.
20. Proceed to Amaxa nucleofection if using a tagged receptor construct, or plate triturated cells directly on poly-D-lysine-coated glass-bottom dishes with fresh DMEM-FBS in tissue culture plate and place in 37 °C/5 % CO_2 incubator (*see* **Note 1**).
21. On the next day, replace the medium with Neurobasal medium.
22. Feed cells every week by removing half the volume of the conditioned media, and add the same volume of fresh Neurobasal medium.

3.3 Imaging of Rat Striatal Neuron Preparation

1. Prepare fluorescently labeled antibodies following the manufacturer's protocol (Alexa Fluor 555-conjugated to ANTI-FLAG M1 antibody in this example).
2. 7–10 days after dissection, pre-warm the imaging medium in the CO_2 incubator for at least 1 h. Additionally, pre-warm fresh Neurobasal medium without phenol red to use to wash the neurons after surface labeling of receptors.

3. Surface label (Flag-δ-opioid) receptor with fluorescently labeled antibody (Alexa Fluor 555-conjugated ANTI-FLAG M1) by adding the antibody solution directly into the dish containing the conditioned medium for 15 min at 37 °C incubator.
4. Wash the surface labeled neurons with pre-warmed Neurobasal medium without phenol red two times, then add 2 mL of pre-warmed imaging medium (*see* **Note 2**).

3.4 Live-Cell Imaging Using Spinning Disk Confocal

All the imaging experiments described in this chapter were performed on a Yokogawa CSU-22 Spinning Disk Confocal on an inverted Nikon Ti microscope with a high NA oil immersion objective (Plan Apo VC 60×/1.40NA).

1. Start the microscope, laser, camera, and the temperature controller for at least 1 h before preparing the neurons for imaging (*see* **Note 3**).
2. Add immersion oil to the high NA oil immersion objective and then place glass-bottomed petri dish on microscope stage. Focus on the neurons using transmitted light and initiate Perfect Focus if available.
3. Select the area of the petri dish to be imaged by searching for neurons that express both (FLAG-δ-opioid) receptor and GCaMP6. It is important that the area has neurons that are as flat as possible on the coverslip. This will ensure that most of the neuronal processes will be in the same focal plane for imaging.
4. Determine the appropriate laser powers and camera exposures for the sample that will be used for acquisition. You want to make sure that you have a good dynamic range of pixel intensity values in your images but that the camera does not become saturated. Pay attention to ensure that the settings are not causing phototoxicity or excessive photobleaching. It might be necessary to reduce the illumination intensity to prevent phototoxicity and ensure that you will be able to monitor the signal for the entire imaging experiment.
5. Once all the parameters for image acquisition are determined, set up the software to run the automated acquisition for imaging. To monitor the agonist-induced activation of neuromodulator receptors and calcium signaling, images for both the labeled receptors (FLAG-δ-opioid receptor labeled with Alexa Fluor 555) and GCaMP6 are collected every 3 s for 2 min without agonist to determine baseline signal. 100 μL of Neurobasal medium containing agonist at the appropriate concentration (here 10 nM of the δ-opioid receptor-specific agonist DPDPE) is added to the sample and images are collected every 3 s for 20 min to measure the internalization and trafficking on the receptor and the dynamics of the calcium response induced by the agonist.

6. All images collected are saved in .TIF file single file image sequence format to be analyzed as described in the next section.

3.5 Analysis

In this protocol, we use freeware ImageJ with McMaster Biophotonics Facility bundle plugins (*see* **Note 4**) to analyze the calcium dynamics. Detailed methodologies of image analysis are beyond this protocol and comprehensive methodologies are addressed elsewhere [20–22]. As with any image analysis, it is critical that the parameters for image acquisition are set correctly for the type of analysis being performed. For example, excessive photobleaching of calcium sensor by a combination of prolonged illumination with short relaxation intervals will make it difficult to interpret the data. Here, we show examples to present the temporal dynamics of calcium response to agonist-induced neuromodulator receptor activation.

1. In Fig. 1a, we show representative time course color-scaled images of GCaMP6 fluorescence dynamics before and after agonist activation of δ-opioid receptor.
   (a). Open GCaMP6 representative files in ImageJ software: *File> Open*.
   (b). Apply color-scale to the images: *Image> Lookup Tables> 16 Colors*.
   (c). Set same value of brightness and contrast for the GCaMP6 representative images: *Image> Adjust> Brightness/Contrast*.
   (d). Insert color-scale bar: *Analyze> Tools> Calibration Bar*.
2. In Fig. 1c, calcium dynamics are represented in graph format using ImageJ plugins, F divided by F_O followed by intensity vs. time measurements.
   (a). Open experiment files in the ImageJ software by selecting: File > Import > Image Sequence.
   (b). Correct for background fluorescence (see Note 5) by using background subtraction plugin: Process > Subtract Background.
   (c). Open the region of interest (ROI) manager: Analyze > Tool > ROI manager. Use the rectangular or oval tool to select regions you want to analyze and add to the RIO manager.
   (d). Process the background corrected GCaMP6 images for fluorescence intensity (F) divided by resting fluorescence (FO) by running the F div by F0 plugin (see Note 6): Plugins > Stack-T-Functions > F div F0. In our analysis, we used an average of the first six frames without agonist addition to calculate $F_{O.}$
   (e). From the new F div F0 window, run the intensity over time plugin (*see* **Note 7**) to obtain the changes in GCaMP6 intensity: *Plugins> Stack-T-Functions> Intensity v Time Monitor*.

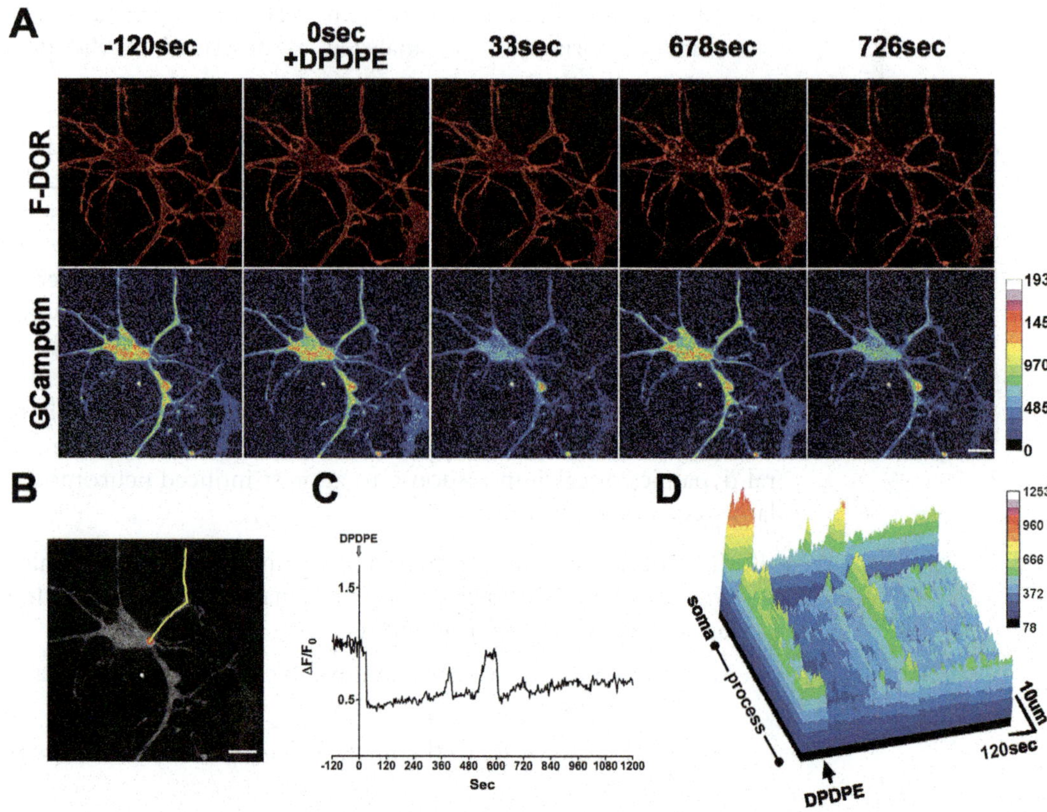

Fig. 1 Activation of δ-opioid receptor decreases calcium in striatal neurons. **(a)** Representative time-lapse spinning disk images of striatal neuron expressing Flag-δ-opioid receptor surface labeled with Alexa-555 conjugated M1 anti-Flag (*top*) and protein calcium sensor, GCaMP6 (*bottom*). GCaMP6 images are color-coded to show changes in cytosolic calcium responses to δ-opioid receptor activation. δ-opioid receptor activation exhibits decrease of GCaMP6 fluorescence approximately 33 s after agonist treatment, followed by bursts of increased GCaMP6 fluorescence at 393 and 678 s. Neurons were imaged for 2 min then treated with δ-opioid receptor specific agonist, 10 nM DPDPE, and continuously imaged for an additional 20 min. Frames are 3 s apart. Color scale is shown to the right. *Scale bar*=10 μm. (**b**) Grayscale image of striatal neuron expressing GCaMP6. *Red circle* shows region of interest used for Fig. 1c and yellow line shows region of interest used for Fig. 1d. *Scale bar*=10 μm. **(c)** The dynamics of cytosolic calcium in the *red circle* region indicated in (**b**) is plotted, $\Delta F/F_0$ (average of first six frames) over 22 min. Arrow indicates the time at which DPDPE was added. **(d)** Color-coded kymograph showing calcium dynamics (3 pixels averaged) of the *yellow line* in (**b**) was converted to surface plot. *Y*-axis indicates regions of neuron and *x*-axis indicates time increasing from *left* to *right*. *Arrow* indicates the time at which DPDPE was added. Color scale is shown to the *right*

3. In Fig. 1d, we use ImageJ plugins kymograph followed by surface plot to illustrate temporal and spatial dynamics of calcium responses to δ-opioid receptor agonist activation.

   (a). Open experiment files in the ImageJ software by selecting: *File> Import> Image Sequence.*

(b). Correct for background fluorescence (*see* **Note 5**) by using background subtraction plugin: *Process> Subtract Background.*

(c). Open the region of interest (ROI) manager: *Analyze> Tool> ROI manager.* Use the straight or segmented line tool to select regions you want to analyze and add to the ROI manager.

(d). Run the kymograph plugin (*see* **Note 8**) to obtain changes in GCaMP6 intensity over time: *Plugins> Kymograph> MultipleKymograph.*

(e). Apply color-scale to the images: *Image> Lookup Tables> 16 Colors.*

(f). From the kymograph result window, run the surface plot (*see* **Note 9**) plugin: *Analyze/ Surface plot.*

4 Notes

1. For any high-performance imaging make sure that # 1.5 coverslips are utilized to help prevent spherical aberration.
2. It is important to pay attention to the composition of the imaging buffer used. Some components normally found in tissue culture media (phenol red, serum, folic acid, and riboflavin) can contribute to background fluorescence. High background fluorescence will decrease your signal-to-noise ratio and make your data analysis more difficult. Preferably use buffers that do not contain Phenol red as a pH indicator. Phenol red can contribute to phototoxicity by high light absorption extinction coefficient [23].
3. Having good temperature stability is important since large fluctuations in temperature can cause problems with focus drift, making it difficult to maintain focus over the time course of the experiments. Adequate equilibration time is needed to help prevent the formation of temperature gradients and to prevent drift in the system. There are several options for incubators to use on microscopes from numerous companies (e.g., Tokai Hit, OkoLab, In Vivo Scientific, Bioscience Tools). There are many pros and cons to the type of incubator that you use but it is important to have stable temperature control.
4. McMaster Biophotonics Facility bundle plugins can be downloaded from http://rsb.info.nih.gov/ij/plugins/mbf/index.html.
5. Detailed description of background correction is covered at http://fiji.sc/Image_Intensity_Processing#Background_correction.

6. Detailed description of F divided by F_O is covered at http://fiji.sc/T-functions#F.C3.B7F0.
7. Detailed description of Intensity over time is covered at http://fiji.sc/mbf/intensity_vs_time_ana.htm.
8. Detailed description of kymograph and required plugins can be found at http://www.embl.de/eamnet/html/body_kymograph.html.
9. Detailed description of surface plotting is covered at http://fiji.sc/mbf/t.htm#t_surfplot.

Acknowledgments

This work was supported by grants from the NIH (NIDA and NIMH). R.W.-Y.C. received a postdoctoral fellowship from the Croucher Foundation, Hong Kong. M.P. received postdoctoral support from the NIMH (T32 MH0889920). Spinning disk confocal imaging experiments were carried out in the UCSF Nikon Imaging Center directed by Kurt Thorn.

References

1. Kenakin T (2004) Principles: receptor theory in pharmacology. Trends Pharmacol Sci 25(4):186–192. doi:10.1016/j.tips.2004.02.012
2. von Zastrow M, Sorkin A (2007) Signaling on the endocytic pathway. Curr Opin Cell Biol 19(4):436–445. doi:10.1016/j.ceb.2007.04.021
3. Sorkin A, von Zastrow M (2009) Endocytosis and signalling: intertwining molecular networks. Nat Rev Mol Cell Biol 10(9):609–622. doi:10.1038/nrm2748
4. von Zastrow M, Williams JT (2012) Modulating neuromodulation by receptor membrane traffic in the endocytic pathway. Neuron 76(1):22–32. doi:10.1016/j.neuron.2012.09.022
5. Irannejad R, Tomshine JC, Tomshine JR, Chevalier M, Mahoney JP, Steyaert J, Rasmussen SG, Sunahara RK, El-Samad H, Huang B, von Zastrow M (2013) Conformational biosensors reveal GPCR signalling from endosomes. Nature 495(7442):534–538. doi:10.1038/nature12000
6. Kotowski SJ, Hopf FW, Seif T, Bonci A, von Zastrow M (2011) Endocytosis promotes rapid dopaminergic signaling. Neuron 71(2):278–290. doi:10.1016/j.neuron.2011.05.036
7. Grienberger C, Konnerth A (2012) Imaging calcium in neurons. Neuron 73(5):862–885. doi:10.1016/j.neuron.2012.02.011
8. Neher E, Sakaba T (2008) Multiple roles of calcium ions in the regulation of neurotransmitter release. Neuron 59(6):861–872. doi:10.1016/j.neuron.2008.08.019
9. Zucker RS (1999) Calcium- and activity-dependent synaptic plasticity. Curr Opin Neurobiol 9(3):305–313
10. Fischer RS, Wu Y, Kanchanawong P, Shroff H, Waterman CM (2011) Microscopy in 3D: a biologist's toolbox. Trends Cell Biol 21(12):682–691. doi:10.1016/j.tcb.2011.09.008
11. Wilson T (2010) Spinning-disk microscopy systems. Cold Spring Harb Protoc 2010 11:pdb top88. doi:10.1101/pdb.top88
12. Graf R, Rietdorf J, Zimmermann T (2005) Live cell spinning disk microscopy. Adv Biochem Eng Biotechnol 95:57–75
13. Wang E, Babbey CM, Dunn KW (2005) Performance comparison between the high-speed Yokogawa spinning disc confocal system and single-point scanning confocal systems. J Microsc 218(Pt 2):148–159. doi:10.1111/j.1365-2818.2005.01473.x
14. Chen TW, Wardill TJ, Sun Y, Pulver SR, Renninger SL, Baohan A, Schreiter ER, Kerr RA, Orger MB, Jayaraman V, Looger LL, Svoboda K, Kim DS (2013) Ultrasensitive fluorescent proteins for imaging neuronal activity. Nature 499(7458):295–300. doi:10.1038/nature12354

15. Looger LL, Griesbeck O (2012) Genetically encoded neural activity indicators. Curr Opin Neurobiol 22(1):18–23. doi:10.1016/j.conb.2011.10.024
16. Tian L, Hires SA, Looger LL (2012) Imaging neuronal activity with genetically encoded calcium indicators. Cold Spring Harb Protoc 2012(6):647–656. doi:10.1101/pdb.top069609
17. Yu YJ, Dhavan R, Chevalier MW, Yudowski GA, von Zastrow M (2010) Rapid delivery of internalized signaling receptors to the somatodendritic surface by sequence-specific local insertion. J Neurosci 30(35):11703–11714. doi:10.1523/JNEUROSCI.6282-09.2010
18. Edelstein A, Amodaj N, Hoover K, Vale R, Stuurman N (2010) Computer control of microscopes using micromanager. Curr Protoc Mol Biol Chapter 14:Unit14 20. doi:doi:10.1002/0471142727.mb1420s92
19. Schneider CA, Rasband WS, Eliceiri KW (2012) NIH Image to ImageJ: 25 years of image analysis. Nat Methods 9(7):671–675
20. Collins TJ (2007) ImageJ for microscopy. Biotechniques 43(1 Suppl):25–30
21. Meijering E, Dzyubachyk O, Smal I (2012) Methods for cell and particle tracking. Methods Enzymol 504:183–200. doi:10.1016/B978-0-12-391857-4.00009-4
22. Waters JC (2009) Accuracy and precision in quantitative fluorescence microscopy. J Cell Biol 185(7):1135–1148. doi:10.1083/jcb.200903097
23. Bogdanov AM, Bogdanova EA, Chudakov DM, Gorodnicheva TV, Lukyanov S, Lukyanov KA (2009) Cell culture medium affects GFP photostability: a solution. Nat Methods 6(12):859–860. doi:10.1038/nmeth1209-859

Chapter 18

Imaging Activity-Dependent Signaling Dynamics at the Neuronal Synapse Using FRET-Based Biosensors

Zohreh Farsi and Andrew Woehler

Abstract

In this chapter, we introduce the combined use of FRET-based biosensors and synaptic markers as an effective tool for studying intracellular signaling pathways in small synaptic terminals of neuronal cells. The approach is based on the unmixing of excitation/emission spectral fingerprints of a FRET donor and acceptor pair, as well as a lipophilic styryl dye, FM1-43, loaded into presynaptic terminals. The destaining of FM1-43 during evoked release provides a map to guide the sampling of fluorescence for FRET analysis. In the example presented here, we measure the temporal dynamics of cAMP at the presynaptic terminal using an intramolecular CFP/YFP-based FRET sensor. However, this methodology can be applied to investigate the spatial and temporal regulation of a variety of signaling processes, as well as dynamic changes in protein–protein interaction.

Key words FRET, Intracellular signaling, Live-cell imaging, Synaptic plasticity, cAMP

1 Introduction

Short-term plasticity (STP) is characterized as a transient change in the amplitude of postsynaptic current in response to depolarization of the presynaptic terminal. Although STP is identified by properties measured at the postsynapse, it is controlled through a wide range of processes that can occur in both the presynaptic and postsynaptic terminals. Within milliseconds of the arrival of an action potential at the presynaptic terminal, depolarization of the plasma membrane opens voltage gated calcium channels. These channels are closely coupled to the active zone and the subsequent calcium influx triggers fusion of neurotransmitter-filled vesicles that have been primed for release. If a second action potential reaches the bouton in close succession, the residual calcium at the active zone results in an increased probability of release and in facilitation of synaptic transmission. On the other hand, if an empty release site has not been refilled or the vesicle priming process has not been completed by the time a subsequent stimulus arrives, synaptic

Alexandros Poulopoulos (ed.), *Synapse Development: Methods and Protocols*, Methods in Molecular Biology, vol. 1538,
DOI 10.1007/978-1-4939-6688-2_18, © Springer Science+Business Media LLC 2017

depression occurs [1]. Facilitation and depression usually only persist for milliseconds to seconds. Sustained high-frequency stimulation at some synapses leads to enhancement of synaptic strength through augmentation and post-tetanic potentiation (PTP), which are longer lasting than facilitation [2]. There are, of course, many intracellular signaling processes that modulate the different types of STP through, for example, the sensitivity of calcium channels, the basal calcium concentration, presence of competing calcium binding proteins, presence of molecules required for priming, the rate at which readily releasable vesicles are replenished and docked to the active zone, and changes in the vesicle pool sizes. Some of these processes arise due to downstream effectors of calcium (i.e., PKC) while others are initiated by other means, such as the auto-activation of presynaptic GPCRs through neurotransmitter spill-over and retrograde signaling [3, 4].

Although a number of intracellular signaling pathways have been identified that modulate synaptic strength, their investigation has usually been limited to electrophysiological techniques with pharmacological perturbation. These approaches are suitable for identification of the key molecules and pharmacological dissection of the pathways involved; however, they are somewhat limited in terms of characterizing the spatiotemporal dynamics of these processes. In order to build a more complete understanding of information processing in the neuronal networks of the CNS, it is important to better understand the molecular signaling pathways and networks that are responsible for modulating short-term synaptic plasticity.

A diverse set of sensors has been developed to monitor many of the processes that are at work in the presynaptic bouton. While these sensors have been used extensively in cell biology, often at the whole-cell scale, the utilization of these sensors to study relatively fast events at small subcellular structures such as the presynapse or postsynapse presents a number of challenges. Among the most pressing is the efficient use of the photon budget provided by a limited number of sensors present in the small subcellular region of interest. With few exceptions, fluorescent reporters eventually undergo irreversible photobleaching limiting the amount of information that can be collected about a given process. Despite these challenges, a large body of work has been conducted to characterize the spatiotemporal dynamics of many kinases, second messengers, and GTPases in dendritic spines using FRET-based sensors [5, 6]. Many of these processes occur on the minute or longer time scales, whereas the processes responsible for STP can occur much faster.

cAMP has been shown to alter synaptic strength through a number of mechanisms at the presynaptic terminal. It has been shown to work through PKA to increase both the release probability and readily releasable pool size [7, 8]. PKA-independent signaling by cAMP at the bouton has also been reported to occur through the direct modulation of hyperpolarization-activated cation channels [9]. In the following we provide an example in which we measure the temporal dynamics of cAMP in response to a train of evoked action potentials at the presynaptic terminal. Neurons expressing an Epac1-based cAMP FRET sensor are first loaded with FM1-43. The axon is identified through colocalization of loaded FM1-43 along fine processes of the transfected neuron. Measurements are performed using a conventional widefield epi-fluorescence microscope equipped with an image splitter to detect fluorescence from the donor and acceptor FRET pair as well as from FM1-43 over four emission channels. A train of action potentials are elicited through field stimulation. The location of presynaptic boutons are identified through destaining of FM1-43 and used to guide the sampling of fluorescence for FRET analysis. Although the example provided in this protocol uses intramolecular FRET sensor, the same approach can be used to measure intermolecular FRET to study protein–protein interaction at the synapse [10].

2 Materials

2.1 Reagents/Consumables

1. Tyrode's Solution: 125 mM NaCl, 2 mM KCl, 2 mM $CaCl_2$, 2 mM $MgCl_2$, 10 mM Hepes, 30 mM Glucose, pH 7.4 with NaOH osmolarity adjusted to 305–310 with sucrose.
2. Imaging Medium: 10 μM 6-cyano-7-nitroquinoxaline-2,3-dione (CNQX), 50 μM DL-2-amino-5-phosphonopentanoic acid (AP5) in Tyrode's Solution.
3. Dulbecco's Modified Eagle Medium (DMEM, e.g., Life Technologies).
4. Neurobasal medium (NBA, e.g., Life Technologies).
5. FM1-43 Solution: 4 μM FM1-43 (e.g., Life Technologies) in Tyrode's Solution.
6. Lipofection reagent (e.g., Lipofectamine 2000, Life Technologies).
7. Fluorescence FRET immersion oil (e.g., Leica).
8. Calibration Dye Mix: Fluorescein, 1 μM Rhodamine B (e.g., Sigma-Aldrich) in water.
9. 18 mm #1.5 coverslips (e.g., Karl Hecht GmbH).
10. Electroporation cuvettes with 2 mm gap (e.g., Thermo Scientific).

2.2 Instrumentation

1. Electroporator (e.g., Bio-Rad Gene Pulser).
2. Inverted epifluorescence (Nikon TE2000 or equivalent).
3. High magnification objective (100 × 1.49NA).

4. Appropriate multiband filter cube (i.e., CFP/YFP) or separate filter cubes to be used in combination with a filter wheel for emission filters (e.g., Chroma Technology, Semrock Inc).
5. EMCCD (e.g., Andor iXon Ultra).
6. Image Splitter (e.g., QuadView QV2, Photometrics) or fast filter wheel with appropriate emission filters.
7. Excitation source: Till Photonics Polychrome IV or comparable.
8. Solution/chamber heating controller (e.g., Warner Instruments).
9. Custom field stimulation imaging chamber with parallel platinum electrodes (comparable chamber offered by Warner Instruments).
10. Master 8 Pulse Stimulator (e.g., A.M.P.I).
11. Stimulus Isolator (e.g., A385, World Precision Instruments).
12. Fast solution exchange system (e.g., Warner Instruments).
13. Six-channel valve control system with six-to-one manifold (e.g., Warner Instruments).
14. Peristaltic pump.
15. Computer for image acquisition and processing (Dual core or better, with Measurement Computing TTL out trigger card).
16. Software for image acquisition (e.g., IQ2, Andor).
17. Software for image processing: ImageJ and Matlab (Mathworks).

3 Methods

3.1 Experimental Setup

A Nikon TE-2000 microscope, equipped with a TILL photonics Polychrome IV monochromator allowing for fast switching between a wide range of excitation wavelengths was used for these measurements. Excitation is focused to the sample and emission collected through a 100× 1.49 NA TIRF objective. Fluorescence emission is then split by a Quadview image splitter and projected onto four quadrants of an EMCCD (Fig. 1). In order to optimize photon collection efficiency multiband filter cubes are not used in the microscope and bandpass filters are excluded from the image splitter. Rather a filter cube with 427/10 excitation, 442 nm dichroic, and 458 LP emission filter for CFP and a cube with a 488/10 excitation, 500 nm Dichroic, and 510 nm LP filters for YFP are used in the microscope and 510, 550, and 600 nm dichroics are used in the image splitter.

The experimental setup is also equipped with a solution superfusion system that allows for rapid exchange of the imaging bath solution, as well as a peristaltic pump to remove solution from the

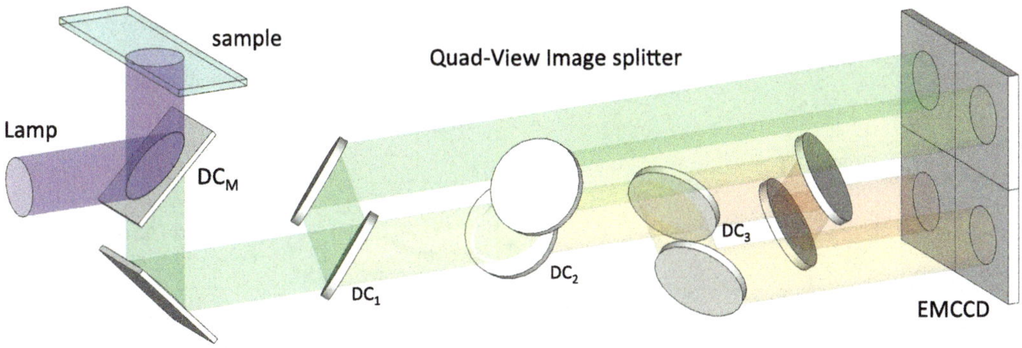

Fig. 1 Optical pathway for multicolor widefield microscopy. Excitation light provided by a xenon lamp monochromator is reflected onto the sample by the main dichroic filter (*DCM*, either 442 nm or 500 nm). Fluorescence emission is then guided to an image splitter composed of three dichroic mirrors and one broadband mirror in series. These mirrors reflect emission less than 510 nm (*DC1*), between 510 and 560 nm (*DC2*), between 560 and 600 nm (*DC3*) and greater than 600 nm in different directions perpendicular to the optical axis. Additional broadband mirrors guide each emission band to optics that project each band onto one quadrant of an EMCCD camera

imaging chamber. This system is controlled manually or by the image acquisition software through a TTL controlled valve system (Fig. 2). In addition to the perfusion inlet and outlet the imaging chamber is equipped with a pair of platinum electrodes for field stimulation. These electrodes are connected to a stimulus isolator that is controlled by a Master 8 pulse generator that is preprogrammed to deliver a given pulse train and is, as well, triggered by the acquisition software.

3.2 Sample Preparation

1. Dissociated hippocampal cultures are prepared from newborn (P0) rats as in Kaech and Banker 2006 [11]. Although we work with hippocampal neurons because of the role of the hippocampus in learning and memory, this protocol is applicable to cultured neurons from other structures (i.e., cortical and DRG).
2. Neurons are transiently transfected with DNA encoding an updated Epac1-based FRET that uses mTurquoise2 as donor and cpVenus as acceptor (henceforth referred to as CFP and YFP, respectively). Transfection is performed at DIV0 before plating via electroporation [12] or at DIV 5–10 using a modified calcium phosphate protocol [13]. The experimental procedure outlined in this work is in no way specific to the sensor demonstrated. There are an ever-growing variety of sensors available for a diverse range of second messengers, cellular

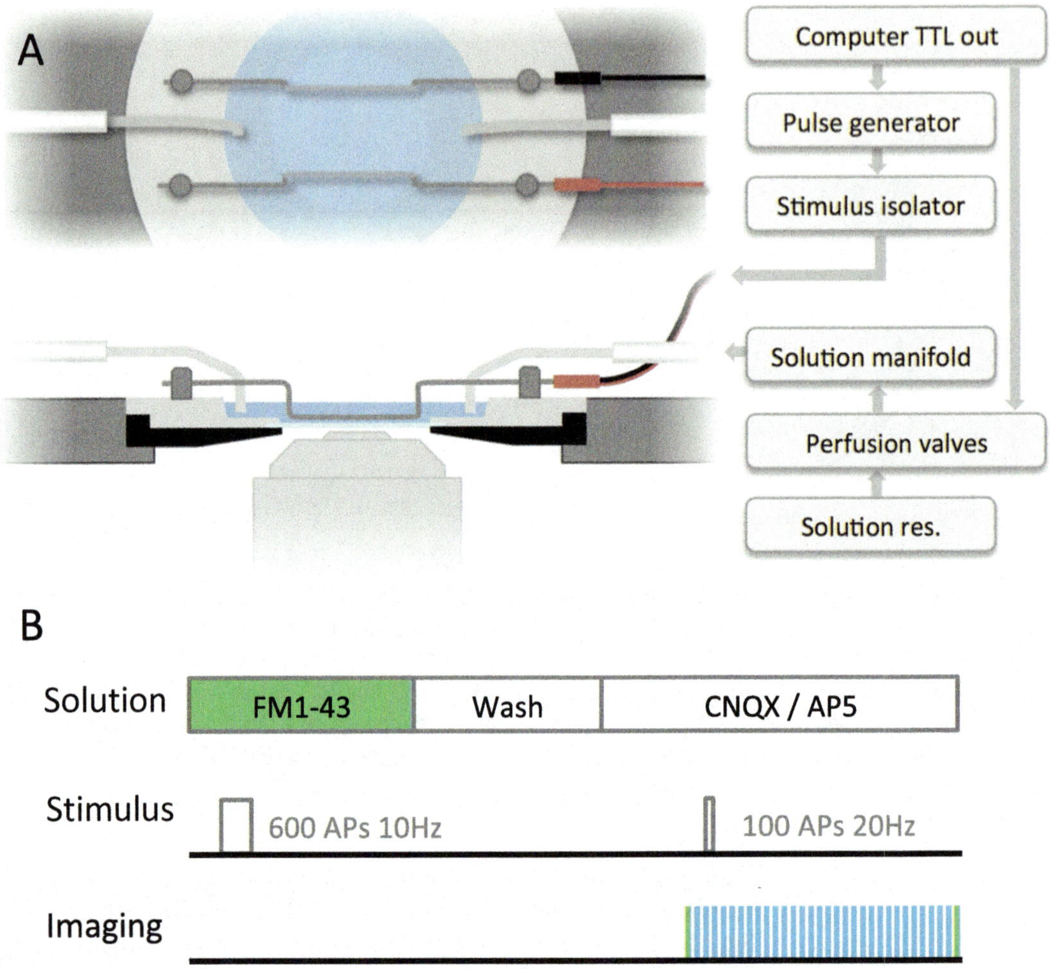

Fig. 2 Systems for automated solution exchange and field stimulation. (**a**) Samples prepared on 18 mm coverslips are mounted in an imaging chamber equipped with parallel platinum electrodes. The imaging chamber is mounted on the microscope stage together with perfusion and suction tubes. Both perfusion and stimulation are controlled via TTL signal provided by the computer running the image acquisition software. Stimulation protocols are programmed into the TTL triggered Master 8 pulse generator, which in turn, triggers the stimulus isolator to which the electrodes are connected. Perfusion is gravity fed from separate reservoirs through a TTL controlled valve system through a manifold to the perfusion tube. (**b**) This system, together with the imaging system and acquisition software, is used to automate the loading of FM1-43, washing of unloaded dye, drug application, and image acquisition before, during, and after stimulation

analytes, posttranslational modifications, and small molecule translocations that can be investigated with this approach (*see* Newman et al. 2011 [14] for a near complete list of genetically encoded sensors).

3.3 Imaging Preparation

1. Before imaging, program imaging protocol and stimulation protocols.
2. On DIV 15–25, wash one coverslip of transfected neurons in a 35 mm dish containing 1.5 ml Tyrode's solution.

3. Mount the coverslip in the field stimulation imaging chamber and add 250 μl of Imaging Medium (which contains CNQX and AP5 to prevent recurrent post-stimulation activity).
4. After applying immersion oil to the 100× objective, mount the imaging chamber to the microscope stage, connect the pulse isolator cables to the field stimulation electrode leads, and mount the perfusion application and suction tubes to the stage.
   (a) If imaging at 37 °C, connect the heating control unit cable assembly to the imaging chamber and place the bead-thermistor in the bath solution. Allow the system to equilibrate for 15 min before starting measurements.
   (b) If imaging at room temperature allow the system to equilibrate for 5 min before proceeding.
5. Locate the correct focal plane in brightfield and switch to epi-fluorescence with donor wavelength excitation (430 nm) to locate a transfected neuron for further imaging. Once a suitable cell has been found turn off excitation.
6. Apply FM 1–43 Solution to the cells. After 1 min incubation, stimulate the cells by executing the loading protocol programmed into the Master 8 Pulse generator (600 APs at 10 Hz). After an additional 5 min incubation in the FM1-43 Solution, wash the cells thoroughly before imaging by superfusing Imaging Medium for 5 min at 1 ml/min.
7. Turn on fluorescence excitation and locate a region of the transfected cell where long, fine processes are labeled with puncta of loaded FM1-43.

3.4 Image Acquisition

In the following steps we outline the sequence in which different images are acquired and events triggered during the experiment. The Andor IQ software allows one to program "protocols" in which different acquisitions settings (excitation wavelength, exposure time, etc.) can be selected and image acquisitions looped for time-lapse imaging during which external events can be controlled via TTL triggers. Similar measurements can be performed with the open source ImageJ-based Micromanager, however, with different hardware to control the monochromator and triggered devices.

1. Acquire one image using direct acceptor excitation at 488 nm, 50 ms exposure time.
2. Immediately start a time series of 20 images acquired at 2 Hz. (200 ms exposure with 430 nm excitation followed by 300 ms with the monochromator set to 300 nm, the default "off" position).
3. Trigger the execution of the stimulation protocol (100 APs at 20 Hz) in the master 8 pulse generator.

4. Immediately start a time series of 200 donor excitation images acquired at 2 Hz.
5. Follow the donor excitation time series with another acceptor excitation acquisition.

3.5 Reference/Calibration Image Acquisition

In order to preform spectral separation of fluorescence emission using linear unmixing one must first measure the individual spectra of each species of interest. This is performed with reference samples expressing either the donor or acceptor or from cells labeled with FM1-43.

1. Prepare cells that are transfected with only donor or acceptor.
   (a) These cells need not be neurons. We usually use more easily transefected cell line (HEK293, Hela, N1E-115, ...) and standard cationic lipid-mediated transfection (i.e., Lipofectamine 2000).
2. Incubate untransfected cells with FM1-43 Solution so that the membrane is labeled.
3. Measure one acquisition of each of the reference samples with the same donor and acceptor acquisition setting as used in the neuron measurements from which reference spectra will be sampled.
4. In epifluorescence microscopy, often times the field of illumination is not uniform. It may be possible that that the image splitter causes nonuniformities in the spatial detection efficiency as well. This can be measured and corrected for by imaging a known uniform sample. To do this place a blank coverslip in the imaging chamber and add the Calibration Dye Mix. Perform the same donor and acceptor acquisition to measure the field of illumination.

3.6 Image Analysis

In the following steps we sample fluorescence from the reference measurements to define reference excitation/emission spectra that are used to unmix the sources of fluorescence in the CFP/YFP/FM1-43 sample measurement. The unmixed FM1-43 images are used to define ROIs from which fluorescence is measured for FRET analysis. If one is only interested in the temporal dynamics of the change in FRET it is possible to simply compute the ratio of unmixed CFP and YFP apparent concentrations or even just the raw fluorescence from the donor and acceptor channels. However, these ratios are instrument dependent and as it is often much easier to calibrate FRET-based sensors in cuvette we prefer to work with an instrument-independent quantity such as FRET efficiency. We perform all image analysis in Matlab; however, all procedures can be implemented using open source options such as ImageJ.

1. To begin, export all images as 16 bit tiff files.
2. For each of the reference image sets (i.e., CFP at 430 nm and 488 nm).
   (a) Load both images into Matlab as a 3D (x, y, ex) stack.
   (b) Split dual or quad-view images to create a 4D (x, y, ex, em) stack.
   (c) Correct for nonuniform field of illumination by dividing each sub-image by the corresponding offset subtracted, normalized field of illumination image.
   (d) Perform image registration of the entire ex/em stack using a subpixel image registration routine [15].
   (e) Measure the background intensity from user-defined ROI and subtract the background value from each pixel for each sub-image in the stack.
   (f) Measure fluorescence from a user-defined ROI and sample from each sub-image of the image stack and save the resulting values as the reference spectra for the given species (CFP, YFP, and FM1-43).
3. Load the "donor excitation" image acquisition from the FRET/FM1-43 sample into Matlab. Apply the same corrections as outlined for the reference images and then perform per pixel linear unmixing using the reference emission spectra from the 430 nm excitation computed above. This will result in time stacks of apparent CFP, YFP, and FM1-43 concentrations (relative to the references).
4. The FM1-43 images can be used to assess loading and destaining and to define ROIs for FRET analysis.
   (a) One can use the intensity of FM1-43 to guide selection of ROIs; however, as staining is often unspecific (Fig. 4c) it is better to use destaining to identify the location of synapses.
   (b) To create a spatial map of destaining fraction subtract the F_{post} (the mean of the 10 images after stimulation) from F_{pre} (the mean of 10 images prior to stimulation) and divide by F_{pre}. *See* Fig. 5a for a spatial map of destaining location and panel B for a graph of typical destaining time-course.
5. Before performing the FRET analysis, the calibration term r^{ex} for donor and acceptor excitation wavelengths must be computed as in Wlodarczyk 2008 [16]. This is then used to compute the spectral fingerprint of FRET between CFP and YFP as $F^{ref}_{ECY} = -F^{ref}_{C} + r^{ex} \times F^{ref}_{A}$ [17].
6. Load the first acceptor-only image together with the first donor-only measurement (Fig. 3d) and perform the same field

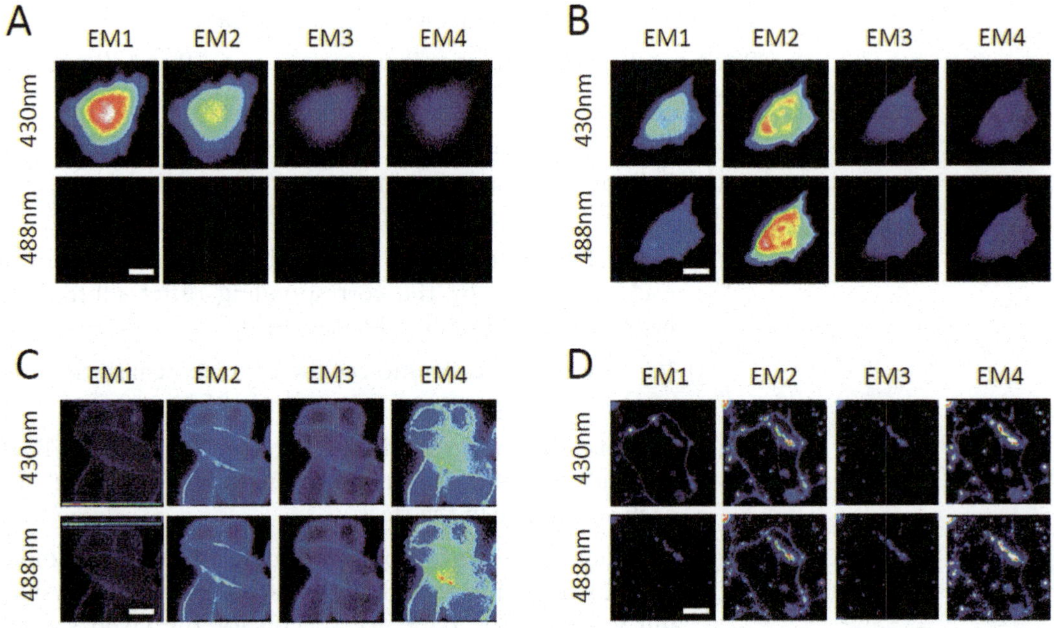

Fig. 3 Measurements of fluorescence spectral fingerprints. Samples selectively labeled with CFP (**a**), YFP (**b**), FM1-43 (**c**) or a combination of the three (**d**), are excited sequentially at 430 nm and 488 nm. Emission with wavelengths less than 510 nm is collected in channel 1, emission between 510 and 550 nm collected in channel 2, emission between 550 and 600 nm collected in channel 3, and emission greater than 600 nm collected in channel 4

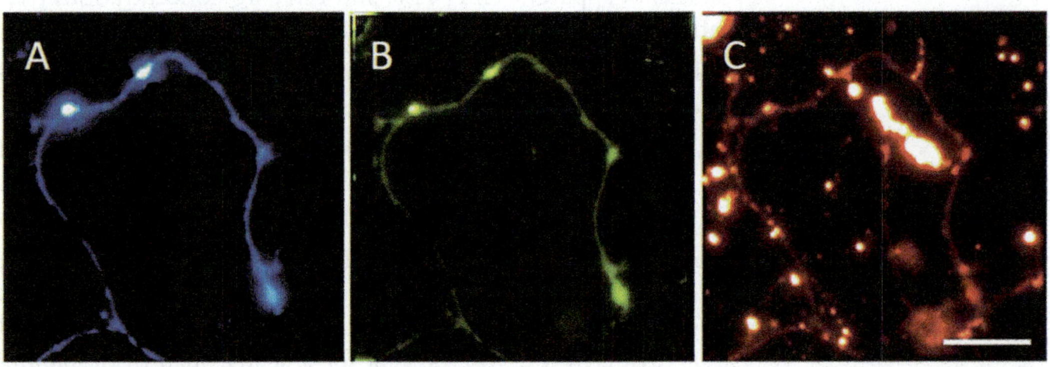

Fig. 4 Unmixing of CFP, YFP, and FM1-43. Using reference spectra measured from ROIs within the reference sample images (illustrated in Fig.3a–c), linear unmixing of fluorescence emission from a sample expressing CFP and YFP and labeled with FM1-43 (**d**) is performed. The resulting images represent the spatial distribution of CFP (**a**), YFP (**b**), and FM1-43 (**c**) within the sample

of illumination correction, image registration, and background subtraction as performed for the reference measurements.

7. Sample the mean fluorescence intensity from the ROI defined in **step 3** for all images in the excitation/emission stack and perform linear unmixing using the reference excitation/emis-

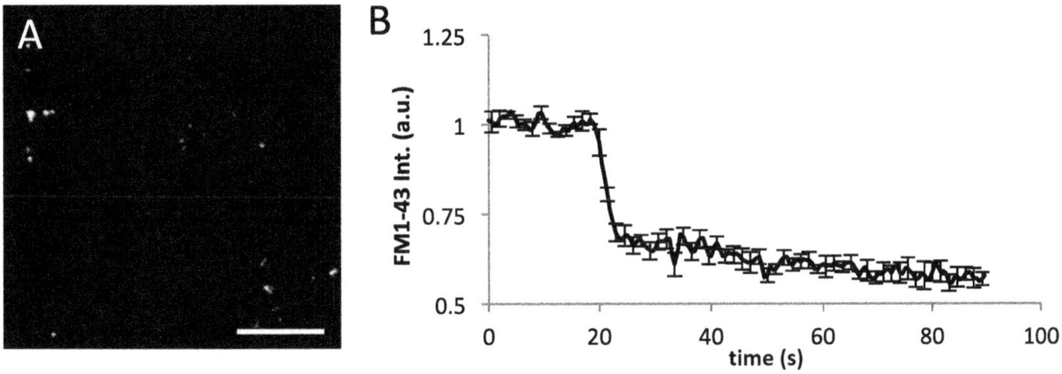

Fig. 5 Destaining of FM1-43. To identify the location of synaptic activity and guide the selection of regions for FRET analysis, the relative difference between the mean of the ten frames before and after stimulation is computed for the FM1-43 apparent concentrations (unmixed images)

sion spectra defined in **step 2** as well as the spectral fingerprint of FRET determined in **step 4**. This results in:

(a) The total FRET corrected CFP concentration, relative to the CFP reference concentration.

(b) The total FRET corrected YFP concentration, relative to the YFP reference concentration.

(c) The total FM1-43 relative to the reference concentration.

(d) The FRET efficiency scaled by the concentration of FRET complexes relative to the CFP reference concentration.

8. Since we are using a fixed 1:1 stoichiometry FRET sensor the concentration of FRET complexes relative to $[C^{ref}]$ is equal to the concentration of CFP relative to $[C^{ref}]$, and thus the mean FRET efficiency within the ROI can be computed by dividing the last product of unmixing by the first.

9. If one assumes that the donor to acceptor ratio will not change within the ROI (a safe assumption for a fixed stoichiometry sensor) one can use the ratio of the total relative concentrations determined in **steps 6a, b** (equal to $[Y^{ref}]/[C^{ref}]$), together with the calibration term defined in **step 4**, to determine FRET efficiency from a single wavelength measurement (Eq. 3a from Woehler et al. 2010 [18]).

(a) Sample the mean of the ROI from the apparent concentrations of CFP and YFP determined in **step 3** for each time point and calculate the FRET efficiency over the duration of the experiment as shown in Fig. 6b.

10. One can obtain an estimate of absolute concentration reported by the sensor during the measurement using a calibration curve as shown in Fig. 6c. If the sensor has been characterized previously, a rough estimate can be obtained by using the K_D and Hill

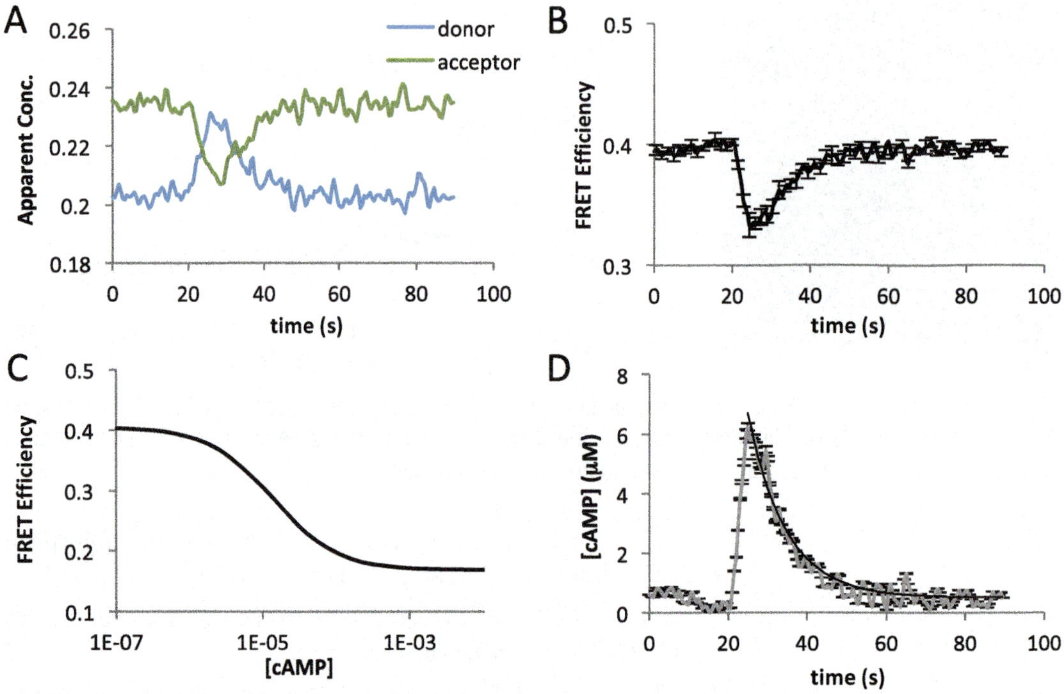

Fig. 6 Measuring dynamic changes in cAMP. The mean fluorescence was measured from ROIs where FM1-43 destaining was observed. Reference spectra were used to unmix the average apparent concentrations of CFP, YFP, and FM1-43 for these ROIs. (**a**) A transient decrease in FRET can be observed as a decrease in the apparent acceptor concentration associated with an increase in the apparent donor concentration upon stimulation with 100 APs at 20Hz starting at 20 s. (**b**) This change in FRET can be quantified using the ratio of apparent concentrations with calibration terms calculated from dual excitation measurements. The mean FRET efficiency of eight ROIs is shown with the error bars indicating the SEM of every third measurement. (**c**) Using the published KD and Hill coefficient of the sensor along with measurements of the FRET efficiency at free and saturating concentrations a calibration curve can be estimated. (**d**) This calibration curve can be used to estimate the transient change in cAMP in terms of absolute concentration and measure a decay time constant of 11.78 +/− 0.86 s (+/− 95 % confidence interval of the fitted parameter)

coefficient originally reported [19], if the response was reported in terms of absolute FRET efficiency. If one wishes to use ratio as a readout, the minimal/maximal responses must be measured on the same instrument as the sample of interest was measured.

11. It is also possible to perform complete in-sample calibrations for many sensors through membrane permeabilization and serial ligand titration [20].

4 Notes

1. Although we demonstrate combined FRET/FM1-43 measurements using an epifluorescence microscope with a four-channel emission image splitter it is possible to apply the methods introduced here to a range of imaging setups.

(a) For information on how the FRET approach has been applied to spectral imaging with a confocal microscope see Woehler 2010 [18].

(b) Although we are imaging three fluorescent species, it is possible to use two-channel (Dual-View or Optosplit) image splitter. This can be accomplished by first measuring the FRET response to stimulation, switching filters, and later loading and destaining FM1-43 to identify synapse location.

(c) It is also possible to perform these experiments without an image splitter. Many instruments are equipped with a filter wheel in the emission pathway that can be used to sequentially image the different emission wavelengths. Of course this decreases the photon detection efficiency as well as the temporal resolution but is applicable to a wider range of instruments.

2. Although we use FM1-43 as a synaptic marker there are alternative approaches to guide ROI selection. There are a number of transgenic mice that express fluorescently labeled synaptic proteins or activity sensors such as synaptopHluoin [21]. There are also several other styryl dyes, such as FM4-64, with fluorescence properties that, depending on the process/sensor of interest, might be more appropriate for multiplexed measurements.

3. Chomatic aberration is the result from the inability of a lens to focus light of different wavelengths to the same point. Although the development of apochromatic lenses has tightened the difference in focus across the visible spectrum, some aberration still persists. When comparing fluorescence emission collected at different wavelengths, as we do for spectrally resolved FRET measurements, it is important to check for and, if necessary, correct for chromatic aberration. This is often performed by imaging diffraction-limited multispectral fluorescence beads with emission spanning the detection channels of a given microscope at different focal depths. By plotting intensity or width of the spot profile for the different emission channels one can evaluate the extent of chromatic aberration. To correct for aberration it is possible to insert lenses into the beam path between the dichroic mirror that splits the emission and the lens that focuses the emission onto the CCD. Both Cairn and Photometrics image splitters provide mounts for emission band pass filters that can also accommodate such lenses.

4. Nonlinear error propagation often occurs in complex image analysis routines. A general rule of thumb is that it is best to sample the mean raw fluorescence from a region of interest and subsequently compute the quantity of interest (i.e., FRET efficiency) rather than sample a ROI of per-pixel computed quantities. Oftentimes error can be skewed by an analysis routine

and lead to the biasing of a sampled mean. For example, converting normally distributed FRET efficiencies measured from a ROI to ligand concentrations using a calibration curve, such as the one illustrated in Fig. 6c, will result in a skewed distribution of ligand concentrations as the mean FRET efficiency deviates from the half-response value.

5. We mentioned several times throughout the protocol that it is possible to use the same approach to measure intermolecular FRET. One thing that is often neglected in investigations of protein–protein interaction using FRET is the dependence of apparent FRET efficiency on the donor-to-acceptor stoichiometry. Apparent FRET is the characteristic FRET efficiency scaled by the fraction of donor molecules interacting with an acceptor (or acceptor labeled with donor, depending on method used). As the donor-to-acceptor ratio is varied, the probability of interaction will also change even if affinity remains constant. It is therefor necessary to measure and ensure similar expression stoichiometry before drawing conclusions about interaction affinity from two intermolecular FRET values. *See* Wlodarczyk et al. 2008 [16] and Renner et al. 2012 [22] for more details.

Acknowledgements

The authors would like to thank KunHan Lin for invaluable comments on the protocol and I. Herfort for the technical assistance. This work was funded in part by the Cluster of Excellence and DFG Research Center Nanoscale Microscopy and Molecular Physiology of the Brain.

References

1. Neher E (2010) What is rate-limiting during sustained synaptic activity: vesicle supply or the availability of release sites. Front Synaptic Neurosci 2:144
2. Habets RLP, Borst JGG (2007) Dynamics of the readily releasable pool during post-tetanic potentiation in the rat calyx of held synapse. J Physiol 581:467–478
3. Fioravante D, Regehr WG (2011) Short-term forms of presynaptic plasticity. Curr Opin Neurobiol 21:269–274
4. Beierlein M, Fioravante D, Regehr WG (2007) Differential expression of posttetanic potentiation and retrograde signaling mediate target-dependent short-term synaptic plasticity. Neuron 54:949–959
5. Murakoshi H, Wang H, Yasuda R (2011) Local, persistent activation of Rho GTPases during plasticity of single dendritic spines. Nature 472:100–104
6. Lee S-JR, Escobedo-Lozoya Y, Szatmari EM et al (2009) Activation of CaMKII in single dendritic spines during long-term potentiation. Nature 458:299–304
7. Kaneko M, Takahashi T (2004) Presynaptic mechanism underlying cAMP-dependent synaptic potentiation. J Neurosci 24:5202–5208
8. Yao L, Sakaba T (2010) cAMP modulates intracellular Ca2+ sensitivity of fast-releasing synaptic vesicles at the calyx of held synapse. J Neurophysiol 104:3250–3260

9. Beaumont V, Zucker RS (2000) Enhancement of synaptic transmission by cyclic AMP modulation of presynaptic I h channels. Nat Neurosci 3:133–141
10. Degtyar V, Hafez IM, Bray C et al (2013) Dance of the SNAREs: assembly and rearrangements detected with FRET at neuronal synapses. J Neurosci 33:5507–5523
11. Kaech S, Banker G (2006) Culturing hippocampal neurons. Nat Protoc 1:2406–2415
12. Karra D, Dahm R (2010) Transfection techniques for neuronal cells. J Neurosci 30:6171–6177
13. Jiang M, Chen G (2006) High Ca 2+-phosphate transfection efficiency in low-density neuronal cultures. Nat Protoc 1:695–700
14. Newman RH, Fosbrink MD, Zhang J (2011) Genetically encodable fluorescent biosensors for tracking signaling dynamics in living cells. Chem Rev 111:3614–3666
15. Guizar-Sicairos M, Thurman ST, Fienup JR (2008) Efficient subpixel image registration algorithms. Opt Lett 33:156
16. Wlodarczyk J, Woehler A, Kobe F et al (2008) Analysis of FRET signals in the presence of free donors and acceptors. Biophys J 94:986–1000
17. Woehler A (2013) Simultaneous quantitative live cell imaging of multiple FRET-based biosensors. PLoS One 8:e61096
18. Woehler A, Wlodarczyk J, Neher E (2010) Signal/noise analysis of FRET-based sensors. Biophys J 99:2344–2354
19. Ponsioen B, Zhao J, Riedl J et al (2004) Detecting cAMP-induced Epac activation by fluorescence resonance energy transfer: Epac as a novel cAMP indicator. EMBO Rep 5:1176–1180
20. Mironov SL, Skorova E, Taschenberger G et al (2009) Imaging cytoplasmic cAMP in mouse brainstem neurons. BMC Neurosci 10:29
21. Kavalali ET, Jorgensen EM (2014) Visualizing presynaptic function. Nat Neurosci 17:10–16
22. Renner U, Zeug A, Woehler A et al (2012) Heterodimerization of serotonin receptors 5-HT1A and 5-HT7 differentially regulates receptor signalling and trafficking. J Cell Sci 125:2486–2499

Chapter 19

Analyzing Structural Plasticity of Dendritic Spines in Organotypic Slice Culture

Mathias De Roo* and Adema Ribic*

Abstract

Understanding the rules of synapse dynamics in the context of development, learning, and nervous system disorders is an important part of several fields of neuroscience. Despite significant methodological advances, observations of structural dynamics of synapses still present a significant experimental challenge. In this chapter we describe a set of techniques that allow repetitive observations of synaptic structures in vitro in organotypic cultures of rodent hippocampus. We describe culturing of slices, transfection with reporter-carrying plasmids, repetitive imaging of dendritic spines with confocal laser scanning microscopy and analysis of spine morphology dynamics.

Key words Organotypic culture, Transfection, Confocal imaging, Dendritic spine, Hippocampus, Plasticity, Synaptogenesis

1 Introduction

In the central nervous system most excitatory synapses are localized on small dendritic protrusions named dendritic spines. These structures appear early after birth during development, but continue to change their presence and morphology, appear, and disappear throughout adulthood serving learning and homeostasis. Additionally, changes in spine density, morphology and dynamics are hallmarks of a number of brain disorders. Studying spine dynamics associated with diverse stimuli, such as long-term potentiation or long-term depression, is thus an important area of investigation toward determining the mechanisms underlying synaptic plasticity. This chapter describes how to perform such studies on an in vitro model of a brain structure involved in learning and memory, namely the hippocampal organotypic slice culture. Originally described by Stoppini and coworkers [1], it allows the study of synaptic and structural plasticity on preserved and functional parts of the

*Authors contributed equally, correspondence may be addressed to either.

Alexandros Poulopoulos (ed.), *Synapse Development: Methods and Protocols*, Methods in Molecular Biology, vol. 1538,
DOI 10.1007/978-1-4939-6688-2_19, © Springer Science+Business Media LLC 2017

hippocampal neuronal network for up to 4 weeks after dissection. The first part describes the procedure to perform organotypic cultures on rat or mouse pups (4–8 days). The second part briefly explains how to transfect these cultures with genes of fluorescent proteins together with other genes of interest such as calcium sensors, receptor subunits or shRNA. The third part describes the steps to follow in order to perform laser scanning confocal microscopy over days or weeks while preserving synaptic integrity. Lastly we describe how we analyze spine dynamics and potential caveats to be considered. This method that we developed has allowed us to successfully describe some important aspects of structural plasticity upon synaptogenesis [2], learning related patterns of activity [3], and pharmacological treatment [4]. Because of easy pharmacological access and relative high-throughput, this procedure, combined with the use of transgenic models of synaptic diseases, provides a useful way to study the mechanisms underlying synapse development and learning in the context of normal physiology or synaptopathies.

2 Materials

Prepare all solutions using sterile ultrapure water (displaying resistance of 18.2 MΩ.cm at 25 °C).

2.1 Hippocampal Organotypic Cultures

1. Laminar flow hood equipped with UV-light.
2. Stereomicroscope with 0.8–3.5× objective and 10× ocular.
3. Chopper or vibratome, compatible razor blades.
4. Incubator CO_2- and temperature-controlled: 5 % CO_2, 37 °C, and 5 % CO_2, 33 °C.
5. White FHLC membrane. Filter type: 0.45 μm.
6. Low profile cell culture inserts. Pore diameter: 0.4 μm, Insert diameter: 30 mm.
7. Flat-bottom tissue culture plates, 6 well, with low-evaporation lid.
8. Vacuum-filtration units, 0.22 μm pore size.
9. MEM 2× stock solution: dilute 9.5 g of MEM Eagle M0268 in 500 ml sterile bidistilled water, add 0.7 g NaCl and 5.96 g HEPES. Filter at 0.22 μm. Store at 4 °C.
10. Culture medium: in bidistilled water, add (v/v) 25 % of MEM 2× stock solution, 25 % heat-inactivated horse serum, 25 % Hank's balanced salt solution (HBSS), 100 unit/ml penicillin–streptomycin, 0.46 % of 7.5 % $NaHCO_3$ stock solution in H_20, 2 mM Tris base. Adjust pH with HCl to 7.2. Filter at 0.22 μm. Aliquot and store at −20 °C.

11. Dissection medium: in bidistilled water, 50% of MEM 2× stock solution supplemented with 100 unit/ml penicillin–streptomycin, 10 mM Tris base, pH with HCl to 7.2. Filter at 0.22 μm and place on ice under the dissection hood before starting the dissection. Prepare fresh every time.
12. Propidium Iodide Solution: 5 μg/ml propidium iodide in culture medium.
13. Dissection tools: curved tip fine tweezers; iris forceps, curved; iris scissors, 10 cm long; scalpel handle and carbon steel blades; bone currete or curved tissue spatula; fine plastic spatulas; Pasteur pipettes, 150 mm.
14. 42.5 mm Whatman filter paper.

2.2 Transfection

1. Plasmid DNA encoding fluorescent protein and other genes of choice.
2. Gene-gun system.
3. 1.6 μm gold micro-carriers.
4. Sonicator.
5. Nitrogen gas cylinder.
6. Helium gas cylinder.
7. $CaCl_2$ Solution: 2 M $CaCl_2$ in bidistilled water. Store at −20 °C.
8. Spermidine Solution: 50 mM spermidine in ethanol. Store at −20 °C.
9. PVP Solution: 50 μg/ml polyvinylpyrrolidone in ethanol. Store at −20 °C.
10. Silica gel or another desiccating agent.
11. 80 μm nylon mesh fixed to a 1 cm-high plastic ring.

2.3 Confocal Imaging

1. Custom chamber made from an observation glass slide and a piece of plastic designed to fit the cell culture inserts.
2. Confocal laser scanning microscope.
3. 4× or 5× objective lens.
4. High NA 40× water-immersion objective lens.
5. Laser source(s) of appropriate wavelength depending on the fluorophore(s) to be imaged.

2.4 Image Analysis

1. Computer (minimal configuration used in our laboratory: Apple Macintosh PowerBook G4; 1.67 GHz; 2 GB RAM).
2. Image Analysis Software: e.g., OsiriX (http://www.osirix-viewer.com), ImageJ (http://imagej.nih.gov/ij/), or Matlab (MathWorks) with appropriate plugins.

3 Methods

3.1 Preparation of Hippocampal Organotypic Cultures

3.1.1 Preparation of Confetti, Pipettes, and Culture Boxes

Prepare the following items on the day before the dissection:

1. Chop 8 mm diameter confetti from white FHLC Membrane. We found that hydraulic or similar industrial hole punchers work well without fraying the confetti edges or producing kinks in the confettis. Wash in 96 % ethanol and sterilize under UV. Store in sterile conditions.
2. Use diamond file to mark the breaking line on a Pasteur pipette tip. Carefully break the tip off at the desired diameter (we use 6 and 2 mm pipettes). Turn on a Bunsen burner and place the tip in the flame until the edges are smooth, taking care that the glass does not change color when extremely hot.
3. Add 1 ml of culture medium to each well of one 6-well culture box (enough for dissection of 3 mice or 1–2 rats).
4. Place a cell culture insert in each well using tweezers.
5. Put four confetti per culture insert using tweezers.
6. Place the box in the 37 °C incubator.

3.1.2 Dissection

1. Just before the dissection starts, prepare dissection medium and sterilize dissection instruments under UV-light for 5 min. Place the dissection instruments on ice after sterilization.
2. For each animal, put 10 ml of cold dissection medium and a filter paper in a 5 cm petri dish.
3. Sacrifice the animal according to your veterinary office policy. Disinfect the animal head by spraying with 70–90 % solution of ethanol and decapitate.
4. Remove the brain and place it on the filter paper in cold dissection medium.
5. Execute a frontal cut with the scalpel to separate the cerebellum from the rest of the brain.
6. Separate the two brain hemispheres with the scalpel.
7. Place the first hemisphere on its medial part and execute a frontal cut to eliminate about 1 mm of the forebrain. *See* **step 11** for alternative method.
8. Reveal the hippocampus as illustrated in Fig. 1.
9. Extract the first hippocampus with the plastic spatula. Avoid as much as possible to touch the hippocampus or the cortex. Under the stereomicroscope, cut the remaining fibers around the hippocampus with the scalpel. Always cut the fibers or tissue instead of tearing out.
10. Repeat **step 6** with the second hippocampus and proceed to producing slices (**step 12**).

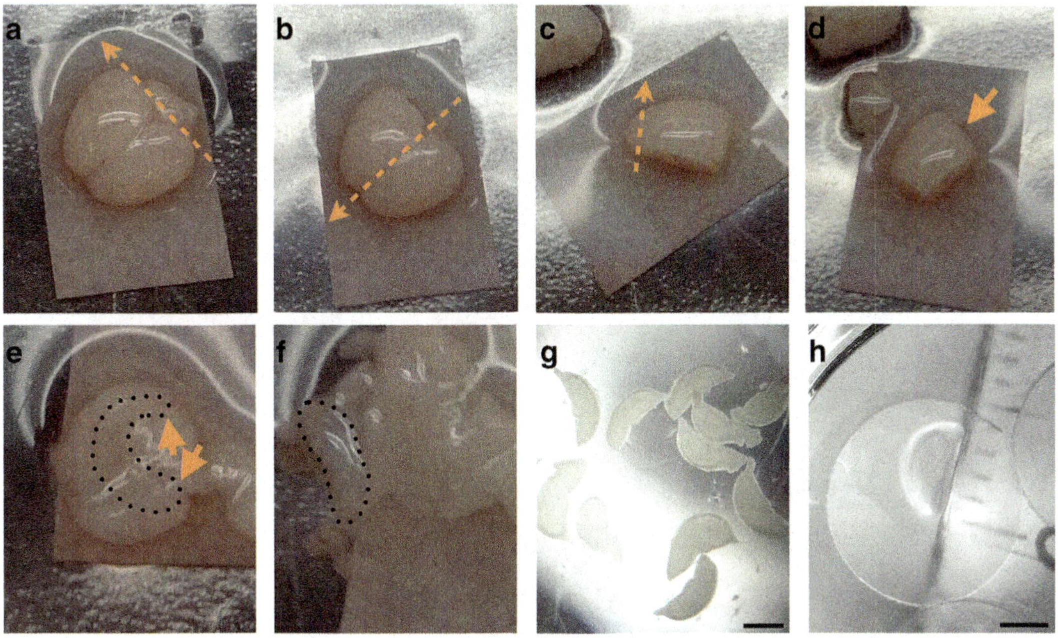

Fig. 1 Dissecting the hippocampus of a 4-day-old rat. (**a**) Execute a frontal cut with the scalpel to separate the cerebellum from the rest of the brain. (**b**) Separate the two brain hemispheres with the scalpel. (**c**) Place the first hemisphere on its medial part and execute a frontal cut to eliminate about 1 mm of the forebrain. (**d**) Put the plastic spatula inside the visible slit (*arrow*) and split the hemisphere to reveal the hippocampus. (**e**) Use the plastic spatula to gently detach the hippocampus extremities (*arrows*) and flip it outside the hemisphere (**f**) Dashed shape highlights the hippocampus. (**g**) Hippocampal slices in dissection medium just after slicing. (**h**) One hippocampal slice in a culture box just after plating on a confetti. A ruler (1 mm per graduation) was placed under the box for illsutration purpose. Scale bars: 2 mm

11. An alternative method for dissecting out the hippocampi is the following: using fine curved forceps, pin the brain gently to the dish along the midline. Using double-ended bone curette or dental spatula, gently scoop the hippocampi from beneath the cortex by slowly displacing the cortex around the hippocampus (Fig. 2). Avoid directly touching the hippocampus. Separate hippocampi from the rest of the brain using fine scissors or tweezers.
12. With the 6 mm fire-polished Pasteur pipette, place the two hippocampi on the chopper, perpendicular to the blade, concave part on top. Remove extra medium around it. Cut 400 μm-thick slices. During cutting, transfer 2 ml of cold dissection medium in a 35 mm petri dish. Drop some cold dissection medium on the slices and use the 2 mm fire-polished Pasteur pipette to transfer them into the petri dish.
13. Select slices containing all cellular regions (DG, CA1-3) from the middle parts of the hippocampi under the stereoscopic microscope and gently separate them with slow back and forth movements with the 2 mm fire-polished Pasteur pipette. Take

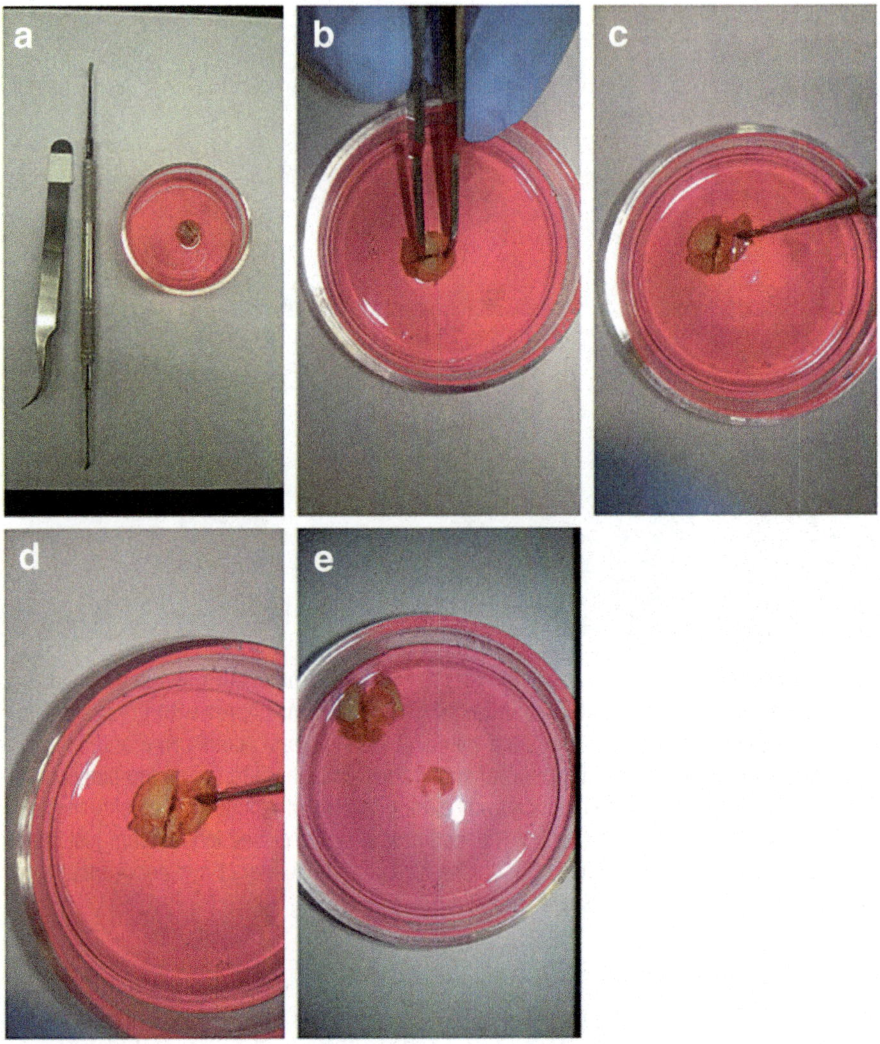

Fig. 2 Alternative method for dissection of hippocampi from rodent pups. (**a**) The tools needed are double ended dental spatula and fine curved forceps. (**b**) Using the curved forceps, gently pin the brain down to the Petri dish along the midline (one end of forceps is in the prefrontal cortex and the other right in front of the cerebellum). (**c**) Using the spatula, lift the hemisphere gently until you see the hippocampus beneath it. (**d**) Gently scoop out the hippocampus by dislocating the cortex around it. Avoid touching the hippocampi directly. (**e**) Separate the hippocampus using either the spoon or fine tweezers. Repeat for another hemisphere. Proceed with slicing as previously described

the culture box out of the 37 °C incubator and place one slice per confetti with the 2 mm fire-polished Pasteur pipette. Use a standard Pasteur pipette to remove extra medium around the slices. Place the culture box back in the 37 °C incubator.

14. Culture health can be assessed after a few days by applying Propidium Iodide Solution that will stain dying cells after a 20 min incubation (*see* **Note 1**).

15. Change the medium in each culture well one day after the dissection, then every 3–4 days. Medium must be preheated it in a water bath then placed in the incubator for 10 min to attain desired temperature and pH. Transfer the cultures to the 33 °C incubator after 3–4 days. Organotypic cultures can be kept under those conditions for 4–5 weeks.

3.2 Transfection

Several methods can be used to visualize dendritic spines in organotypic slice cultures. Doing organotypic culture from transgenic mice expressing YFP or similar reporter may be the simplest option. We describe here the gene-gun method, which allows a rapid transfection of one or several genes in rat or mice organotypic culture. Originally developed to transfect plants, several labs working on dissociated or organotypic cultures have adapted this method for their needs [5–8].

In our hands, gene-gun transfection can be done after 1 day and up to 3 weeks after the dissection following these steps:

3.2.1 Preparation of Cartridges

1. Cut 80 cm of tubing (e.g., Bio-Rad) and wash it with 100% dehydrated ethanol using a syringe. Place the tubing into the tubing prep station (Bio-Rad), switch on the Nitrogen knob set to 0.3 on the tubing prep station scale and rotate for 15 min to allow the tubing to dry. Then switch off the Nitrogen knob.
2. Put 5 mg of gold micro-carriers in a 1.5 ml microfuge tube.
3. Put gloves on. Add 1 ml of spermidine Solution, vortex at high speed for 10 s, and sonicate in a bath for 5 s.
4. Add 5–20 μl of each of the plasmids containing the DNA of interest stored at −20 °C at 1 $\mu g.\mu l^{-1}$ (*see* **Note 2**).
5. Gently add 25 μl of $CaCl_2$ solution while vortexing at slow speed. Wait for 10 min.
6. Remove the supernatant and dilute in 500 μl of 100% ethanol. Centrifuge for 15 s at 500 × *g*. Repeat this step twice.
7. Gently mix the pellet in 1.5 ml of PVP Solution.
8. Take out the tubing from the tubing prep station. Attach the tubing to a 2 ml syringe and aspirate the preparation while slowly vortexing the microfuge tube to maintain the gold micro-carriers in suspension. Stop aspiration when the preparation has been fully transferred into the tubing.
9. Put back the tubing with the syringe into the tubing prep station and immediately start the tubing rotation. Rotate for 1 min with the syringe still attached to the tube to avoid leakage.
10. Wait for 1 min for the micro-carriers to attach to the tubing, then use a pen to mark the part of the tube containing the preparation.

11. Aspirate the preparation with the syringe very slowly. Detach the syringe and rotate the tubing for 2 min.
12. Switch on the Nitrogen knob set to 0.3 on the tubing prep station scale and rotate for 15 min.
13. Cut the tubing with the Bio-Rad device to get 1 cm-long cartridges.
14. Hermetically store the cartridges with silica gel/desiccant at 4 °C. Wait for at least 24 h before use.

Gold beads in cartridges can be observed under the microscope with a 10× lens (*see* **Note 3**). Cartridges can be used for up to 4 months.

3.2.2 Shooting

Shooting must to be done under a laminar or horizontal flow hood to ensure sterile conditions.

1. Take the vial containing the cartridges out of the refrigerator and wait for a few minutes before opening it to avoid condensation.
2. Place the desired number of cartridges into to gene-gun. Count one cartridge per cell culture insert containing at maximum four organotypic cultures. Put the vial containing cartridges back into the refrigerator. Take the culture box out from the incubator.
3. Place the nylon mesh at 1 cm from the top of the insert containing the cultures to be transfected. Shoot at 100–120 PSI. Do the same with the remaining inserts and place the culture box back into the incubator.

3.3 Confocal Imaging

Time-Lapse repetitive imaging sessions start 3–4 days after transfection, once gene expression is sufficiently high (depending on the promoter used). We routinely use eGFP or mRFP to image dendritic spines. Thus we excite these fluorophores with 488 and 568 nm laser sources, respectively.

1. Check the transfection efficiency: Put 1 ml of culture medium in the observation chamber. Transfer the insert containing the cultures on the observation chamber and gently cover each culture with medium until the insert is filled with 1 ml of culture medium. Place the chamber under the confocal microscope with a 4× objective lens and scan each culture at minimum laser power sufficient to detect transfected neurons. Under sterile hood, transfer the confetti containing slices-of-interest on single culture inserts. Place the box back into the incubator. Wait for at least 30 min for the confetti to attach to the culture insert.
2. Perform repetitive imaging (Fig. 3): Put 1 ml of culture medium in the observation chamber. Transfer the insert containing a selected culture on the observation chamber and cover the cul-

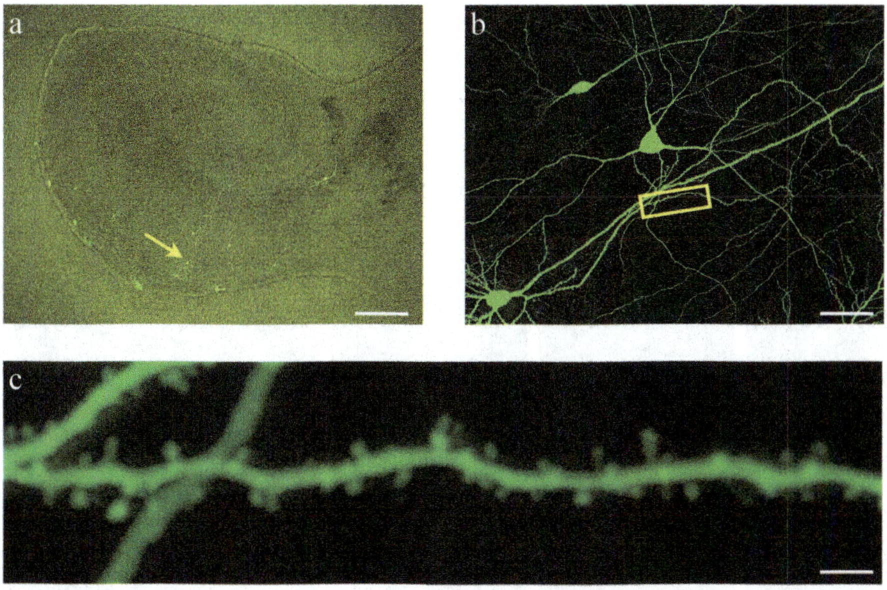

Fig. 3 Imaging of an eGFP transfected neuron. (**a**) 5×-image of an eGFP transfected organotypic slice culture obtained with a confocal microscope and a 488 nm laser. Arrow indicates a transfected neuron in CA1. Scale bar: 400 μm (**b**) Maximum Intensity Projection (MIP) Image of the neuron indicated in (**a**), obtained with the 40× objective lens. Box shows a dendrite of interest. Scale bar: 50 μm. (**c**) MIP of the dendrite-of-interest observed with the 40× objective lens with 10× additional zoom. *Scale bar*: 2.5 μm. Images are presented as raw from the acquisition software

ture with medium until the insert is filled with 1 ml of culture medium. Place the chamber under the confocal microscope with a 4× objective lens and scan the culture. Center on the neuron-of-interest and switch to the 40× objective lens. Always use minimal laser power as possible (*see* **Note 4**). Perform a complete 3D scan of the neuron (3 μm steps) in order to evaluate its health and to select a portion of a dendrite (*see* **Note 5**). Perform a 3D scan of the chosen dendritic segment (0.4–0.6 μm-steps). Put the culture back in the incubator.

3.4 Spine Turnover Analysis

Spine turnover analysis is theoretically possible by automated methods. Nevertheless, several parameters can change from one set of experiments to the other depending on spine density and image quality. It is also very difficult to write a code powerful enough to resolve dubious cases. Lastly, cases that are unexpected would be missed by automated analysis. We therefore prefer to perform non-automated analysis. For this, it is essential to have a software where Region-Of-Interest (ROI) management is well designed. We have chosen to use OsiriX, a free and open source imaging program. In addition, we developed a plugin to export ROI data for further statistical analysis. We have also successfully used the MatLab environment, as well as ImageJ with available plugins that facilitate such analyses. A typical analysis procedure is as follows (Fig. 4 and **Notes 6** and **7**):

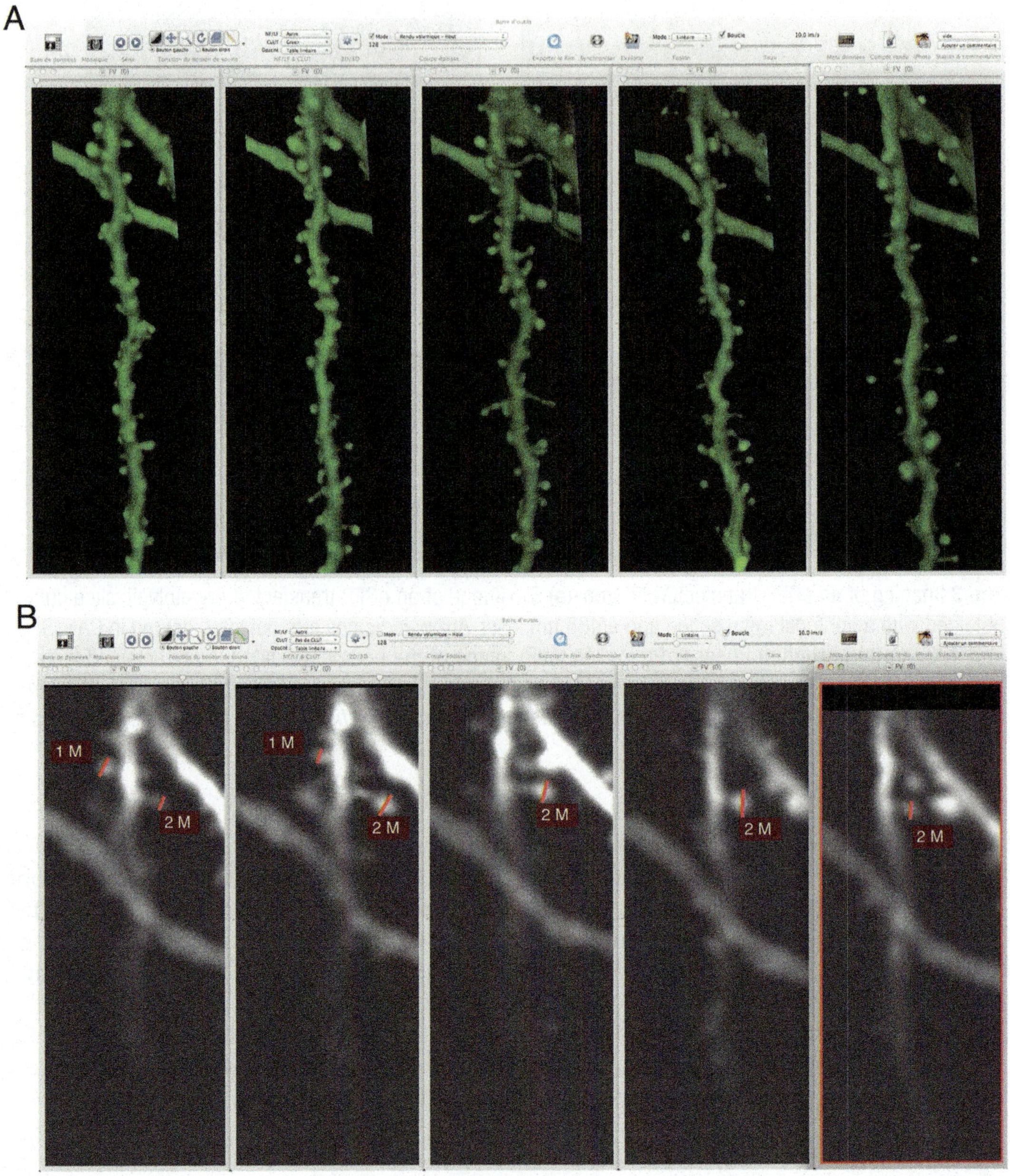

Fig. 4 Analysis of spine dynamics. (**a**) Screen capture of the analysis software showing the dendrite imaged in Fig. 2 at time-points 0, 5, 24, 48, and 72 h. Images are MIP with volume rendering. (**b**) Individual *z*-axis optical sections of the top part of the dendrite shown in (**a**) during analysis of spine dynamics. The software allows fast navigating through the *z*-axis to identify and label individual spines. ROI labels are shown for two spines. Note that the spine labeled 1 M is lost at 24 h, whereas the spine labeled 2 M is conserved until the end of the experiment (72 h-time-point). ROI labels are made bigger for illustration purposes

1. Open each stack of images corresponding to the observation time-points and position them chronologically. Scroll across the stack of images and place the first ROI (n°1) on the spine that is closest to a dendritic branching point. Add another tag to the ROI if needed (e.g., spine type: M for "Mushroom"). Depending on the parameters measured, ROI can be a line or a circle (*see* **Note 8**).
2. Scroll across the next stack of images and place an ROI of the same label number if the spine is still present at the same place. Do the same for each following stack of images.
3. Repeat the operation for the next spines until the end of the dendritic segment is reached.
4. Repeat **steps 1–3** but starting with the second stack of images in order to label spines that are not present at the first session of images ("new spines"). Repeat again these steps for the next stacks of images until all spines have been labeled.
5. OsiriX creates a matrix of ROI names and measurements that can be exported to any spreadsheet for quantifications (*see* refs. [2, 3] for examples).

4 Notes

1. Health of the organotypic culture is an essential determinant of transfection efficiency and high quality imaging. Take particular care during dissection and culturing. Always use new blades for each culture and replace the dissection medium for each brain. After cutting, select slices with no apparent damages, with dentate gyrus and CA1-CA3 regions clearly visible. One can use fine paint brushes (#1) to aid the separation of slices.
2. Up to three different plasmids can potentially be co-transfected in the same cells, although the level of transfection may decrease. In this case, DNA concentration may need to be adapted.
3. Gold micro-carrier coating should appear homogenous. Cartridges with agglomerates of beads may cause culture death and thus should be eliminated. To ensure homogenous distribution and coating of tubing with micro-carriers and efficient DNA binding, make sure to replace PVP and spermidine solutions every 4 months.
4. Always use minimum laser power when you image transfected neurons in order to avoid phototoxicity. We set our laser power at minimum (1 %) which corresponds to power between 0.5 and

5 µW under the 40× objective, depending on laser source. We have never observed photobleaching or phototoxicity using these parameters. Keep imaging session short (less than 15 min). If imaging session requires more time, use a CO_2 and temperature-controlled imaging chamber, or a perfusion of warm oxygenated ACSF. In our hands, perfusion of warm oxygenated ACSF increased risk of premature death and contamination, whereas short imaging sessions at room temperature with culture medium stored in a petri dish in the culture incubator as described allowed daily imaging of the same dendritic branches for up to 2 weeks without any obvious signs of toxicity.

5. Choose carefully the dendrite of interest. It has to be long enough to bear enough spines to allow proper statistics. We typically image 40 µm-long dendritic portions, which present at least 45 spines. At the same time, it has to be short enough to allow short imaging sessions. Include the branching point of the dendrite in the scanned image to use it as a 100% reliable landmark for the next imaging session.
6. Spine turnover analysis is possible only under certain conditions. Choose your time-points depending on the turnover rate of your preparation and the spine density before running whole sets of experiments. Turnover analysis may be impossible with an overly spine-dense dendrite observed over long intervals of time.
7. Spine analysis is time consuming and the experimenter may come across dubious cases. In particular, a spine aligned with the *z*-axis may cause difficulties. For some cases, a 3D reconstruction or a simple volume rendering may help the experimenters depending on their own capability to visualize 3D objects in space. Blind analysis made by two independent researchers is an essential step to estimate the average imprecision of analysis. As a rule-of-thumb we consider differences in scores of new, lost, and stable spines less than 3% for the same set between two experimenters to be acceptable. A training period may be required.
8. In addition to spine density and turnover, experimenters may be interested in getting other parameters such as spine-neck length, spine-head diameter or volume. A deep understanding of the imaging system used for image acquisition is an absolute prerequisite for those analyses. For length, note that the acquisition software often provides more pixels per µm than the actual maximum resolution that is achieved by the imaging set with the given experimental conditions. Thus an analysis program that counts the number of pixels may provide different lengths for two objects of the same size. Second, even if excitation and detection parameters are kept the same

along imaging sessions of a given dendrite, the level of expression of the fluorophores can vary sufficiently across time to create errors in the object's size. To overcome these problems when studying the size variation of a spine on images taken at different time-points, first normalize the size variation with the size of a transfected object that has a constant size. In a mature preparation, the diameter of the dendrite itself can be used. Second, rather than comparing raw measurements, perform a qualitative estimation of size variation based on a cut-off greater than the best detection capability of the imaging system used. For example, classify a spine as a growing spine only if its diameter variation is greater than 0.2 μm. Otherwise consider the spine as non-growing or false negative. Spine volume is difficult to estimate with classical single photon confocal microscopy. Although deconvolution may improve the quality of the estimation if parameters are wisely chosen, some aberrations in the Point-Spread Function generate signal loss that cannot be corrected by post hoc image processing. Therefore, estimations of spine volume variation should be performed with extreme caution.

This work is dedicated to Professor Dominique Muller.

References

1. Stoppini L, Buchs PA, Muller D (1991) A simple method for organotypic cultures of nervous tissue. J Neurosci Methods 37:173–182
2. De Roo M, Klauser P, Mendez P, Poglia L, Muller D (2008) Activity-dependent PSD formation and stabilization of newly formed spines in hippocampal slice cultures. Cereb Cortex 18:151–161
3. De Roo M, Klauser P, Muller D (2008) LTP promotes a selective long-term stabilization and clustering of dendritic spines. PLoS Biol 6:e219
4. De Roo M, Klauser P, Briner A, Nikonenko I, Mendez P, Dayer A, Kiss JZ, Muller D, Vutskits L (2009) Anesthetics Rapidly Promote Synaptogenesis during a Critical Period of Brain Development. PLoS One 4:e7043
5. Yoshida Y, Kobayashi E, Endo H, Hamamoto T, Yamanaka T, Fujimura A, Kagawa Y (1997) Introduction of DNA into rat liver with a hand-held gene gun: distribution of the expressed enzyme, [32P]DNA, and Ca2+ flux. Biochem Biophys Res Commun 234:695–700
6. O'Brien JA, Lummis SC (2006) Diolistic labeling of neuronal cultures and intact tissue using a hand-held gene gun. Nat Protoc 1:1517–1521
7. Woods G, Zito K (2008) Preparation of gene gun bullets and biolistic transfection of neurons in slice culture. J Vis Exp pii:675
8. Arsenault J, O'Brien JA (2013) Optimized heterologous transfection of viable adult organotypic brain slices using an enhanced gene gun. BMC Res Notes 6:544

Part V

Revealing Patterns of Synaptic Connectivity

Chapter 20

Using Fluorescent Markers to Estimate Synaptic Connectivity In Situ

Mrinalini Hoon*, Raunak Sinha, and Haruhisa Okawa*

Abstract

Labeling fixed brain tissue with fluorescent synaptic and cellular markers can help assess circuit connectivity. Despite the diffraction-limited resolution of light microscopy there are several approaches to identify synaptic contacts onto a cell-of-interest. Understanding which image quantification methods can be applied to estimate cellular and synaptic connectivity at the light microscope level is beneficial to answer a range of questions, from mapping appositions between cellular structures or synaptic proteins to assessing synaptic contact density onto a cell-of-interest. This chapter provides the reader with details of the image analysis methods that can be applied to quantify in situ connectivity patterns at the level of cellular contacts and synaptic appositions.

Key words Immunofluorescence, Synaptic puncta, 3D appositions, Dot-finding, Volume estimation, Correlation analysis

1 Introduction

This chapter addresses different image analysis techniques that can be adopted to estimate the distribution of synaptic connectivity patterns. The techniques discussed in this chapter are centered on the analysis of neural circuitry using fluorescent markers in fixed tissue preparations. We discuss methods to estimate the location and number of synaptic contacts in intact tissue preparations depending on the signal-to-noise ratio of the fluorescent markers. These markers can either be dyes used as cell-fills or antibodies against known cellular or synaptic proteins of the circuit. Connectivity can be assessed both at the cellular and at the synaptic level depending on the available markers for the circuit of interest. We do not cover fluorescence assessments of in vitro culture systems or measures of fluorescence intensities to quantify the amount of protein at synaptic sites.

*Correspondence may be addressed to either author.

Alexandros Poulopoulos (ed.), *Synapse Development: Methods and Protocols*, Methods in Molecular Biology, vol. 1538,
DOI 10.1007/978-1-4939-6688-2_20, © Springer Science+Business Media LLC 2017

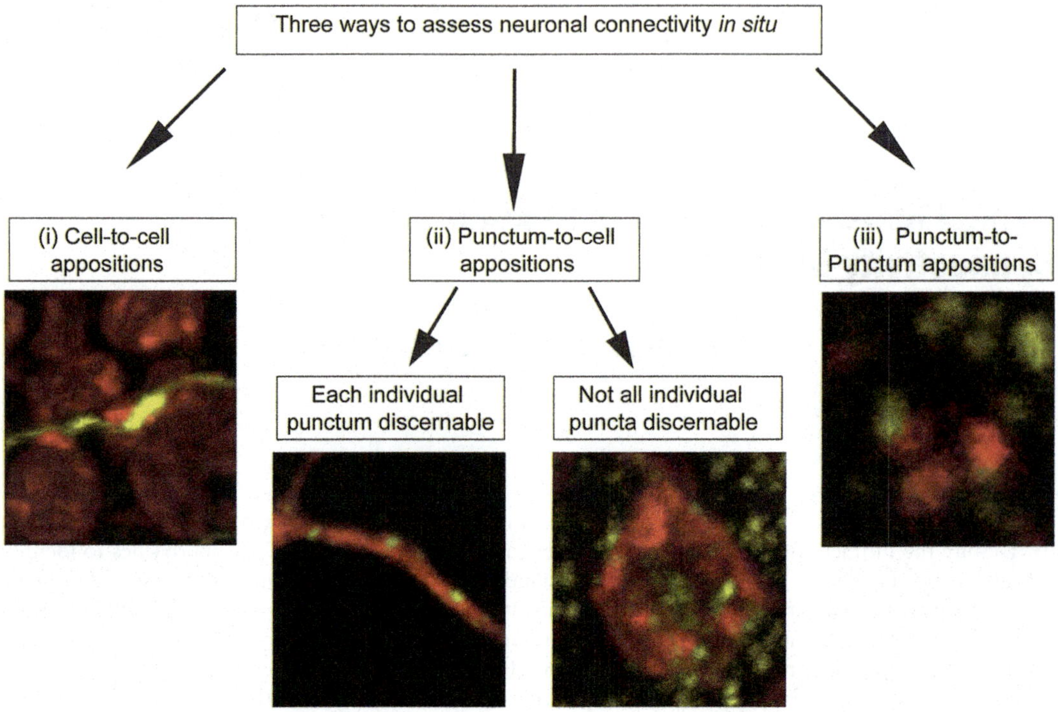

Fig. 1 Three ways to identify neuronal connectivity in situ. Neuronal connectivity can be examined for (*i*) Cell-to-cell appositions, (*ii*) Punctum-to-cell appositions, (*iii*) Punctum-to-punctum appositions. Shown are examples of images that can be quantified in each case. (*i*) Cell–cell appositions: The dendrites of a dye-filled retinal interneuron (*yellow*) contacting the axon terminals of retinal bipolar cells labeled with a cell-specific marker (*red*). To analyze cell-to-cell connectivity, 3D appositions with volume overlap between the two labeled cells can be estimated. (*ii*) Punctum-to-cell appositions: Punctum-to-cell apposition analysis can be used to assay contacts onto a cell-of-interest. For this connectivity assessment two methods can be adopted depending on the signal-to-noise ratio of the labeled signal for the synaptic protein. If individual synaptic puncta are clearly discernable, as illustrated for PSD95 clusters (*yellow*) on the dendrites of a retinal ganglion cell (*red*), then a dot-finding routine can be used. Alternatively, if the labeling of the synaptic protein does not allow the identification of individual puncta, as in the case for $GABA_A$ receptor (*yellow*) labeling on bipolar cell axon terminals (*red*), a volume estimation of the synaptic signal colocalized with the cell-fill signal is recommended instead. (*iii*) Punctum-to-punctum appositions: To establish the relationship between two synaptic proteins, for example the inhibitory presynaptic marker glutamic acid decarboxylase (*red*) and the inhibitory postsynaptic $GABA_A$ receptor (*yellow*) in the inner retinal synaptic layer, a correlation analysis can be utilized

We cover three approaches to assess connectivity in situ as outlined in Fig. 1. The first approach can be applied to estimate connectivity between two cells in cases where cellular markers are available to identify these cells. This level of analysis estimates cell-to-cell contacts in a given circuit [1]. The second approach is applicable in cases where a synaptic marker (presynaptic or postsynaptic) needs to be quantified together with the corresponding (post- or presynaptic) cell-fill label (punctum-to-cell appositions). This can be achieved using one of two methods depending on the signal-to-noise ratio of the synaptic label. If synaptic labeling is clearly punctate and easily discernable from background noise, a "dot-finding" routine [2–6] can be used to map the number and distribution of sites-of-apposition between the synaptic marker and the cell-fill label. If, however, synaptic labeling is not discernably punctate, a number estimation would prove erroneous. In such cases "volume-estimation" [7–9] can be applied to capture the volume of the synaptic pixels colocalized with the cell-fill label. The third image analysis approach we discuss in this chapter is applicable to cases where the relationship between two synaptic puncta (punctum-to-punctum appositions) needs to be established [5, 10]. Most often this method can be used to determine the correlation between presynaptic and/or postsynaptic markers. For each of the image analysis methods outlined here we will discuss the advantages and the limitations intrinsic to each approach such that the reader in search of an image analysis routine can weigh the respective pros and cons of each.

2 Materials

The materials needed to estimate synaptic connectivity using immunohistochemical methods fall into three categories: (1) the biological sample labeled for cellular or synaptic proteins, (2) a microscope to image the fluorescence, and (3) image analysis software.

2.1 Samples

Samples should be prepared such that the cell or synapse of interest is brightly labeled. Depending on the circuit being addressed, different cell type-specific markers or synapse-specific antibodies can be used. In the retinal tissue used as a model system in this chapter, cells of interest can be labeled with biolistic techniques [11] or electroporation [12]. For the study of most synaptic circuits in situ, a list of techniques and immuno-markers can be readily gathered from a survey of the available literature. The first step would be to determine which markers or techniques for cell or synapse labeling are most suitable and work reliably for the circuit of interest. The fluorescence signal of the cell or synapse of interest should be relatively photostable allowing repetitive imaging. For antibody labeling, most Alexa dye-conjugated secondary antibodies (e.g., Molecular Probes or Jackson Immunoresearch) are quite photostable. A note of caution should however be added for the use of far-red Alexa dyes, as these tend to have higher photobleaching rates.

2.2 Imaging Microscope

Acquisition of images can be carried out in any multichannel confocal microscope (Olympus FV1000 used here) allowing the simultaneous acquisition of presynaptic and postsynaptic cells or markers of the circuit. It is imperative to assess the alignment of all lasers of the confocal module, and to correct for any potential chromatic aberration of the microscope objectives as any shift of the channels during acquisition will drastically impair one's ability to accurately determine appositions and connectivity. This can be addressed by imaging multispectral beads [13] and checking the alignment of all the channels prior to imaging the sample of interest. During image acquisition it is important to acquire a z-stack of all the planes containing the region of interest. In general, images should be acquired satisfying the Nyquist criterion of sampling with at least twice the spatial resolution in both xy (lateral) and z (axial) dimensions. The xy and z sampling distance following the Nyquist criteria are a function of the numerical aperture of the objective and the emission wavelength of the fluorophore being imaged [14]. For imaging of a fluorophore with an emission wavelength around 500 nm using a 60× oil immersion objective with a numerical aperture of ~1.3–1.4, the xy pixel size following the Nyquist criteria should be at most ~0.1 μm/pixel and a z-step no larger than ~0.3 μm/pixel [14]. For the quantification of single synaptic puncta however, a degree of oversampling is recommended to achieve a smoother profile for each individual punctum to enable robust detection. The single-synapse images depicted in the figures of this chapter have been acquired with a pixel size of 0.1 or 0.051 μm for xy and 0.2–0.3 μm for z.

2.3 Image Analysis Software

The software needed for post-acquisition analysis depends on the question being addressed. The image analysis methods outlined in this chapter rely on the following software:

Cell-to-cell apposition analysis

1. MetaMorph (Molecular Devices).
2. Fiji (or ImageJ, freely downloadable online) [15].
3. Amira (FEI Visualization Sciences Group).

Punctum-to-cell apposition analysis

1. MetaMorph (Molecular Devices).
2. Amira (FEI Visualization Sciences Group).
3. Fiji (or ImageJ, freely downloadable online) [15].
4. Imaris (Bitplane Oxford instrumentations company).
5. Matlab (Mathworks) with the image processing tool box.

Punctum-to-punctum apposition analysis

1. MetaMorph (Molecular Devices).
2. Fiji (or ImageJ, freely downloadable online) [15].
3. Amira (FEI Visualization Sciences Group).
4. Matlab (Mathworks) with the image processing tool box.

3 Methods

3.1 Basic Image Processing

Before outlining the different analysis methods (Fig. 1) that can be used to determine cellular and synaptic connectivity through estimating the distribution and density of contacts, it is worth describing in brief some general image analysis steps, which are applicable to the raw files generated by confocal microscopes. A convenient method to generate stacks is by using the MetaMorph software. Using this software each individual image of the 3D stack can be "median filtered". The option to median filter an individual image-plane or a complete stack can be found under the *Process* icon on the MetaMorph toolbar. Median filtering of images is a way to remove the noise generated by detectors, i.e., photo-multiplier tubes, while preserving the borders of the cellular structure or synaptic puncta, which is crucial to the analysis methods outlined in this section. Image stacks generated by MetaMorph and saved as .stk files can be opened using the Amira software for the 3D visualization of the cellular or synaptic circuit. Fiji can also open image stacks, median filter images and generate a Z or 3D projection of all planes. The options to generate the Z or 3D projection of the image planes are listed under the Stacks option found by selecting the *Image* icon on the Fiji toolbar.

3.2 Cell-to-Cell Apposition Analysis

The assessment of cell-to-cell contacts can be used in cases when both presynaptic and postsynaptic cells can be labeled with markers and the number of contacts between cells needs to be estimated. A good example for the utilization of this analysis method in the retinal circuit comes from the estimation of the number or density of presynaptic amacrine cell contacts onto a population of postsynaptic bipolar cell terminals [1], both of which can be easily identified with cellular markers. This analysis maps in 3D the sites of apposition (volume overlap) between the two cells-of-interest and precludes the need for a synaptic marker at each apposition site.

3.2.1 The Analysis Routine

1. Generate image stacks of the cell-fill signal in different channels. It is advisable to use MetaMorph to open individual tiff files and generate .stk files as described earlier for each of the cell-fill labels. While generating the .stk files remember to median filter the stacks before saving.
2. Open and display the .stk files in Amira. To estimate the contact sites of cells labeled with marker 1 (channel 1, Fig. 2a) on processes labeled with marker 2 (channel 2, Fig. 2a top panel), create a mask around the signal in channel 1 that is apposed to the channel 2 processes. Amira has a *Labelfield* function, which can be selected under the *Labelling* category for any channel and allows signal segregation into distinct "labels". The *Labelfield* function on a specific channel allows the user to

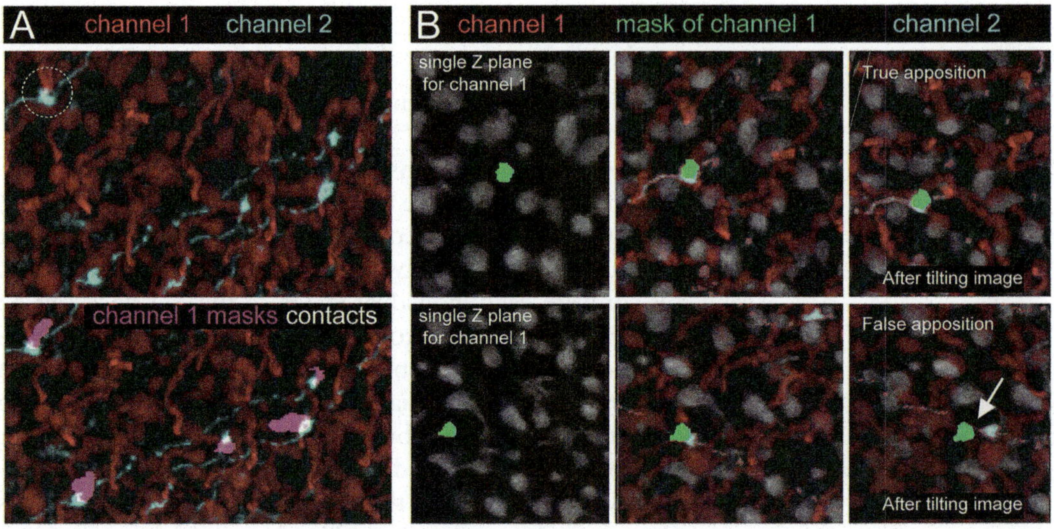

Fig. 2 Identifying putative synaptic contact sites between two cellular markers. (**a**) Contact sites (*white*) of cellular marker in channel 1 (*red*, masked profiles of channel 1 in *magenta*) that need to be mapped onto processes labeled in channel 2 (*cyan*) by a separate marker. (**b**) To map the contacts sites (shown for the circled contact in the top panel of A), the signal in a single plane of channel 1 (*gray*) can be masked (or isolated in 3D, *green*) and the mask tilted in 3D to assess a true overlap with the channel 2 terminal (*cyan*). Tilting of the image helps distinguish overlap of the observed signals from random overlap of the channels. *Arrow* points to a false apposition as revealed by tilting the image

manually select the fluorescence signal of interest with which to create a "mask".

3. When a particular region of channel 1 is highlighted (as in Fig. 2b), tilt the image in 3D to assess whether or not there is a site of apposition between the two markers. For each site of apposition between the two channels create a contact mask as a separate *material*, which can then be edited and colorized as needed. The 3D rotation is crucial to estimate a true apposition.
4. The mask for channel 1 that is created for the different apposition sites on channel 2 can then be colorized as desired (shown in magenta in Fig. 2a bottom panel). This enables an estimation of the total number of apposition sites between the two marker channels and one can further assess the total number of contact sites for a given stretch of neuronal process.

3.2.2 Controls Needed for this Analysis

While carrying out this analysis it is imperative to assess each apposition by tilting the image in 3D. A preliminary 3D visualization of the two channels to assess contacts can often be misleading as a false contact might display itself as the signal from the two channels overlapping in one but not in all planes—there could be several points that look like potential appositions, but when rotated the overlap between the two channels disappears (Fig. 2b bottom panel).

3.2.3 Strengths and Limitations of this Analysis

+ This analysis allows easy estimation of potential contact density and distribution between two cellular profiles, provided the two cells can be differentially labeled.
- It only provides potential contact number and distribution of the contact profiles between two labeled cellular partners. This implies that other groundwork needs to be done to ascertain whether the apposition between the two cellular partners has both presynaptic and postsynaptic substrates.
- This analysis routine will not reveal which of the cellular partners are presynaptic or postsynaptic. These details would need to be assayed by other experiments using synaptic markers.

3.3 Punctum-to-Cell Apposition Analysis

Cases where synaptic contacts onto a cell-of-interest need to be determined, the labeling for the synaptic protein may be punctate or be contaminated with background fluorescence, which hampers the delineation of each and every synaptic punctum. For either of these cases a separate method of analysis can be utilized. The first step before starting quantification is to determine the quality of the immunoreactive signal and assess whether the signal is punctate or not, as that will determine which analysis method is most appropriate (*see* **Note 1**). For a clear punctate synaptic signal a "dot-finding" analysis can be used to quantify the number of synaptic puncta onto a particular cell. However, if the labeling for the synaptic protein is not clearly punctate and puncta are difficult to resolve from background, then an estimation of the volume of the pixels representing synaptic contacts onto the cell-of-interest is more appropriate.

3.3.1 Clear Punctate Signal: Dot-Finding Analysis to Count Individual Synaptic Puncta

If labeling for either presynaptic or postsynaptic protein is punctate with each individual punctum easily discernable, one can proceed to quantify the puncta number. Automatic detection of puncta can be done using commercial software or custom-written programs. Different programs use varied algorithms to identify fluorescent puncta. In this section, we describe the principles of a custom-written Matlab-based routine for punctum identification [4]. We refer to this routine as the "dot-finding" analysis. The goal of the routine is to be able to detect isolated puncta independent of their intensity distribution, separate overlapping puncta and eliminate background signal that appear as isolated dots, but are much smaller than individual puncta (Fig. 3).

As an example (Fig. 4), we show how to identify fluorescent protein-tagged postsynaptic density protein 95 (PSD95-FP) expressed in retinal neurons after biolistic labeling [11], which allows sparse labeling of cells such that the arbors of individual cells are highlighted. To label the dendritic arbor of the cell-of-interest, add a second cytosolic (cell-fill) fluorescent label. The advantage of expressing PSD-95-FP compared to immunolabeling for endogenous PSD95 is that there is no need to determine which PSD95-FP puncta belong to the cell-of-interest, as only the co-labeled cell-of-interest displays fluorescent puncta.

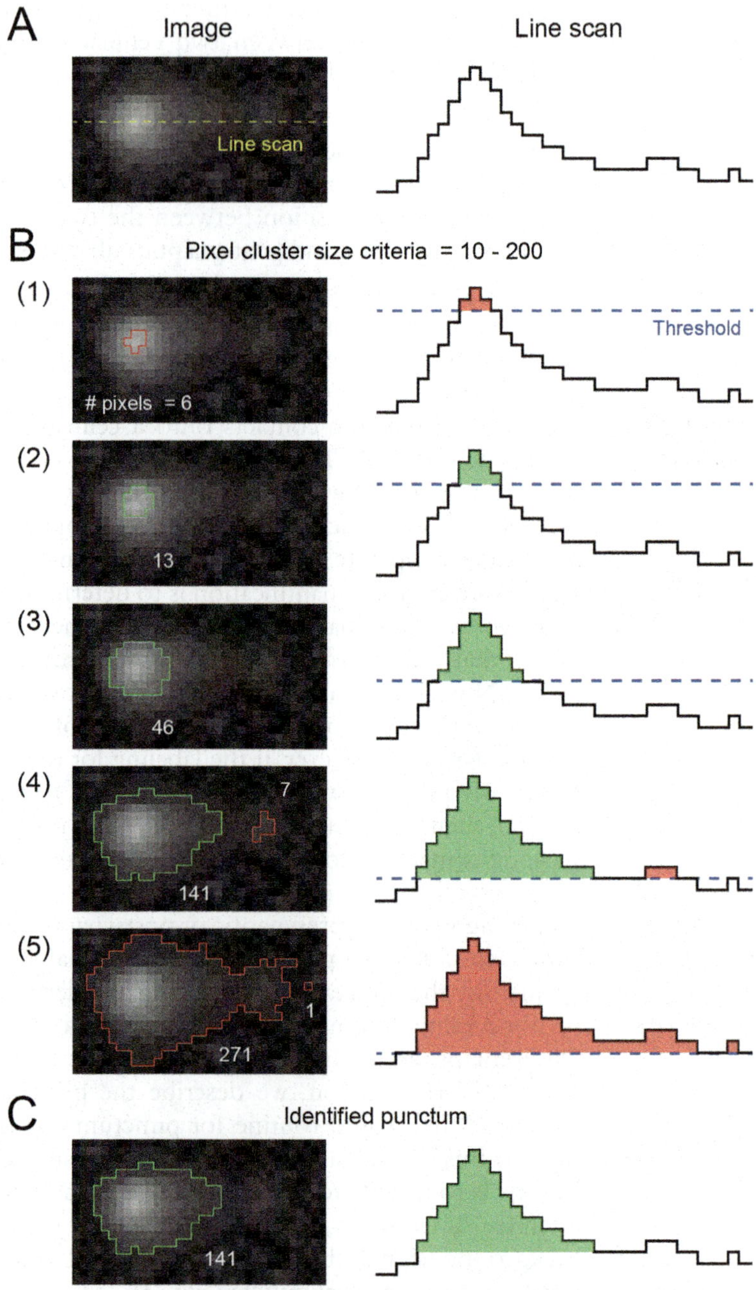

Fig. 3 Basic principle of the dot-finding program. (**a**) Left column shows the image of a hypothetical punctum and the right column shows the line scan of pixel intensity along the dotted line in the image. (**b**) The program iterates thresholding from the maximum pixel intensity (255 for 8-bit) down to zero. If the number of adjoining pixels above the threshold falls into a pixel cluster size criterion that defines the adequate size of a punctum (*see* text), the pixel cluster is isolated as a punctum. For simplicity, we set the pixel cluster size criterion between 10 and 200 (upper and lower limit can be determined using the largest and smallest puncta in the image respectively) in the 2D image although this routine is normally run on a 3D image stack. (*1*) The threshold first reaches the intensity of the brightest pixels in a punctum. Until the total number of adjoining pixels reaches the lower limit of the pixel cluster size criterion (10 in this example), the pixel cluster is not considered as a punctum. (*2*) The threshold moves down. Once the number of adjoining pixels above the threshold exceeds

1. The dot-finding routine requires the establishment of two main criteria, i.e., the lower and upper limit of the number of adjoining pixels that can be considered as a punctum. The lower threshold limit works to eliminate background noise during detection whereas the upper limit works to separate individual puncta from background fluorescence. Any punctate structure that does not fall within these size criteria will not be considered a single punctum. We will refer to this kind of thresholding as the "pixel cluster size" thresholding (for a comparison of the dot-finding program with simple intensity thresholding techniques *see* **Note 2** and Fig. 9).
2. The program repeats the pixel cluster size thresholding for all puncta in an image, starting from the highest intensity pixel value (255 for a 8-bit image) marching down to the lowest value of zero. To hasten this process the user can always identify the lowest and highest pixel values of the image being analyzed and set those relevant limits for the program to execute the pixel cluster size thresholding. An easy way to estimate the lower and upper bound of pixel intensities in the image of interest is by plotting all the gray values of the image.
3. Defining the minimum and maximum puncta dimensions in number of pixels can be effectively done by looking at the maximum intensity projection of the PSD95 channel. An initial estimate of these parameters is achieved from a line scan along a dendritic stretch, which maps the intensity profile of the puncta along that stretch. The line scan provides an idea of the intensity distribution of puncta as a function of their spatial extent, i.e., size in pixels (Fig. 4b). Isolated puncta have a clear peak in their intensity profile and give a fairly reasonable measure of the puncta dimensions, which can then be used as starting values for the iterative fitting procedure, which is done in 3D. This step eliminates background noise below the minimum puncta dimension and gives a plot where each punctum is represented by number of iterations instead of intensity. After detecting puncta that fit within the range of defined dimensions, the dot-finding program returns a set of properties/parameters associated with each detected punctum such as the maximum number of iterations (Max IT#), cumulative sum of the number of iterations (Sum IT#), volume, circularity, together with a few others (examples in Fig. 4b).

Fig. 3 (continued) the lower limit of the pixel cluster size criterion, the pixel cluster is marked as a punctum. (*3–4*) When the threshold moves further down, more pixels are included as a part of the punctum. (*5*) As the threshold approaches the background level, the number of adjoining pixels above threshold increases by a large amount. When the number surpasses the upper limit of the pixel cluster size criterion (200 in this example), the pixel cluster is no longer registered as a punctum. Any signal with a pixel cluster size less than the lower limit (in this case 10) is also not registered as a punctum. (**c**) The final result showing the identified punctum (*green*) with 141 pixels. Note that the signals with 1 or 7 pixel cluster sizes are not identified (*red*) by the program as puncta

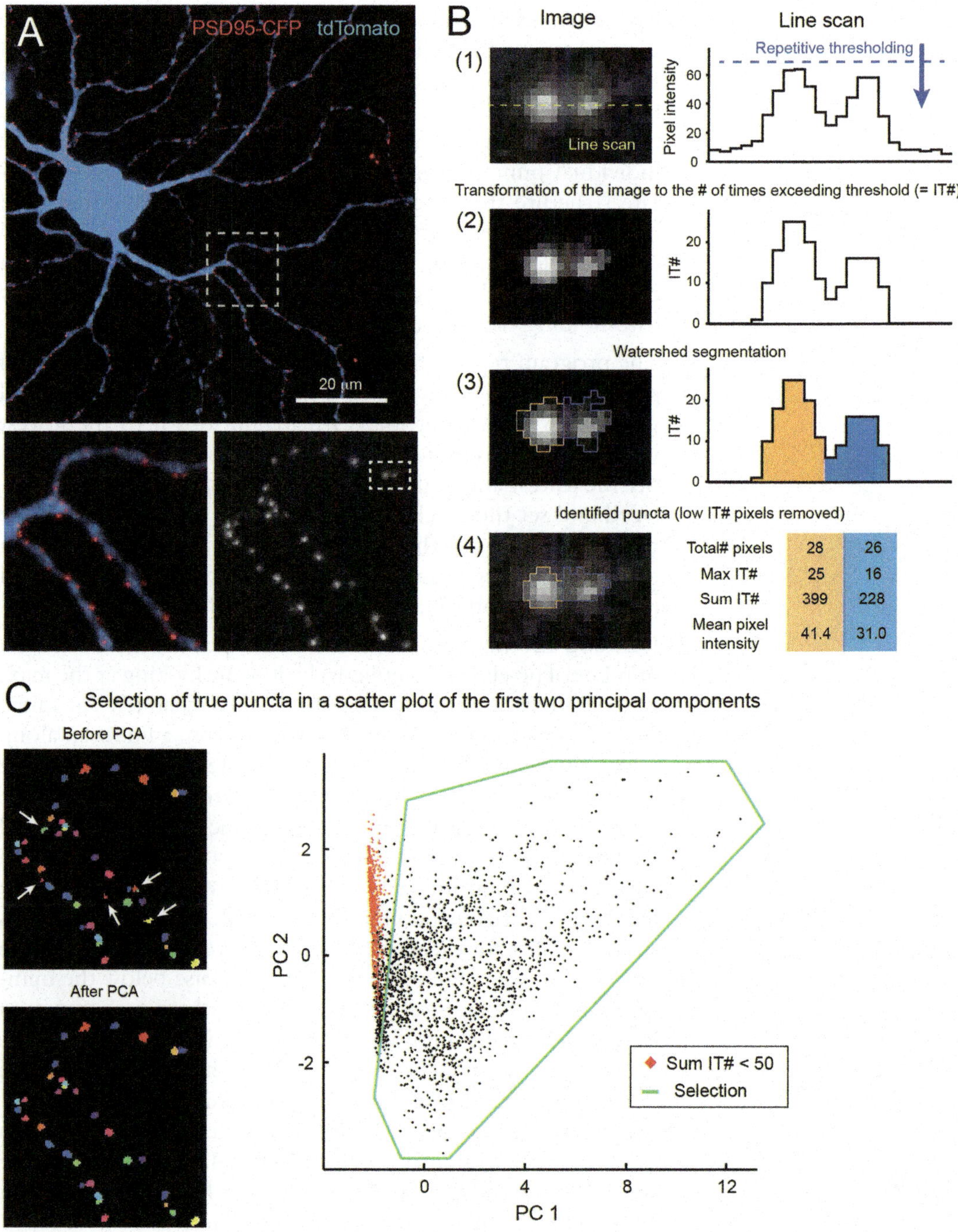

Fig. 4 Example of the dot-finding routine. (**a**) An example retinal ganglion cell biolistically labeled with PSD95-CFP and cytosolic tdTomato. The enlarged images are shown in the bottom panel and are used for demonstrating the process of the dot-finding program in (**b**). (**b**) The first stage of dot-finding program: (*1*) Image of two adjacent PSD95-CFP puncta (marked by the dotted box in **a**, *lower panel*) and the line scan of their pixel intensity distribution. As explained in Fig. 3, the first part of the dot-finding routine repeats thresholding of the same image, starting from the highest intensity value marching down to the lowest intensity. If a cluster of pixels above a certain intensity meets the user-defined pixel cluster size criteria, those pixels are counted as a punctum. For this example, the intensity interval was set to 2 (i.e., thresholding iterates in descending steps,

4. Let us consider the example of a single punctum with a range of pixel intensities. Now if we set a lower limit of 2 pixels and an upper limit of 6 pixels for the pixel cluster size, the program will not consider a punctum as "detected" until at least 2 pixels can be detected and allocated to it. At the upper end the program will stop allocating any more pixels to this punctum once 6 pixels are accumulated. When the program iterations begin, the pixel on the punctum with the brightest intensity will be the first to be included. This captured pixel is still not considered as a possible punctum because 1 pixel does not surpass the criterion of the lower limit. When the next level of intensities are sampled, pixels with the second highest intensity adjacent to the first brightest pixel will pass threshold and a punctum will be designated. As the threshold marches down to lower intensities, more pixels are detected for the punctum and the detected size (pixel cluster) for the punctum correspondingly increases until the pixels allocated to this punctum reach the maximum of 6. Fluctuations within the background fluorescence of an image may contain puncta-like structures, potentially causing false-positive puncta assignment. Many of these structures can be eliminated by the criterion of the lower limit in the pixel cluster size thresholding.
5. Setting the upper limit of puncta dimensions does not preclude detection of overlapping puncta and such clusters often get counted together. This is better illustrated in the line scan profile where two neighboring puncta have overlapping intensity distributions but distinct peaks. To detect such instances of nearby dots of similar or dissimilar intensities, the dot-finding program includes a region-based image segmentation tool called watershed transformation (*see* **Note 3** and Fig. 4b).

Fig. 4 (continued) i.e., from an intensity of 60 to 58 to 56 and so on). (*2*) Each pixel within an identified punctum is detected a certain number of times during the repetitive threshold sampling. For example, the brightest pixel in the left punctum was detected 25 times at different intensity thresholds. The image of pixel intensity is now transformed into an image giving this detection number (here we call it IT#). (*3*) Next, a watershed algorithm segments all the puncta, allowing adjoining pixels consisting of multiple puncta to be split into individual puncta. (*4*) Final identified punctas with all their detection properties. For each identified punctum, pixels with the lowest 10–20 % IT# can be removed because these low IT# pixels tend to expand the representation of the punctum at its transition to the background noise. Each punctum has several parameters associated. Here, four of those parameters are listed. Although the dot-finding program is preferred on 3D image stacks, the parameters calculated here are based on the 2D image shown in the left column. (**c**) The second stage of the dot-finding program: Principal component analysis (PCA) is performed to reduce the dimensionality of puncta parameters shown in (*B4*). A scatter plot is created using the first two principal components. Each data point in the scatter plot represents a single identified punctum obtained from the image. *Red dots* show puncta with their sum of IT# smaller than 50, indicating the cluster of false-positive puncta. *Green polygon* shows the manual selection of true puncta. *Arrows* in the image indicate false-positive puncta that are successfully removed by application of this manual selection process

6. The dot-finding program so far can eliminate diffuse background noise and distinguish individual PSD95 puncta in a cluster. But there is still one caveat that remains in such images which is the presence of a small fraction of background signal in the shape of dots often much smaller than individual puncta themselves. The iterative size algorithm detects these isolated spots, since they are slightly bigger than the typical minimum punctum dimensions. To prevent these spots from being counted, parameters associated with each punctum, such as the brightness or volume, can be used for defining the boundary between true puncta and false-positive puncta. One way to define the boundary is to select some of these parameters and set a threshold for each. The other way is to include all the parameters and perform principal component analysis (PCA) (*see* **Note 4** and Fig. 4c).

Based on the combination of these key steps the dot-finding program provides a platform for a reliable semi-automated detection of synaptic contacts onto a neuron. Once the dot-finding program has registered all the dots in 3D space, one can make statistical estimates such as number of puncta, their relative distance to each other, and volume of each punctum. In case of multiple synaptic protein labeling, the dot information retrieved for each label by the dot-finding program allows the comparison of such statistics between two synaptic protein labels, for example comparing sites of excitatory and inhibitory receptors on a neuronal arbor.

Additional Considerations for this Analysis

1. After the dot-finding routine has detected the puncta for a particular image, a further refinement step can be carried out to remove falsely identified puncta. Each detected punctum has its associated parameters such as brightness, volume and contrast. These parameters can be used to define a good border between a true punctum and a falsely identified one. This process can be enhanced by uploading all the detected puncta and their associated parameters in Imaris and displaying them onto the cell-of-interest in 3D.
2. Cases where synaptic labeling is not restricted to the cell-of-interest (such as when immunolabeling of endogenous synaptic markers is used), an additional step is required either before or after running the dot-finding routine to filter for synaptic puncta belonging to the cell-of-interest while omitting other puncta in the vicinity.
   (a) One way to do this before starting the dot-finding routine is by using Amira to *Labelfield* the processes of the cell-of-interest via the cell-fill signal (*see* Subheadings 3.2 and 3.3.2.) and thereafter isolate the puncta within the cell-of-interest using the *Arithmetic* function in Amira (*see* Subheading 3.3.2.).

(b) The Imaris software can alternatively be used to skeletonize and separate the cell-of-interest. The information of skeletons (called *Filament* in Imaris) can be exported to Matlab, which allows users to create a digital mask around the skeletonized processes.

(c) One can also isolate puncta specific to a particular cell after running the dot-finding routine by assessing how much of the cell-fill signal is associated with each detected punctum. After all the detected dots are overlaid on the cell-fill signal that labels the soma and dendrites of the cell-of-interest, set a threshold for the brightness of the cell-fill signal overlapping with each punctum. This threshold should be able to delineate the puncta only associated strongly with the cell-fill signal. Check that the selected intensity threshold is reasonable for the particular cell-of-interest, and apply it to each identified synaptic punctum to filter puncta specific to the cell-of-interest.

Controls Needed for this Analysis

1. When running the dot-finding routine for detecting puncta expressed exclusively by the cell-of-interest using biolistic expression techniques, several control experiments need to be performed to verify that each expressed punctum is correctly localized at the synapse. For such control experiments labeling with synaptic partners (for example using a presynaptic excitatory marker to verify that the PSD95-FP signal represents bona fide synaptic puncta) and assessing the apposition between the expressed synaptic-fusion protein and endogenous immunolabeled marker is imperative [4]. The synaptic location of a punctum at the level of the light microscope further needs to be verified at the level of the electron microscope. For biolistic expression of synaptic proteins by retinal neurons a correlative light and electron microscopic analysis has recently revealed that more than 70 % of fluorescently expressed puncta display correct synaptic localization in electron micrographs [5]. Therefore for retinal neurons expressing these fluorescently tagged proteins one can assume that the majority of puncta are synaptic. However, in other cases this assumption needs to be assessed by correlative electron-microscope analysis.
2. When running the dot-finding routine on immunolabeling for synaptic proteins in a particular region not limited to a cell-of-interest, there can be some percentage of puncta that are falsely assigned to the cell owing to the resolution limit of light microscopy. One way to obtain the number of false positive synaptic puncta identified within the cell is to flip one channel relative to the other, and examine how many puncta still get identified within the cell using the same thresholds. That number of puncta would then serve as the number of puncta erroneously

ascribed to the cell and can be subtracted from the total number of detected puncta to give the actual number of synaptic puncta associated with the cell-of-interest. A note of caution should be added for the analysis of random interactions by channel rotation for cases in which the cell-of-interest occupies a considerable volume together with immunolabeling for a synaptic protein, which is robustly expressed. If the dendritic processes of the cell-of-interest are wide and thick then rotating either the cell-fill channel or the immunolabeled synaptic protein channel will cause a high percentage of dots to be associated. This does not mean that a high percentage of synaptic puncta is erroneously being ascribed to the cell-of-interest, but rather reflects a constraint imposed by the neural circuit being assessed by the light microscope. A better way to estimate the percentage of random associations for such a case would be by immunolabeling for a synaptic protein known to not be associated with the cell-of-interest and then determining the number of associated dots. Preferably, such a "negative control" synaptic marker should be as densely expressed as the synaptic marker of interest. Otherwise, the percentage of random association has to be corrected for by the difference in their density.

Strengths and Limitations of this Analysis

+ The dot-finding routine not only quantifies the number of puncta across a cell-of-interest, it can also provide information about the volume occupied by each identified punctum (number of voxels in a punctum multiplied by the voxel volume). The volume information can be useful, for example, when two populations of puncta (contacts made by two different presynaptic cell types, for example control vs. experimentally manipulated) have synaptic differences in lengths or density. These differences however have to be large enough to be resolved at the level of the light microscope.

– Use of the dot-finding program is limited to situations when the synaptic labeling is clear and punctate. As labeling becomes less punctate, the results of the dot-finding routine become more erroneous. An example of such a case is shown in Fig. 5. In such cases, it is recommended to instead estimate the volume occupied by the synaptic protein within the cell-of-interest as described in the next section.

3.3.2 Individual Puncta Cannot be Discerned: Volume Estimation of Synaptic Signal

For quantifying labeling of synaptic proteins in cases where individual puncta cannot always be discerned it is advisable to carry out a volume estimation of the immunofluorescent signal rather than using the dot-finding routine. For example if one wants to assess the amount of $GABA_A$ receptors (Fig. 6a, in yellow) located on retinal bipolar cell terminals (Fig. 6a, red) one cannot run a dot-finding routine as $GABA_A$ clusters often appear aggregated (Fig. 6a,

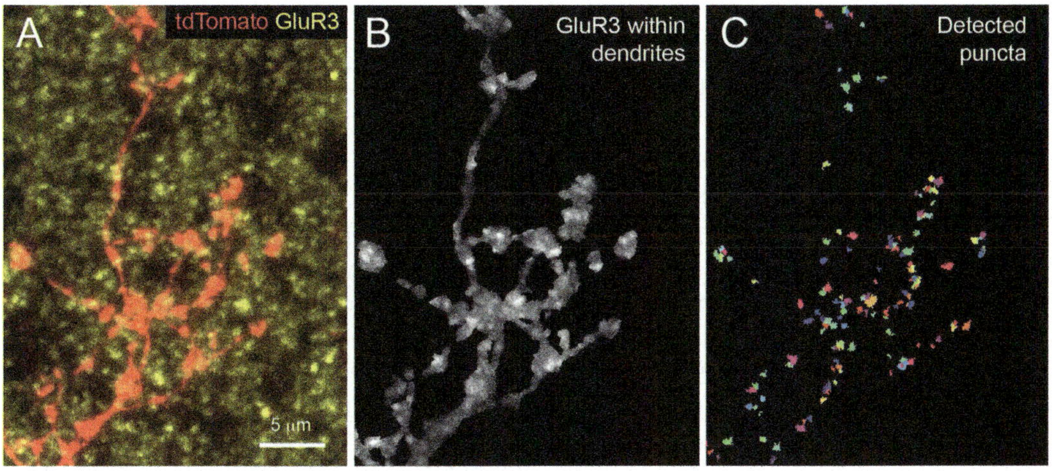

Fig. 5 Limitations of the dot-finding program. (**a**) The dendrite of a retinal ganglion cell biolistically labeled by tdTomato and immunostained for a glutamate receptor subunit, GluR3. The image shows the maximum z-projection of a confocal image stack. (**b**) GluR3 signal is isolated within the dendrite 3D mask. The distinction between synaptic signals and background is relatively ambiguous. (**c**) Dot-finding program is performed on the entire 3D volume and detected puncta within the dendrite are shown. Individual detected puncta are pseudocolored

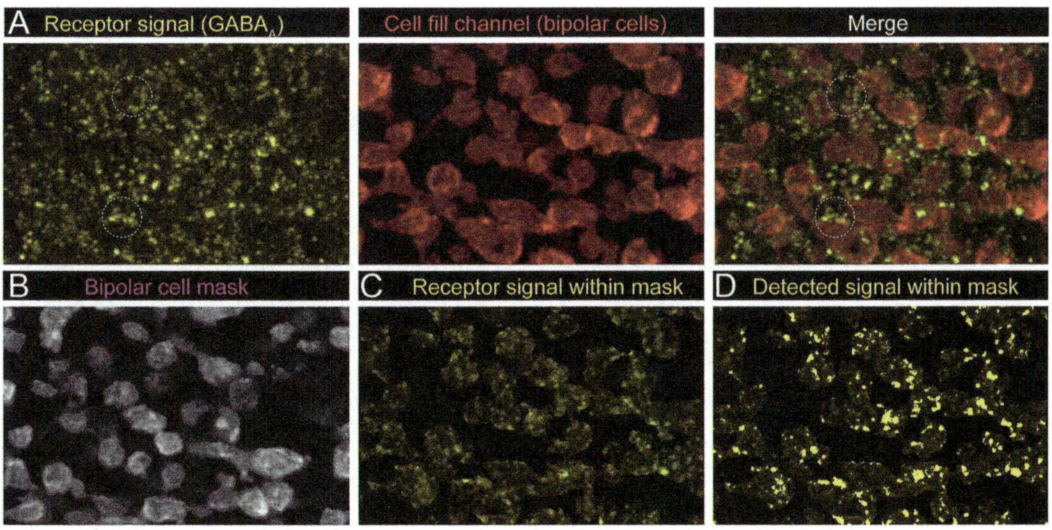

Fig. 6 Representing the volume occupied by synaptic proteins on a cell-of-interest. (**a**) Labeling for GABA$_A$ receptors (*yellow*) to estimate the density of these receptor clusters on retinal bipolar cell terminals (*red*). White circles depict the difficulty in resolving separate GABA$_A$ clusters. This makes it impossible to count the number of GABA$_A$ clusters on the bipolar cell terminals. Thus an estimation of the total pixel volume occupied by GABA$_A$ clusters on bipolar cell terminals is recommended instead. (**b**) To isolate the GABA$_A$ receptor signal within bipolar cell terminals, a 3D mask is generated first by encircling (*purple*) the labeled bipolar cell terminals. (**c**) After the bipolar cell mask has been generated it can be multiplied with the GABA$_A$ receptor channel to isolate the receptor signal specifically within the mask (*yellow*). (**d**) Thereafter, a threshold for the GABA$_A$ receptor signal can be applied to eliminate any background pixels within the bipolar cell terminal mask and identify "receptor pixels" within the 3D mask (digitized in *yellow*)

white circle) making it difficult to resolve individual puncta. In such cases one can isolate the signal of the labeled protein ($GABA_A$ receptors in Fig. 6a) within the cell-of-interest (bipolar cell terminals in Fig. 6a) and quantify the total pixel volume of the synaptic signal within the cell-of-interest by assessing the volume of the immunofluorescent pixels within the cellular structure of interest.

1. The first step for the volume estimation of any immunolabeled synaptic protein is the isolation of the immunolabeled signal within the cellular structure of interest. For the example depicted in Fig. 6a it would be the isolation of the $GABA_A$ signal within the terminals of the bipolar cells. In order to isolate the $GABA_A$ signal within the bipolar cell terminals, create a 3D mask encompassing the terminals of the bipolar cells. The software Amira allows this kind of masking of immunolabeled structures using the *Labelfield* function on the cell-fill channel (also *see* Subheading 3.2.). This function allows the selection of the cell-fill signal to be incorporated within the mask and all encircled structures (shown in purple, Fig. 6b) represent the signal that is added to the 3D mask.
2. Once this mask has been generated, carry out an *Arithmetic* function in Amira (listed under the *Compute* category) to multiply the channel containing the receptor signal (input A) with the 3D mask (input B). The result of this multiplication (A × B) yields the receptor signal exclusively within the cell-fill mask (Fig. 6c).
3. Determine an intensity threshold for the receptor channel that is best for the isolation of the immunofluorescent signal above background fluorescence (*see* **Note 5** and Fig. 10).
4. Create a mask (*label* in Amira) for each voxel of the receptor channel above the determined threshold. This voxel label for the receptor channel can be applied through the *LabelVoxel* option listed under the *Labelling* category for the receptor channel. The *LabelVoxel* option allows a threshold value to be set as input and when applied generates a mask of each receptor pixel above the threshold.
5. Multiply the receptor labels (set as input A) with the 3D cell-fill mask (set as input B) using the *Arithmetic* function described above to reveal the detected receptor pixels (Fig. 6d) within the cell-fill mask.
6. It is advisable to express the volume of the receptor pixels as a % occupancy compared to the volume of the cell-fill channel, which can be estimated by the volume of the 3D mask. The volume of the 3D mask can be obtained from Amira using the *TissueStatistics* function listed under the *Measure* category for each label.

7. The pixel volume (both the volume of the cell-fill masks and of the detected receptor pixels) can alternatively be estimated using Matlab. Export the label files in a binarized tiff format from Amira and then using Matlab open the tiff files and convert to .mat format. Once in .mat format, Matlab can then import the binarized matrix (of zeros and ones representing noise and signal) and get a count of the voxels representing "signal" (estimation of the number of ones in the matrix) and thus the total number of voxels in the 3D mask. After importing the voxel size of the image into Matlab the total volume of the signal can be estimated and thereafter the % occupancy of the receptor signal within the cell-fill mask.

Controls Needed for This Analysis

1. The controls needed for this analysis are similar to those discussed for the dot-finding routine in Subheading 3.3.1. If the cell-of-interest has thin dendrites that do not occupy a large volume in 3D then flip the cell-fill channel relative to the immunolabeled synaptic protein channel to assess random volume occupancy.
2. Alternatively, if the cell-of-interest has wide and thick dendrites such that flipping the cell-fill channel relative to the immunolabeled synaptic protein always yields a high volume occupancy, then label for synaptic proteins known to not associate with the cell-of-interest and ascertain their percentage occupancy on the cell-of-interest. Using a synaptic protein that is not supposed to contact the cell-of-interest is beneficial to set a lower limit for the volume occupancy analysis.

Strengths and Limitations of This Analysis

+ There is a high range of immunolabeled synaptic protein signal that can be quantified by this analysis. Most often labeling for synaptic proteins in situ does appear with considerable background signal, which prevents the reliable identification of every individual punctum. With the volume estimation analysis one can work with non-punctate synaptic labeling to estimate density or strength of synaptic contacts onto a cell-of-interest.
- It cannot provide the number of synaptic puncta associated with a cell-of-interest. The volume estimation analysis can only provide the volume of the cell that is occupied by the labeled synaptic protein.

3.4 Punctum-to-Punctum Apposition Analysis

In cases where two synaptic proteins are immunolabeled it becomes possible to ask about the spatial association between the two synaptic puncta types. By this we mean an assessment of either the colocalization of two synaptic puncta types (presynaptic or postsynaptic) or a quantification of the appositions between presynaptic and postsynaptic puncta. In both these cases a cross-correlation analysis between the two synaptic markers can prove beneficial.

Alternatively, one can utilize a manual technique for establishing the association between two synaptic puncta types by defining regions of interest (ROIs) around the synaptic puncta of one marker and then examining these ROIs in the other synaptic puncta channel to quantify the percentage colocalization or apposition between the two synaptic puncta types.

3.4.1 Correlation Analysis Between Synaptic Puncta

If both presynaptic and postsynaptic markers are labeled, one way to assess the colocalization between the two markers is to calculate the pixel intensity correlation between the two channels [5, 10]. In principle, if two markers are completely nonoverlapping, the correlation coefficient should be near zero. On the other extreme, if one takes two identical images, the correlation coefficient is one. Presynaptic or postsynaptic markers that are colocalized or apposed should give correlation coefficients somewhere between zero and one. Labeling of two presynaptic or two postsynaptic markers should however provide higher correlation values compared to the association between a presynaptic and postsynaptic label. Figure 7 illustrates how one can visualize the correlation between synaptic protein markers. The example image is of two immuno-markers that target the same synaptic protein. The correlation coefficient of these two markers, as expected, is high and is near 0.8, which represents the realistic upper bound of the correlation analysis. As one shifts one channel relative to the other in x and y directions, the correlation quickly falls off. The width of the correlation peak provides an estimate of the spatial relationship between the two synaptic markers. For example, if two markers target presynaptic and postsynaptic proteins that are known to be apposed, the peak correlation coefficient will be smaller than the example where two immuno-markers target the same synaptic protein, but the width of the peak will be broader because the two puncta are located on either side of the synapse and occupy a bigger space in 3D compared to the spatial extent of the labels marking the same protein. Note that if the size of the puncta is considerably large then the width of the correlation peak will be correspondingly broad. Taken together this analysis plots the cross-correlation of the intensity between two channels of an image.

Figure 8 compares three examples of cross-correlation analysis between different synaptic markers in a retinal slice preparation. Figure 8a is the same image from Fig. 7, labeling for markers of the same excitatory presynaptic protein (RIBEYE) using two different antibodies: anti-CtBP2 and anti-RIBEYE. As noted above, the correlation value generated for these two markers represents the upper bound of correlation. As shown in the rightmost panel of Fig. 8a, upon rotation, the correlation coefficient becomes flat and nears zero confirming that the observed correlation between the two markers is not a result of random association between the two markers. Figure 8b shows correlation between an inhibitory

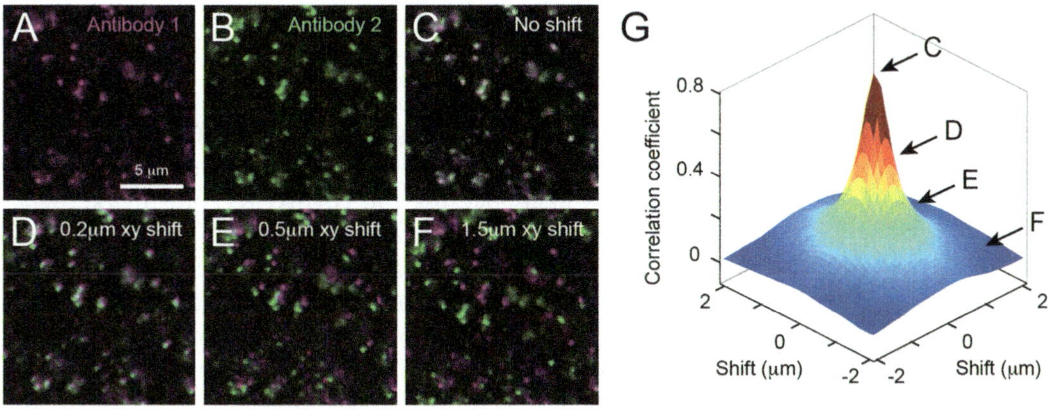

Fig. 7 Cross-correlation analysis. (**a**) A retinal slice immunostained for a synaptic protein with Antibody 1. (**b**) The same protein immunostained with Antibody 2. C: Overlay of the two channels. (**d–f**) The image of Antibody 2 was shifted relative to the image of Antibody 1 both in x and y directions by 0.2 μm (**d**), 0.5 μm (**e**), or 1.5 μm (**f**). (**g**) Correlation coefficients plotted for all X and Y displacements. Coefficients estimated at the shift positions in (**c**) to (**f**) are indicated by *arrows*

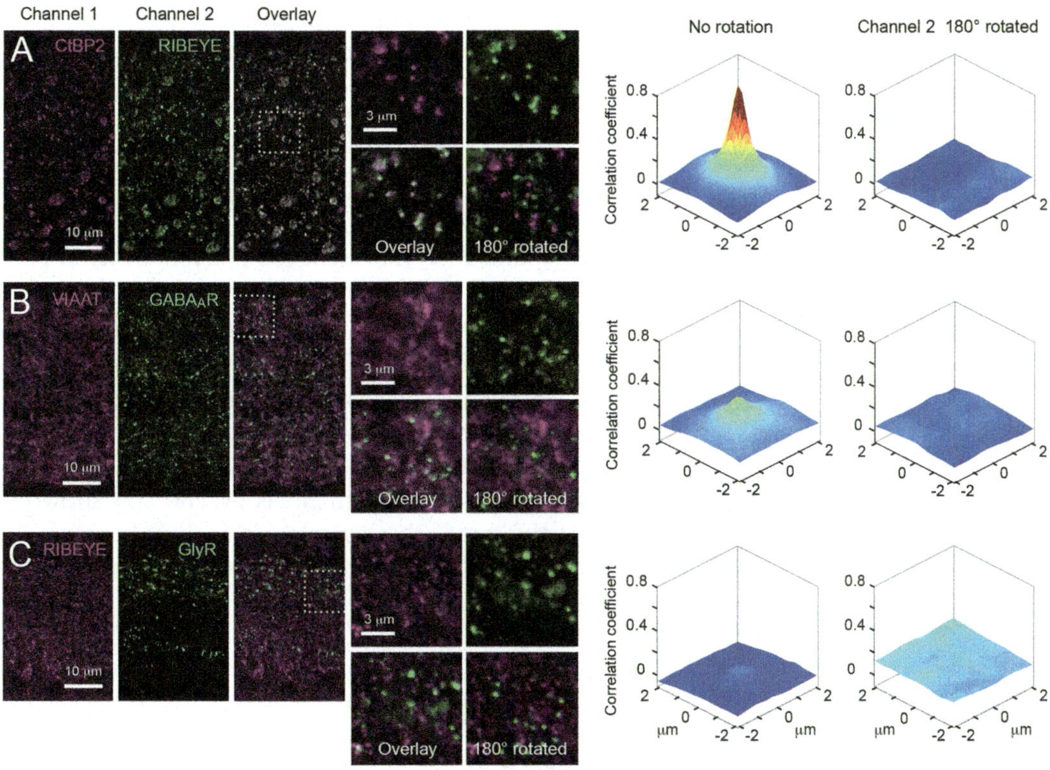

Fig. 8 Cross-correlation analysis between different synaptic labels in the inner synaptic layer of the retina. (**a**) An example of cross-correlation between two channels that label the same synaptic protein. Both anti-CtBP2 antibody and anti-RIBEYE antibody label synaptic ribbons, found in excitatory presynaptic terminals in the retina. (**b**) An example of cross-correlation between presynaptic and postsynaptic proteins expressed at the same type of synapse. Antibody against vesicular inhibitory amino acid transporter (VIAAT) labels inhibitory presynaptic sites whereas antibody against $GABA_A$ receptor labels a subset of inhibitory postsynaptic sites. (**c**) An example of cross-correlation between presynaptic and postsynaptic proteins expressed at different synapse types. Anti-RIBEYE labels all excitatory presynaptic sites whereas the antibody against Glycine receptors labels a subset of inhibitory postsynaptic sites. All the images are single confocal planes, and magnified views of boxed regions are shown. Images at the same locations after rotating channel 2 by 180° show poor colocalization

presynaptic marker (vesicular inhibitory amino acid transporter or VIAAT) and an inhibitory postsynaptic marker ($GABA_A$ receptor). These markers show a degree of correlation that is, however, not higher than 0.4 likely because the $GABA_A$ receptor marker labels only a subset of GABAergic synapses in the retina whereas VIAAT is expected to label all GABA- and glycinergic retinal presynapses. Figure 8c shows the correlation between an excitatory presynaptic marker (RIBEYE) and an inhibitory postsynaptic marker (Glycine receptor). As expected, the correlation coefficient for these unrelated markers is near zero, providing the lower bound in the correlation analysis. In each case controls were performed, by rotating one channel (180°) relative to the other.

1. For the cross-correlation analysis use Fiji or MetaMorph to median filter and crop the image.
2. Use Matlab to perform the analysis between the two channels. Correlation coefficient (r) of pixel intensity between two channels (ch1 and 2) can be calculated by the following equation:

$$r(\mathrm{ch1},\mathrm{ch2}) = \frac{\left(P_{i,\mathrm{ch1}} - \overline{P_{\mathrm{ch1}}}\right)\left(P_{i,\mathrm{ch2}} - \overline{P_{\mathrm{ch2}}}\right)}{\sqrt{\left(P_{i,\mathrm{ch1}} - \overline{P_{\mathrm{ch1}}}\right)^2}\sqrt{\left(P_{i,\mathrm{ch2}} - \overline{P_{\mathrm{ch2}}}\right)^2}}$$

   Where $Pi_{,\mathrm{ch1}}$ and $Pi_{,\mathrm{ch2}}$ represent the pixel intensity at pixel position i in channel 1 and 2 respectively and $\overline{P_{\mathrm{ch1}}}$ and $\overline{P_{\mathrm{ch2}}}$ represent the average pixel intensity of channel 1 and 2. Matlab function *corrcoef* can calculate this equation.
3. Obtain cross-correlation by displacing one channel over the other in x and y directions, and repeating the calculation.

Controls Needed for this Analysis

Rotate one of the immunolabeled channels by 180° (as shown in Fig. 8) and repeat the same cross-correlation analysis between the two markers to estimate any correlation that results from a random overlap between the two labels.

Strengths and Limitations of this Analysis

+ Cross-correlation analysis does not require a threshold for detection (for more information on cross-correlation analysis *see* refs. [16, 17]). Thus, compared to other methods that require thresholds to separate synaptic signals from background, cross-correlation analysis is more objective.
+ Because there is less need of manual operations for this analysis, it is faster and less laborious than other methods.
– Although the cross-correlation analysis is robust for comparing colocalizations across different synaptic markers or the same pair of markers between different experimental conditions, the absolute number you get as the correlation coefficient is less intuitive. For example, the peak correlation coefficient in

Fig. 8b (~0.3) does not provide information about the percentage of the presynaptic labels that colocalize with the postsynaptic labels or vice-versa. Because the presynaptic marker in Fig. 8b is supposed to label all the inhibitory presynaptic sites, and the postsynaptic marker labels a subset of inhibitory postsynaptic sites, theoretically all the postsynaptic labels should be apposed with the presynaptic labels. However, since only a fraction of presynaptic labels is apposed to postsynaptic sites, the correlation analysis shows a reduced correlation coefficient without differentiating between the reciprocal relationships of the synaptic markers.

- If you label for a certain synaptic protein using a method that detects only a fraction of the total distribution in the tissue, such as expressing fluorescent protein-tagged synaptic proteins in a subset of cells, the correlation with other synaptic proteins will be underestimated. In such a case, the alternative method would be to isolate the fluorescent protein-tagged synaptic labels as described in previous sections (dot-finding program using Matlab or *LabelVoxel* function in Amira) and limit the analysis to those labels.

3.4.2 Manual Methods for Assessing Association Between Two Synaptic Puncta Types

Apart from the correlation analysis outlined in Subheading 3.4.1., one can also estimate the association between synaptic markers by defining ROIs for synaptic puncta in one channel and then superimposing those ROIs onto the synaptic protein labeled in the other channel to count how many puncta show colocalization or apposition.

1. ROIs can be defined or drawn in multiple ways (*see* **Note 6**).
2. Count appositions by scoring for colocalization between two presynaptic or two postsynaptic markers [18], or by scoring for 3D appositions between a presynaptic and postsynaptic label [8].
3. While assessing appositions it is imperative to check for associations in 3D and when scoring for the number of colocalized puncta or ROIs it is advisable to set a lower bound for association by quantifying the volume of the ROI (defined in channel 1) being occupied by the label in the other channel (channel 2). A good way is to quantify the volume occupancy of the ROI by the other channel and then delineate a lower bound of at least 20% or more occupancy for a punctum to be scored as "colocalized". This way the experimenter can objectively set a lower bound for identification of a punctum as colocalized or not (for more details on colocalization of fluorescent puncta *see* refs. [19–23]).

Controls Needed for this Analysis

While assessing the association between two synaptic labels using defined ROIs, an easy control is to flip the synaptic puncta channel horizontally by 180° and then assess the percentage of ROIs that are still apposed or colocalized [18]. This would give the percentage of random associations between the two puncta channels as synaptic puncta are small relatively well defined structures occupying a definite space in 3D.

Strengths and Limitations of this Analysis

+ The advantage of this method over the cross-correlation analysis is that the experimenter can assess the relationship for the protein-of-interest even if it labels only a subset of synapses as one is directly looking for associations with the selected ROIs in the channel of interest. Therefore, one can ask about the relationship of one marker relative to the other which is not permitted by the cross-correlation analysis.
– A key limitation is that this manual method is time-consuming thus restricting the number of puncta-puncta appositions that can be sampled.

4 Notes

1. While selecting from the image analysis routines described in this chapter, the signal to noise of the image being processed is the most important factor that determines which image analysis method is most suited. The importance of this cannot be stressed enough as improper image acquisition and processing can distort the analysis and lead the experimenter to misleading interpretations. Once an analysis method is selected to be most appropriate it is imperative to perform the described adequate controls to ensure an optimal in situ connectivity analysis. Most often an experimenter does not understand the limitations of the acquired image and erroneously applies an imaging routine, which generates a flawed connectivity estimate. An example of such an erroneous application is using the dot-finding routine for the analysis of a synaptic protein label that is not clearly punctate. The described image analysis routines in this chapter are the routines currently used in our Lab that facilitate connectivity estimates for fixed tissue samples at the light microscope level. However, as more image analysis software and tools develop we should soon see an enhancement of the image analysis routines an experimenter can apply for a quick and easy assessment of synaptic connectivity.
2. The brightness of PSD95-FP signals can often show a gradient from the soma towards the dendritic tips as illustrated in Fig. 9 for the expression of PSD95-CFP. If one wants to quantify the distribution of such puncta along the length of a dendrite from the soma to the periphery, a simple intensity threshold method for isolating puncta will not be able to accurately capture all

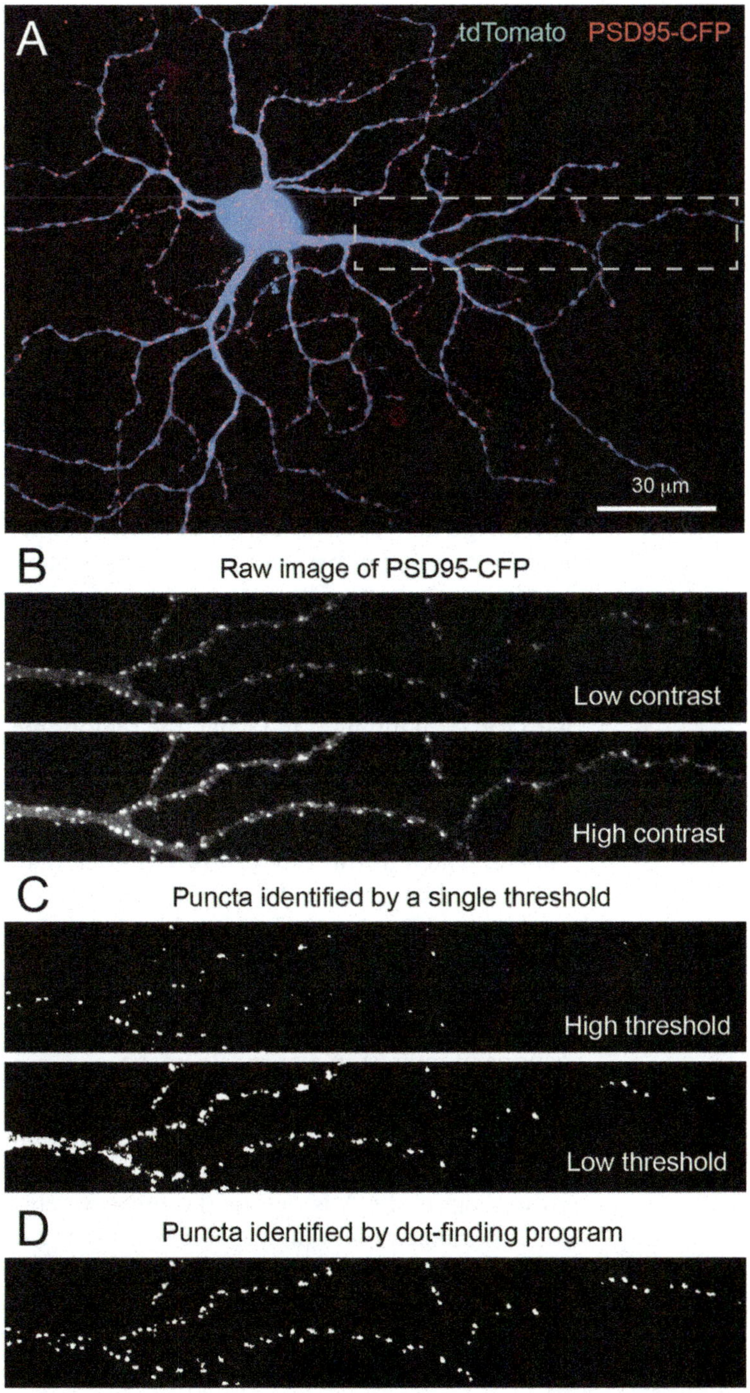

Fig. 9 Comparison of dot-finding program with simple thresholding in isolating punctate synaptic labels along cellular processes. (**a**) A retinal ganglion cell biolistically transfected with CFP-tagged postsynaptic density protein 95 (PSD95-CFP) together with tdTomato that fills the dendrites. (**b**) The raw image of PSD95-CFP in the boxed region of (**a**) is shown at two different contrast levels. (**c**) PSD95-CFP puncta identified by a single threshold. A high threshold set (top panel) for isolating puncta in the proximal dendrite fails to detect dim puncta in the distal dendrite. A low threshold set (bottom panel) for isolating dim puncta in the distal dendrites fails to reject background in the proximal dendrite. (**d**) A custom-written dot-finding program isolates dim puncta in the distal dendrites without overestimating puncta in the proximal dendrites

the puncta along the dendrite (Fig. 9c). Intensity thresholds adjusted for isolating bright puncta in the proximal dendrite will fail to detect the dim puncta located at distal dendrites, whereas thresholds that allow detection of dimmer puncta will cause brighter proximal puncta to fuse into large unsegmented fluorescent foci. A similar problem could arise for immunolabeling of synaptic proteins in a relatively thick tissue volume, as it is common to have variability in the strength of the signal as one images deeper into the tissue. The dot-finding program has proven to be efficient in both these cases (Fig. 9d) because it is designed to isolate each individual punctum locally from the surrounding background signal.

3. To detect instances of nearby dots of similar or dissimilar intensities, the dot-finding program includes a region-based image segmentation tool called watershed transformation. In brief, any image can be considered as a topographic surface with regions of interest that have local intensity maxima such as the overlapping puncta. The watershed transform essentially determines the boundaries of the puncta based on a principle in geography that a drop of water at the point of separation between two regions of interest will flow down to the region with the local minima, and when the regions merge, i.e., water level rises and flows into the other region, lines of separation called the watershed lines can be estimated [24, 25]. This analysis provides closed contours thus distinguishing overlapping puncta, which in this case can be considered as regional maxima instead of minima. Using this technique with predefined markers such as the pixel dimensions helps to refine the contours and better detect puncta location. Matlab offers a built-in function called "watershed" in its image processing toolbox, which can locate the lines/boundaries and the separated regions of interest in a grayscale image.
4. Using PCA analysis, the dot-finding program can restrict puncta selection to mostly synaptic puncta and reduce the number of false positives that may arise from isolated background spots. PCA is one of the most widely used statistical tools in image processing capable of reducing dimensions and revealing relationships between data components [26, 27]. The basis of this technique is to linearly transform a set of likely correlated variables into a smaller set of new variables called principle components. These principle components represent the directions over which the data is most spread out and has the largest variance. The first two principal components usually account for roughly 90% of the total variance in the dataset. Thus, it is recommended to plot all the puncta within the space of the first two components. The outliers in this plot represent the isolated background spots and can be deselected from the

plot by drawing a region around the cluster that represent the actual puncta. However, often these two populations are not well separated in the scatter plot. To help define the cluster of false-positive puncta in such a distribution, the puncta with low values of Sum IT# (the cumulative number of iterations for each detected puncta) are represented in different colors because empirically Sum IT# seems like a good indicator to differentiate true puncta from false-positive puncta. The cluster of puncta with low Sum IT# values can then be omitted from the selection (Fig. 4c).

5. An important determinant of the volume estimation analysis relies on where the threshold is set for the receptor signal channel. Where the threshold is set for pixel detection depends greatly on the background (intensity of the non clustered pixels) signal observed in the channel of interest. As such it is helpful to plot the distribution of pixel intensities from all the pixels of the image and determine the peak of the "noise" signal observed in the channel, which is a good quantitative estimate of the background fluorescence of the channel. Thereafter, one can set the threshold for pixel detection well above the noise (background) peak, the value for which can be determined by fitting the pixel-gray value curve with a gamma function (Fig. 10a). Figure 10b illustrates what signal can be captured in the receptor channel as one selects thresholds at increasing gray values, i.e., from 1 standard deviation (SD) above the noise peak to a limit of four standard deviations above the noise peak. Which exact threshold is ideal for pixel detection is dependent on the signal to noise ratio observed in the respective channel but for the example discussed in Fig. 10, a threshold above 3 SDs of the noise peak would prove to be ideal as it successfully captures all the signal of the image without including the background fluorescence. If however, the background signal is higher than what is shown in the Fig. 10 example, one would need to increase the threshold to above 4 SDs. Immunolabeling can vary from sample to sample and thus one does need a quantifiable defined threshold that can be applied consistently for the volume analysis of synaptic proteins. As depicted in Fig. 10b as one increases the threshold the volume of the signal captured correspondingly decreases. It is thus advisable to choose a rather conservative threshold (3 or 4 SDs above the noise peak), as this would decrease variability across and within samples. Selecting a lower threshold however, would display much more variability across and within samples as more of the background "noise" pixels of the image get captured with lower thresholds.
6. ROIs can be defined or drawn in multiple ways. One could use a strict threshold as described earlier in **Note 5** or use the

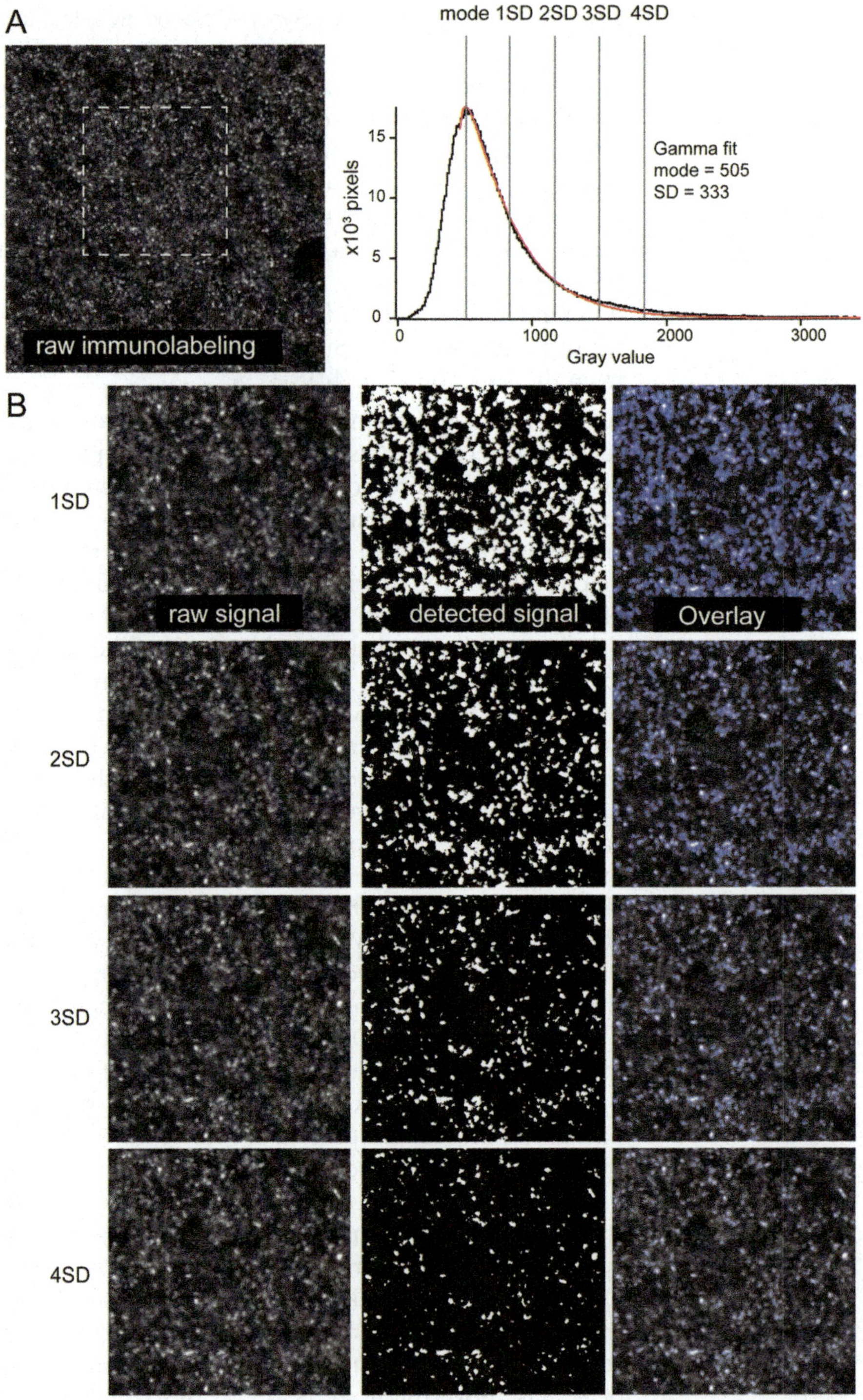

Fig. 10 Determining the threshold during synaptic protein volume estimation analysis. (**a**) Raw image of a synaptic protein immunolabeling (*left*). All the pixels of the image can be plotted against the spread of the gray value intensities (*right*). This plot can be fitted with a gamma function (*red curve*) to determine the value of the noise peak (mode of the distribution) and the values of the standard deviation (SD) above the noise peak. (**b**) Example of the raw signal (*dotted box* in **a**) before and after the application of increasing thresholds. The middle panel illustrates the binarized accepted pixels. Note that at lower thresholds set at 1 SD above the noise peak a lot of the background signal gets erroneously included

Labelfield option in Amira to outline ROIs. Alternatively, one can simply draw a border around pixel intensities deemed to be above background by eye. ROIs do not necessarily need to be drawn in Amira or Metamorph other image analysis software like Fiji, Photoshop etc, can be used to define ROIs around fluorescent puncta, which can be easily superimposed on the other synaptic protein channel.

Acknowledgments

The dot-finding routine was developed by Josh Morgan and further refined by the collaborative efforts of past and present Rachel Wong Lab members. Credit for the dot-finding routine goes to the concerted efforts of Josh Morgan, Daniel Kerschensteiner, Adam Bleckert, and Haruhisa Okawa. Daniel Kerschensteiner, Florentina Soto, Josh Morgan, and Adam Bleckert together developed the correlation analysis. We are grateful to Rachel Wong and Adam Bleckert for critical feedback and suggestions during preparation of this chapter. This work is supported by NIH grants (EY10699, EY17101, and EY14358 to R. Wong and a Vision Core grant EY01730); a Human Frontier Science Program grant (RGP0035 to L. Lagnado, F. Schmitz, and R. Wong); a Human Frontier Science Program Long term fellowship (R. Sinha); and a Knights Templar Eye Foundation career starter grant (M. Hoon). The immunolabeling for Ribeye and $GABA_A$ receptors shown in the Figures has been possible due to antibodies generously provided by Frank Schmitz and Jean-Marc Fritschy.

References

1. Grimes WN, Hoon M, Briggman KL, Wong RO, Rieke F (2014) Cross-synaptic synchrony and transmission of signal and noise across the mouse retina. Elife 3:e03892
2. Okawa H, Della Santina L, Schwartz GW, Rieke F, Wong RO (2014) Interplay of cell-autonomous and nonautonomous mechanisms tailors synaptic connectivity of converging axons in vivo. Neuron 82:125–137
3. Morgan JL, Soto F, Wong RO, Kerschensteiner D (2011) Development of cell type-specific connectivity patterns of converging excitatory axons in the retina. Neuron 71:1014–1021
4. Morgan JL, Schubert T, Wong RO (2008) Developmental patterning of glutamatergic synapses onto retinal ganglion cells. Neural Dev 3:8
5. Bleckert A, Parker ED, Kang Y, Pancaroglu R, Soto F et al (2013) Spatial relationships between GABAergic and glutamatergic synapses on the dendrites of distinct types of mouse retinal ganglion cells across development. PLoS One 8:e69612
6. Kerschensteiner D, Morgan JL, Parker ED, Lewis RM, Wong RO (2009) Neurotransmission selectively regulates synapse formation in parallel circuits in vivo. Nature 460:1016–1020
7. Schubert T, Hoon M, Euler T, Lukasiewicz PD, Wong RO (2013) Developmental regulation and activity-dependent maintenance of GABAergic presynaptic inhibition onto rod bipolar cell axonal terminals. Neuron 78:124–137
8. Newkirk GS, Hoon M, Wong RO, Detwiler PB (2013) Inhibitory inputs tune the light response properties of dopaminergic amacrine cells in mouse retina. J Neurophysiol 110:536–552
9. Dunn FA, Della Santina L, Parker ED, Wong RO (2013) Sensory experience shapes the development of the visual system's first synapse. Neuron 80:1159–1166

10. Soto F, Bleckert A, Lewis R, Kang Y, Kerschensteiner D et al (2011) Coordinated increase in inhibitory and excitatory synapses onto retinal ganglion cells during development. Neural Dev 6:31
11. Morgan JL, Wong RO (2008) Ballistic labeling with fluorescent dyes and indicators. Curr Protoc Neurosci Chapter 2:Unit 2.11
12. Swartz M, Eberhart J, Mastick GS, Krull CE (2001) Sparking new frontiers: using in vivo electroporation for genetic manipulations. Dev Biol 233:13–21
13. Zucker RM, Rigby P, Clements I, Salmon W, Chua M (2007) Reliability of confocal microscopy spectral imaging systems: use of multispectral beads. Cytometry A 71:174–189
14. Scriven DR, Lynch RM, Moore ED (2008) Image acquisition for colocalization using optical microscopy. Am J Physiol Cell Physiol 294:C1119–1122
15. Schindelin J, Arganda-Carreras I, Frise E, Kaynig V, Longair M et al (2012) Fiji: an open-source platform for biological-image analysis. Nat Methods 9:676–682
16. Wu Y, Eghbali M, Ou J, Lu R, Toro L et al (2010) Quantitative determination of spatial protein-protein correlations in fluorescence confocal microscopy. Biophys J 98:493–504
17. Comeau JW, Kolin DL, Wiseman PW (2008) Accurate measurements of protein interactions in cells via improved spatial image cross-correlation spectroscopy. Mol Biosyst 4:672–685
18. Hoon M, Bauer G, Fritschy JM, Moser T, Falkenburger BH et al (2009) Neuroligin 2 controls the maturation of GABAergic synapses and information processing in the retina. J Neurosci 29:8039–8050
19. Mokin M, Keifer J (2006) Quantitative analysis of immunofluorescent punctate staining of synaptically localized proteins using confocal microscopy and stereology. J Neurosci Methods 157:218–224
20. Mokin M, Keifer J (2004) Targeting of GLUR4-containing AMPA receptors to synaptic sites during in vitro classical conditioning. Neuroscience 128:219–228
21. Fish KN, Sweet RA, Deo AJ, Lewis DA (2008) An automated segmentation methodology for quantifying immunoreactive puncta number and fluorescence intensity in tissue sections. Brain Res 1240:62–72
22. Ippolito DM, Eroglu C (2010) Quantifying synapses: an immunocytochemistry-based assay to quantify synapse number. J Vis Exp pii:2270
23. Dunn KW, Kamocka MM, McDonald JH (2011) A practical guide to evaluating colocalization in biological microscopy. Am J Physiol Cell Physiol 300:C723–742
24. Meyer F (1992) Color image segmentation, in *Proceedings of the IEE International Conference on Image Processing and Its Applications*, Maastricht, the Netherlands, IEE, London, UK, pp. 303–306
25. Vincent L, Soille P (1991) Watersheds in digital spaces: an efficient algorithm based on immersion simulations. IEEE Trans Pattern Anal Mach Intell 13:583–598
26. Busse B, Smith S (2013) Automated analysis of a diverse synapse population. PLoS Comput Biol 9:e1002976
27. John Wiley & Sons, Inc is the publisher which has multiple locations worldwide so cannot specify a single location with certainty but if one must absolutely be used use New York

Chapter 21

Dual Anterograde and Retrograde Viral Tracing of Reciprocal Connectivity

Matthias G. Haberl, Melanie Ginger, and Andreas Frick

Abstract

Current large-scale approaches in neuroscience aim to unravel the complete connectivity map of specific neuronal circuits, or even the entire brain. This emerging research discipline has been termed connectomics. Recombinant glycoprotein-deleted rabies virus (RABV ΔG) has become an important tool for the investigation of neuronal connectivity in the brains of a variety of species. Neuronal infection with even a single RABV ΔG particle results in high-level transgene expression, revealing the fine-detailed morphology of all neuronal features—including dendritic spines, axonal processes, and boutons—on a brain-wide scale. This labeling is eminently suitable for subsequent post-hoc morphological analysis, such as semiautomated reconstruction in 3D. Here we describe the use of a recently developed anterograde RABV ΔG variant together with a retrograde RABV ΔG for the investigation of projections both to, and from, a particular brain region. In addition to the automated reconstruction of a dendritic tree, we also give as an example the volume measurements of axonal boutons following RABV ΔG-mediated fluorescent marker expression. In conclusion RABV ΔG variants expressing a combination of markers and/or tools for stimulating/monitoring neuronal activity, used together with genetic or behavioral animal models, promise important insights in the structure–function relationship of neural circuits.

Key words Connectome, Neural circuits, Sparse labeling, Pseudotyping, Axonal arbor, Dendritic spines, Projections, Rabies virus, Neuroanatomy

1 Introduction

Connectomics—the study of neuronal connectivity of circuits or entire brains—is an important emerging discipline in modern neuroscience. This is evidenced by the substantial investment in large-scale brain-mapping projects, which promise to extend our knowledge of the connectome in model species as well as humans. Such comprehensive knowledge of connectivity matrices would provide scientists with a road map to test or discard predictive hypotheses regarding the function of defined neural circuits or cell types and to perhaps understand their dysfunction in human disease. These ambitious studies require a range of innovative methods, including approaches to label defined neurons within a circuit or brain area to permit their

Alexandros Poulopoulos (ed.), *Synapse Development: Methods and Protocols*, Methods in Molecular Biology, vol. 1538, DOI 10.1007/978-1-4939-6688-2_21, © Springer Science+Business Media LLC 2017

fine-scale reconstruction. Such morphological reconstructions help to identify neuron types and their projections, to predict their functional role, and to assess changes in these properties following plastic or pathological modifications.

A number of criteria must be fulfilled to enable reliable and complete reconstruction of all morphological features of neurons—such as axons and dendritic spines: (1) These structures must be intensely labeled to allow the use of high-resolution light-microscopy approaches (super-resolution-, multi-photon-, or confocal microscopy) for visualization and automated post hoc analysis approaches; (2) the neuronal structures must be labeled in their entirety (this may be a particular issue for axonal projections which extend over long distances throughout the brain); (3) ideally the labeling method can be applied to individual or few neurons (sparse infection) or to neuronal populations (bulk infection).

Many of these requirements can be fulfilled by the use of recombinant viral vectors expressing fluorescent markers. However, for tracing reciprocal projections—namely those arising from neurons projecting both to and from any region of interest—two vectors with closely matched properties must be used. To address this problem, we turned to the use of recombinant glycoprotein-deleted rabies virus (RABV ΔG) derived from the highly attenuated Street Alabama Dufferin (SAD) B19 laboratory strain of rabies virus. Rabies virus has been used for years as a transneuronal tracer of synaptically connected neurons. However, advances in the understanding and manipulation of its genome, and in the ability to reconstitute the virus from recombinant DNA have led to an important diversification of its use as a viral expression vector and non-synaptic tracer of neurons. The finding that the envelope glycoprotein is essential not only for the infectious properties of the virus but also for transsynaptic spreading led to the development of glycoprotein-deleted variants incapable of transsynaptic spreading. Pseudotyping this glycoprotein-deleted variant with its native glycoprotein created a retrograde, single-cycle viral vector [1] that infects axon terminals at the site of injection effectively labeling neurons that project to a site of interest

We recently developed a novel recombinant RABV ΔG variant that enables the infection of neurons at their cell body/dendrites [2] rather than their axon terminals. Subsequent expression of a soluble fluorescent marker fills the neuron, including all its processes, leading to the type of labeling classically achieved with anterograde (non-transsynaptic) chemical or biochemical tracers. The high level expression of the encoded transgene, however, results in a very detailed morphological labeling of superior quality—similar to that reported previously for retrogradely tracing RABV ΔG forms. To do this, we "repackaged" the virus with a chimeric envelope protein (VSV G^{RtmC}) containing sequences derived from the glycoproteins of both the vesicular stomatitis virus and rabies virus. Importantly, although exhibiting a very different tropism for infection (*see* **Note 1**), these two variants have the same temporal window of application, diffusion characteristics, and high-level expression of target proteins. Here, we discuss the combined use of anterograde and retrograde RABV ΔG for the labeling of reciprocal projections (Fig. 1). This approach now

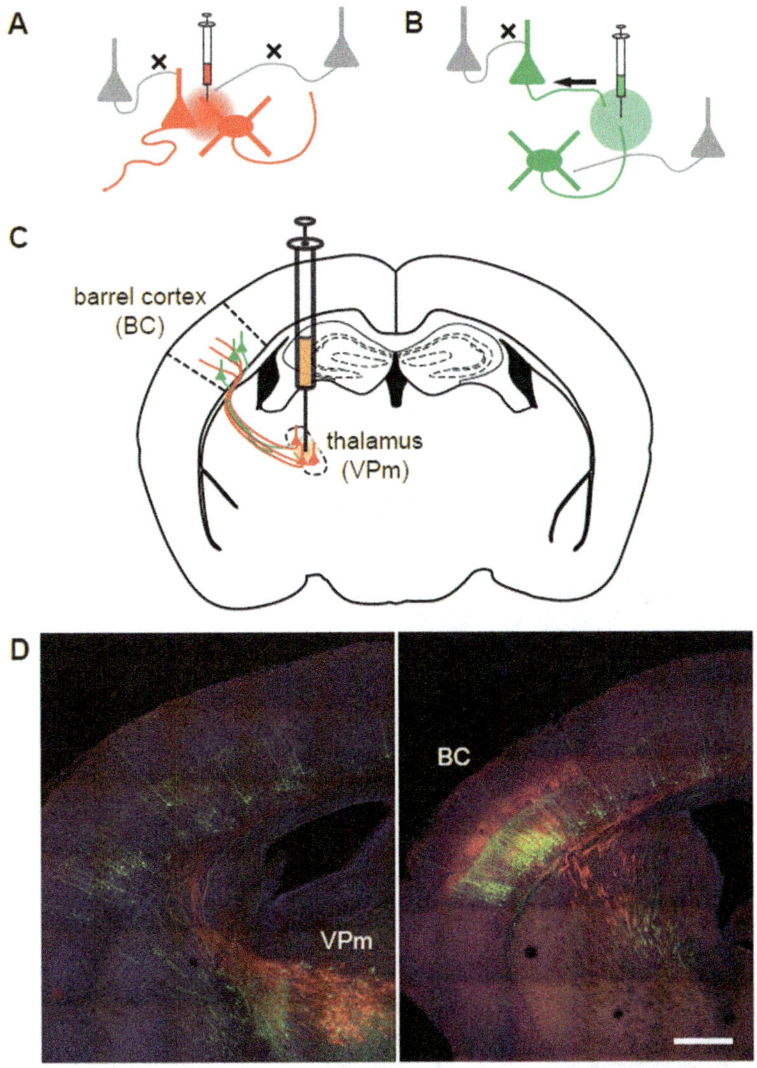

Fig. 1 Combination of RABV ΔG-mediated anterograde and retrograde tracing. (**a**) Schematic of anterograde tracing using RABV ΔG(VSV G^{RtmC})-mCherry. (**b**) Schematic of retrograde tracing using RABV ΔG(SAD G)-eGFP. (**c**) Combined injection of anterograde and retrograde rabies virus injection into the VPm region of the thalamus, which projects to, and receives input from the barrel cortex. (**d**) Reciprocal thalamocortical/corticothalamic connections are labeled after injection of anterograde and retrograde RABV ΔG. Here, the anterogradely labeled projections are descending from the injection site, the VPm region (*left panel*), and arriving more anterior (*right panel*) in the barrel cortex. While the strongest innervation of thalamo-cortical projections terminates in layer 4, the retrogradely labeled cells are highly concentrated in layer 5 and 6, layers known to provide feedback to the thalamus. Scale bar 500 μm. Figure modified from ref. [2]

allows the reconstruction of all key morphological features of neurons: dendrites (Fig. 2), spines, boutons (Fig. 3), and importantly long-ranging axons extending throughout the brain [2]. Infection can be sparse (single or few neurons) or neuron population-wide (hundreds). We demonstrate its usefulness for a number of experimental questions, using proprietary as well as custom-written reconstruction approaches.

Lastly, the ability of these vectors to encode additional transgenes [3] and to rapidly express them at high levels provides an attractive and often superior alternative to lentivirus and adeno-associated virus-based expression vectors [1, 2]. This method promises to greatly enhance the toolbox for the structure–function analysis of neuronal circuits [4, 5].

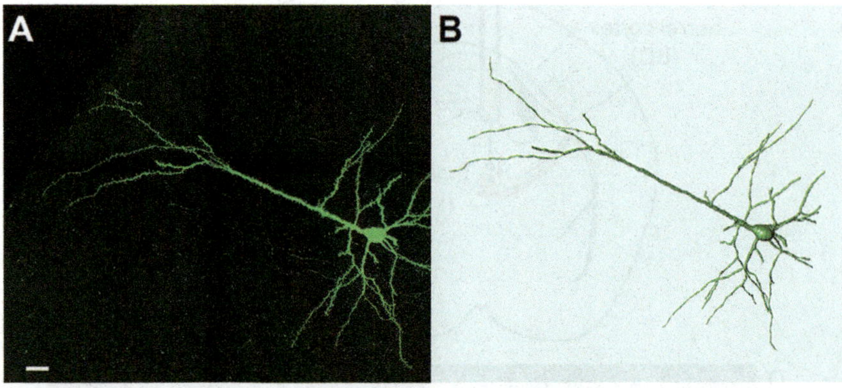

Fig. 2 Reconstruction of the dendritic tree. (**a**) Confocal stack of an individual neuron labeled by RABV infection. Scale bar 20 μm (**b**) Automated reconstruction of the dendritic tree using APP2 algorithm (of the VAA3D software) applied to the confocal stack

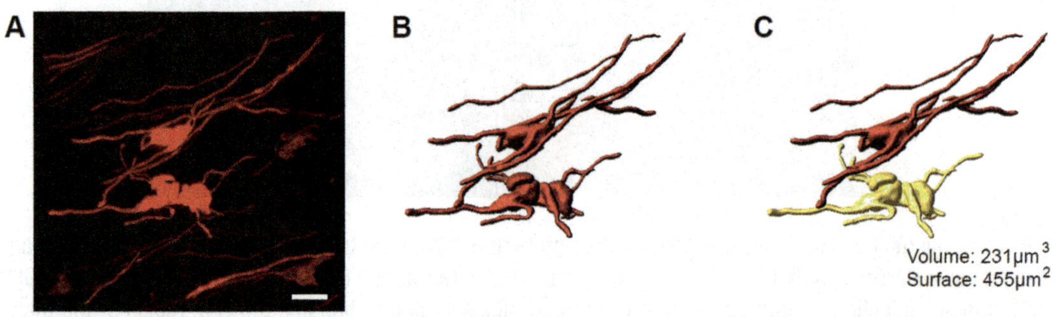

Fig. 3 Volume reconstruction of RABV ΔG(VSV G^{RtmC})-mCherry labeled axon tracks and mossy fiber boutons. (**a**) Confocal stack showing the axonal tracks of **RABV ΔG** (VSV G^{RtmC})-mCherry infected dentate gyrus cells projecting to the CA3 region. (**b**) Volume reconstruction of individual boutons and axonal projections. (**c**) Readout of the parameters of the volume reconstruction of one individual bouton (*yellow*). Scale bar 5 μm. Bouton Volume: 231 μm^3. Bouton Surface: 455 μm^2

2 Materials

2.1 Viral Vectors

Our protocol uses a combination of two RABV ΔG variants expressing fluorescent markers for the visualization of morphological features. For retrograde tracing, we employ RABV ΔG (expressing eGFP) psedudotyped with the glycoprotein of the CVS (challenge virus standard) strain of RABV. For anterograde tracing, we use RABV ΔG (expressing mCherry) pseudotyped with a hybrid glycoprotein (VSV G^{RtmC}; [2]). This chimeric protein contains the extracellular domain of the glycoprotein of vesicular stomatitis virus and the transmembrane and cytoplasmic domains of the SAD glycoprotein [2] (*see* **Note 2**).

We utilize ultracentrifugation for purification of the RABV ΔG variants to remove potential contaminants that likely cause inflammatory reactions when injected into the brain and for obtaining a high titer stock for these in vivo applications. Typically titers of 10^6–10^8 infectious units per ml are achieved. These stocks are stored in small aliquots at −80 °C (in low protein-binding tubes) and thawed the day of use. Repeated freeze-thaw cycles should be avoided as they reduce the titer. Unlike other viral vectors that are readily available from viral production cores, RABV ΔG-based vectors are only available from a limited number of facilities and labs (*see* **Notes 3** and **4**).

2.2 Virus Handling (See Note 5)

1. Noncorrosive surface disinfectant (e.g., Surfa'safe, Anios).
2. Bleach solution: 2% bleach in water (freshly prepared from pastilles).
3. Hanks buffered salt solution (HBSS).
4. Pipette tips with low protein binding and filter barrier.

2.3 Surgery

1. Sterile saline, e.g., from a veterinary supplier, plus cartridges delivering an individual dose if available (may be purchased from a pharmacy).
2. Calibrated glass capillaries with 1–5 μl graduations (e.g., Blaubrand® IntraMARK).
3. Nontoxic rapidly curing silicon sealant (e.g., KWIKSIL, World Precision Instruments).
4. Iodine-based skin disinfectant.
5. Syringes with needles: 1 ml and 20 ml tuberculin syringes and 23G, 25G, and 26G needles.
6. Absorbents: Cotton swabs (Q-tips® or similar), absorbent swabs (e.g., Sugi ®, Kettenbach GmbH).
7. Ready-to-use sutures (appropriate for the species).
8. Disposable scalpels.

9. Eye cream.
10. General anesthetic that is appropriate for the species in question (e.g., isoflurane).
11. Local anesthetic for skin application. The use of bupivacaine may be preferable over lidocaine as it has a longer duration of action.
12. Analgesic appropriate to species (e.g., buprenorphine; 0.03 mg.ml^{-1} in sterile saline, prepared from a 0.3 mg.ml^{-1} stock solution; *see* **Note 6**).

2.4 Histological Approaches (See Note 7)

1. Ringer's solution: 135 mM NaCl, 5.4 mM KCl, 1.8 mM $CaCl_2$, 1 mM $MgCl_2$, 5 mM HEPES; pH 7.4.
2. Heparin: 5000 units.ml^{-1} pharmaceutical grade heparin.
3. Fixative solution: 4% paraformaldehyde (PFA) in 0.1 M phosphate buffer (PB), freshly prepared and pH adjusted to 7.4, cooled to 4 °C prior to use (or alternative fixative solutions as appropriate).
4. Collection buffer: 0.02% sodium azide in 0.1 M PB (alternatively 0.02% thimerosal in PB).
5. DAPI stock solution: 5 mg.ml^{-1} 4′,6-diamidino-2-phenylindole dihydrochloride (DAPI) in dimethylformamide, stored in dark at −20 °C.
6. Coverslips with thickness 0.17±0.01 mm (e.g., Hecht-Assistant reference# 1014.
7. Mounting medium (e.g., Prolong Gold, Invitrogen; *see* **Note 8**).
8. Antibodies for post-processing if necessary (e.g., Anti-GFP, Roche, mouse IgGk clones 7.1 and 13.1, reference# 11814460001; anti-DsRed, Clontech, rabbit polyclonal reference# 632496).
9. Nail polish.

2.5 Instrumentation

1. Glass pipette puller (e.g., P-97, Sutter).
2. Gas anesthesia apparatus.
3. Precise stereotaxic frame (we use Kopf model 940 with digital display and 10 μm resolution), mouse head-holder and non-rupture ear-bars.
4. Electrode holder adapted to stereotaxic frame, modified to accept a glass pipette (e.g., Kopf 1771).
5. Ultraprecise injection pump (e.g., Ultramicropump; World Precision Instruments) plus controller and V-mount for attachment to stereotaxic frame.
6. Glass syringe 10 μl capacity (e.g., Nanofil syringe; World Precision Instruments) with injection needles (e.g., 34 gauge beveled Nanofil needle, internal diameter 85 μm, outer diameter 185 μm; World Precision Instruments).

7. Regulatable temperature controlled heating pad plus controller and rectal probe e.g., FHC reference: 40-90-8D, AD Instruments reference: ML295, World Precision Instruments, reference: ATC2000).
8. Stereoscope (with good optics, suitable magnification and working distance).
9. Cold-light with flexible goose neck fitting (e.g., Photonic reference: PL3000).
10. Hand-held micro drill (e.g., Ideal micro drill, World Precision Instruments).
11. Stainless steel drill burrs 0.7 mm diameter (e.g., Fine Science Tools).
12. Electric razor (optional).
13. Heating blanket for recovery following surgery.
14. Glass bead sterilizer for surgical tools (optional).
15. Peristaltic pump equipped with Tygon® tubing of appropriate size for the species e.g., ~3 mm (inner diameter) for rat and 1.3 mm (inner diameter) for mouse. Tubing is fitted with a switchable Y-shaped connector to permit the sequential perfusion of two different solutions.
16. "Feeding" cannula with rounded tip (e.g., Dutscher reference:061310).
17. Surgical tools for perfusion (forceps, bone-cutting scissors, Dumont forceps, various hemostats, bone-cutters, spatula, cover glass forceps). Tools for perfusion should be kept separated from other tools.
18. Vibratome (e.g., VT1200, Leica).

2.6 Software and Digital References for Analysis

1. The Mouse Brain in Stereotaxic Coordinates (CD-ROM version) [6].
2. ImageJ (free software; http://imagej.nih.gov/ij/download.html) and NDPI plugin (http://www.imnc.in2p3.fr/pages-perso/deroulers/software/ndpitools/).
3. Imaris Bitplane 7 or later (Demo Version available at http://www.bitplane.com/).
4. Adobe Illustrator CS3 or later (30 day trial Version available at https://creative.adobe.com/products/download/illustrator).

3 Methods

3.1 Experiment Design

The essential feature of our approach is the co-utilization of a retrograde and an anterograde viral tracer, which are matched in their temporal expression and diffusion characteristics. Such vectors permit the strong morphological labeling and/or expression of gene of interest in neurons located at the site of injection, as well as in neurons defined by their projection characteristics. These neurons are targeted either by coinjection of a 1:1 mixture of retrograde and anterograde tracers into the same site (as described in ref. [2]) or by sequential injection into two sites, defined by the projection or target area of a given population of neurons. This procedure necessitates the selection of coordinates for stereotaxic injection, brain fixation and post-processing, imaging by conventional or specialized microscopy approaches (*see* **Notes 9** and **10**) and finally image analysis.

The Mouse Brain Atlas [6] is a good starting point for the selection of appropriate stereotaxic coordinates and reference points. However, it is often very useful to consult the literature concerning a specific target region. The Allen Brain Atlas (http://mouse.brain-map.org/), for example, provides improved maps based on the incorporation of data from gene expression studies. Other online atlases include the Mouse Brain Library (http://www.mbl.org/), the Brainmaps (http://brainmaps.org/), and the Scalable Brain Atlas (SBA) (http://scalablebrainatlas.incf.org/WHS12). Several of those initiatives (e.g., Allen Brain and SBA) not only provide a map of the Mouse Brain but also of other species (rat, marmoset, macaque, or human).

3.2 Stereotaxic Injection

As an essential step in any tracing experiment, stereotaxic injection permits the correct spatial targeting and introduction of the vector(s) of interest into the CNS. Several protocols have previously described this procedure in detail (e.g., [7]). Here we provide the key points for the successful stereotaxic injection into a defined brain region.

Surgery should take place in a dedicated room with appropriate lighting and surfaces that are readily disinfected. For anesthesia we prefer gas-based-, over injectable (e.g., ketamine/xylazine) anesthesia, due the rapid recovery times, high survival rate, and reliability over a range of ages and transgenic mouse models.

1. Preparation of virus. Thaw virus aliquot the day of the injection. Briefly spin down tube in a benchtop centrifuge to remove droplets from the lid of the tube (to prevent the formation of aerosols). Store virus on ice until needed (*see* **Note 11**).
2. Disinfect bench surface prior to beginning and ensure that sterile tools and all other materials are at hand.
3. Prepare injection syringe. Rinse syringe and needle 5–10 times each with sterile distilled water and then HBSS.
4. Initiation of anesthesia. Place mouse in the induction chamber and anesthetize with a 4 % isofurane mixture in air (delivered at a flow-rate of 0.8 l min^{-1}). It is important to closely monitor

changes in the breathing pattern and tonus of the mouse. Do not leave mouse unattended as prolonged exposure to this isoflurane concentration is lethal.

5. Once anesthetized, head-fix mouse in a stereotaxic frame and reduce flow-rate and isoflurane concentration to levels sufficient to maintain a surgical plane of anesthesia (in our hands 1.5 % isoflurane, delivered at a rate of 0.4 l min^{-1}). The head of the mouse is stabilized by placing the teeth into the mouthpiece and the earbars into the post-glenoid foramen (a small cavity in the skull above the ear canal). The earbars have a scale to verify that the head is correctly placed. Use a small pair of blunt forceps to gently pull the tongue out of the mouth to avoid breathing problems/suffocation. Tighten the nosepiece slightly, without applying pressure to the snout. Verify the absence of a reflex by gently puffing air on the eyes and pinching the toes (both front and hind paw). Apply eye cream to prevent corneal drying during surgery. Place the mouse on an adjustable heating pad and maintain at 37 °C.
6. Shave the top of the head (for instance by clipping the hair with a pair of scissors), and anesthetize the scalp by subcutaneous injection of approximately 30–50 μl of local anesthetic and disinfect with iodine solution.
7. Open the scalp with a single vertical cut using a scalpel. Use clamps or "custom-made" hooks prepared from 25G needles to hold the skin aside and keep the skull accessible.
8. Clean the top of the skull using sterile saline (~22 °C), a cotton swab and bone scraper if necessary. This is necessary to clearly visualize the suture lines and intersection points of the skull bones. We discourage the use of H_2O_2 as it oxidizes the bone and surrounding tissue. Perform a visual check to make sure that the head is positioned as horizontal as possible—verify with respect to bregma and lambda—and that there is no lateral movement.
9. Load the syringe by placing a droplet of virus onto a small piece of Parafilm and gently aspiring the virus into the syringe using a slow regular movement. Discard the Parafilm and tips in fresh bleach solution. Place the syringe in the pump assembly (*see* **Note 12**).

   The next steps are performed using the stereoscope to visualize the skull and brain surface.

10. Position the needle over the appropriate reference point (e.g., lambda or bregma). Set the anterior/posterior (A/P) and lateral/medial (L/M) coordinates to zero, move to the target location using the digital display, and mark the target with a fine-point indelible marker.

11. Perform the craniotomy by using a hand-held drill to thin the skull. Be careful not to pierce the bone or to cause undue heating of the skull surface. The skull may be washed using sterile saline—individual cartridges that deliver the saline in a drop-wise manner are ideal for this purpose. Use a "blunt" 26G needle to break through the thinned bone. A clear view of the brain surface is essential for the next step since we determine the injection depth with respect to the pia (rather than the top of the skull, which may vary in thickness due to the age of the animal). Using a fresh 26G needle, gently scratch the dura to provide access for the needle. Any blood should be washed away with sterile saline and blotted with absorbent swabs.
12. Make sure that the syringe is correctly charged and that the tip is not blocked by ejecting a small amount of virus, then blot dry with laboratory paper and advance the needle until the tip touches the brain. Set the dorsal/ventral (D/V) coordinate to zero using the digital display and then descend until the desired location is reached, advance another ~30 μm, then step back up to the desired target position.
13. Inject virus at a rate of 50 nl per min until the desired volume is ejected. We typically use a total volume of 100–500 nl; however, this will depend on the titer of the virus and the number of neurons one wishes to target (*see* **Notes 13** and **14**). After injection the needle remains in place for an additional 5 min to prevent "tracking" as it is withdrawn.
14. Withdraw the needle slowly, cover the craniotomy with a small drop of KWIKSIL sealant, close the skin with three or four stitches and place the mouse on an absorbent towel in an empty cage on top of a heating blanket. When the mouse awakens, inject ~300 μl sterile saline (i.p.) and analgesic (e.g., buprenorphine 0.01–0.05 mg/kg s.c.). Place the mouse in a clean cage and monitor for several days to ensure the absence of infection or discomfort.

3.3 Fixation of Injected Brains and Post-Fixation Processing

The purpose of this step is to fix the brains and to prepare samples for subsequent microscopic analysis. Here, we describe a generalized procedure for perfusion-based fixation, sectioning, and mounting of slices (*see* **Note 15**). Perfusion is performed using a peristaltic pump equipped with tubing joined with a Y-shaped connector to permit sequential perfusion of two different solutions. The use of a pump permits the fixative to be delivered at a constant rate over a prolonged period. The tubing is fitted with a 23G needle (for mice) and "feeding" cannula (for rats). For most applications we use a pH-equilibrated solution of 4 % paraformaldehyde in 0.1 M PB (pH 7.4) (*see* **Note 15**). Since these procedures are fairly standard in neuroscience (for example, *see* ref. [8]), we describe only the major steps, and details that we consider to be specific to our procedure.

Animals are typically sacrificed 6–10 days after the injection of RABV ΔG particles (*see* **Note 16**), but the time period may need to be defined empirically for the species and structure of interest.

1. Intracardiac perfusion. Pre-charge each "arm" of the tubing with room temperature Ringer's solution (supplemented with heparin at dilution of 1:1000) and Fixative solution (separated by Y-shaped connector). Deeply anesthetize animals with an injection of a lethal quantity of sodium pentobarbital. Once the animal has stopped breathing, and after ensuring the complete absence of any reflex, open the abdomen and chest cavity, clamp the rib cage in an upward position using a curved hemostat. For rats, clamp the descending aorta (optional for mice). Introduce a small cut into the right atrium. Start the pump at a slow rate—the Ringer's solution is perfused first and should already start flowing before introduction of the needle or cannula to prevent the introduction of bubbles. The needle or cannula is then introduced into the left ventricle (for the latter a small cut must be made first and then the cannula is kept in place by clamping with a hemostat). Perfuse Ringer's solution until the effluent fluid is clear. The lungs, tongue, and liver (if you have not clamped aorta) will also become pale. Now switch to fixative solution and perfuse for approximately 10–15 min. Dissect out the brain and postfix it overnight in the same fixative solution.
2. Histology (*see* **Note 9**). Cut 50-μm-thick coronal slices using a standard vibratome and collect them in Collection buffer. Vibratome sectioning is preferable over cryostat sectioning, which introduces freezing artifacts that may complicate subsequent reconstruction approaches. Process slices either for immunofluorescence, or immediately counterstain them with DAPI (*see* **Note 17**).
3. The DAPI staining step is performed as follows:
   - Prepare a working solution of DAPI by diluting the DAPI stock solution 1:10000 in 0.1 M PB. The remaining concentrated stock solution can be stored at 4 °C.
   - Incubate and gently shake slices for 15 min in this working solution at room temperature, then wash three times in 0.1 M PB to remove excess dye.
   - Place slices in a 2 × 2 configuration in the center of a microscope slide (for rat brain slices, mount a single slice per slide), remove excess moisture using filter paper, mount slices in appropriate mounting medium and coverslip. Seal the edges of the coverslip with nail polish, allow to dry, and store samples at 4 °C until imaged.

3.4 Imaging Labeled Neurons and Brain Areas

The resulting samples can be analyzed in much the same way as samples resulting from the injection of conventional chemical/viral tracers, with the caveat that we prefer to use fluorescence detection due to its compatibility with high-resolution imaging and 3D reconstruction approaches (*see* **Note 9**). The type of imaging strategy will depend on the question under consideration, determining the desired spatial resolution of the images, volume of tissue that needs to be imaged, speed of acquisition etc.

We typically start by acquiring a low-resolution overview of the labeled brain areas using a rapid image acquisition approach, such as an automated slide scanner (e.g., the Nanozoomer by Hamamatsu). Images are acquired with three color channels to permit not only the visualization of the anterogradely and retrogradely labeled neurons, but also of the underlying anatomical features (DAPI stain). This approach permits the acquisition of more that one z-planes–or a single z-plane for faster scanning. For focussing, the DAPI channel is selected and a number of focus points covering the area of interest are selected. Once the parameters and the scan area have been defined, the acquisition proceeds in an automated manner (*see* **Note 18**).

To acquire higher-resolution images required for morphological reconstruction or volume analysis, we use laser scanning confocal microscopy. The acquisition parameters will depend on the desired downstream analysis, but we typically image with bidirectional scanning at 400–700 Hz, at 12 or 16 bit depth, and for most applications at the best resolution achievable (*see* **Note 19**).

3.5 Analysis

As always for imaging, the type of image analysis and the choice of analysis software will largely depend on the experimental question under consideration (for a range of potential applications, *see* **Note 20**). Here we present two concrete examples demonstrating procedures for (1) mapping of labeled neurons with respect to a brain atlas and (2) volume reconstruction of any neuronal process e.g., boutons (Fig. 3) (*see* **Notes 21** and **22**).

3.5.1 Cell Counting from Large-Scale Scans of the Brain Using Adobe Illustrator and ImageJ

1. NDPI format reader. The first step is the conversion of files from NDPI format to TIFF format (*see* **Note 23**). For first time users: install the NDPI-Tools ImageJ plugin to permit the import of NDPI files into ImageJ. The toolbox can be found at (http://www.imnc.in2p3.fr/pagesperso/deroulers/software/ndpitools/). Plugins can be added by simply moving the file/s into the Plugin folder of ImageJ.
2. NDPI single file opening. Open ImageJ. Click "Plugins"—"NDPI tools"—"Open NDPI file".
3. NDPI batch extraction. Extracting a batch of Hamamatsu NDPI slide scans into tiff files is achieved using the "Convert directory to Tiff" function. The TIFF files are automatically saved.

   A complete folder can be extracted either as complete files (at different magnifications or planes) or further segmented into smaller mosaics (to decrease the resulting file size). Large slide scans at 20× magnification in the tiff format can cause significant problems due to substantial memory requirements (in our case PC with 18 GB). Therefore, we split the scans (that cover 6 or 8 slices) as early as possible into individual images (*see* below).

4. Split slide images into individual images for each slice. In effect, the original image contains information for all slices mounted on the same slide. This can be done manually or, for higher throughput, using a macro that splits files with objects of a given size. Save as individual TIFF files.
5. Creating the overlay. Open the sequential images of the brain sections using ImageJ. Go to the digital version of the Mouse Brain Atlas [6] to assign a coordinate to each brain slice (i.e., anterior/posterior A/P with respect to atlas). For the visual orientation use clearly visible morphological landmarks such as the ventricles, the hippocampus, and/or the corpus callosum. Open one individual image of a brain slice using Adobe Illustrator (AI). Open the pdf-file of the Mouse Brain Atlas with AI and chose the page corresponding best to the A/P coordinates. Create an overlay, by using the outlines of the structures or by reducing the opacity of the image on top (*see* **Note 24**). The two images can be rotated and shrunk until they match, but the x/y ratio should be kept unaltered. Save the overlay as a TIFF file.
6. Cell counting. Open the overlay TIFF file with ImageJ. Start built in "cell counter" plugin of ImageJ. Click "Initialize". Select a counter type (e.g., chose one type per brain area to differentiate the neurons) and place one marker on each neuron. Transfer cell count to an Excel File for analysis. The marker file is saved for documentation and can be reopened and edited on the same background image.

3.5.2 Confocal Volume Reconstruction Using Imaris Bitplane (Fig. 3)

Volume reconstructions can be performed in a semiautomated manner using the reconstruction wizard (**steps 2–4**) provided that the signal-to-noise ratio is sufficient. In a stepwise manner the surface and volume of the neuronal structures in the image are reconstructed. The source data is an image stack obtained by confocal microscopy (Fig. 3a).

1. File import. Open an image stack in Imaris Bitplane. A range of file types including Leica (.lif) files can be readily imported using the Imaris Bitplane software. Note that Leica files also import the pixel-/voxel size (whereas tiff images/stacks do not encode this information and therefore it has to be manually entered into the image properties once a tiff file is loaded).
2. Surface creation. Click "Add new surface" in the main menu on the left side of the window. Tick "Segment only a region of

interest" in the appearing window (lower left), to speed up the subsequent process, since treating the entire stack can increase the processing time substantially. If the loading of a large image stack already causes the computer to slow down, individual regions of interest (from this stack) can be imported instead, to reduce the processing time for the surface reconstruction.

3. Selecting an ROI. Select sequentially the region-of-interest (ROI) of your image stack, the source channel, the level of detail (based on the image resolution). Choose "thresholding based on local contrast", which provides a finer level of detail of the reconstructed object.
4. Removal of unwanted objects. Unwanted objects are removed based on their size from the final reconstruction by setting a minimum and maximum threshold.
5. Post reconstruction editing. Imaris provides a number of readouts from the reconstructed objects, globally and from each individual object, e.g., volume and surface of the objects (Fig. 3c). If objects, which are initially separate are merged in the reconstruction, they can be manually split in the reconstruction by using the scissor tool. Select the newly created surface (in the main menu), click "Edit", (Set Pointer as Select), press Shift and use the mouse to set a marker at the place where the object should be split. Click "Cut Surface".

4 Notes

1. *Tropism.* While capable of transducing both neurons and glial cells, RABV ΔG pseudotyped with VSV G^{RtmC} has a strong tropism for excitatory neurons [2]. The tropism depends on the brain region under investigation and should thus be determined empirically.
2. *Expression of target proteins.* Although we focus on neuroanatomical applications in this protocol, several RABV ΔG constructs encoding sensors or activators/inhibitors of neuronal activity are available (e.g., [9, 10] discussed in [4] and [5]. These vectors may be pseudotyped (as described here) for dual anterograde and retrograde tracing and integrated into a similar experimental design. Likewise, constructs expressing synaptic markers [11] have also been recently described and would greatly complement the approaches presented here.
3. *RABV ΔG production.* The production of infectious viral particles from cDNA has previously been described in detail (e.g., [12–14]. This is a lengthy and challenging process and it is preferable to obtain a starter stock from a collaborator, and then to amplify and pseudotype it yourself rather than starting

from scratch. Nonetheless, the production of pseudotyped vectors typically takes four weeks to complete and requires specialized training in cell culture or virology. The pseudotyping plasmids expressing the glycoprotein of the CVS strain of RABV ΔG and VSV G^{RtmC} fusion glycoprotein are available from Prof. Klaus Conzelmann (LMU Munich). Test aliquots of "in vivo" quality CVS G and VSV G^{RtmC} pseudotyped RABV ΔG are available from either the authors or Prof. Klaus Conzelmann.

4. *Safety regulations.* The use of recombinant viral vectors requires institutional approval and appropriate safety measures. It is recommended to consult the respective institutional office prior to commencing experiments to determine the necessary safety/administrative measures for your situation.
5. *Inactivation of virus.* All disposable materials that have come into contact with the virus are immediately inactivated by treatment with freshly prepared 2 % bleach solution (prepared from pastilles if possible). Personal protective equipment includes the use of a FFP2 mask, lab-coat and gloves offering the appropriate level of protection (e.g., gloves corresponding to European Norm EN374-2003 and possessing an AQL value of 1.5 or better). Gloves should be changed every 20 min and discarded in a container for infectious biological waste. Stereotaxic frame, surgical instruments and surrounding bench space can be disinfected with a noncorrosive surface spray such as Surfa-safe (ANIOS) followed by 70 % ethanol (or similar product containing a mixture of a detergent and ethanol). Rabies virus is inactivated by ethanol (for further reading *see* ref. [15]). The injection needle/syringe assembly is rinsed copiously with distilled H_2O and disinfected in 70 % ethanol overnight. Prerinsing with water is necessary to prevent fixation of the virus in the barrel of the needle and subsequent clogging.
6. *Animal use.* All experiments involving the use of live animals require institutional/ethical permission. We advise you to contact the appropriate institutional authority. Anesthetics are restricted substances and their use must be meticulously recorded.
7. *Osmolarity of solutions.* For certain applications (e.g., clearing approaches for ultramicroscopy) it may also be desirable to check the osmolarity of solutions and adjust as necessary.
8. *Mounting media.* The choice of mounting medium will depend on the microscope objective lens used for imaging. For glycerol immersion objectives, we use Slow-fade (Invitrogen). For oil immersion objectives, we routinely use Prolong Gold mounting medium (Invitrogen), which has refractive properties similar to immersion oil. In our hands, samples retain their fluorescence for years when mounted in this medium.

9. *Super-resolution microscopy or permanent staining.* RABV ΔG vectors are capable of conferring intense labeling of all neuronal structures, revealing even morphological fine-details. The intensity of labeling is ideal for post hoc reconstruction procedures. Nevertheless, it may sometimes be necessary to adapt the protocol for use of super-resolution techniques such as stimulation emission depletion microscopy, or non-fluorescence-based detection microscopy. For these approaches, a combination of antibodies against eGFP and DsRed (indicated in **Materials**) may be used. For detection of eGFP alone, we have successfully used a rabbit polyclonal antibody (Invitrogen reference A-11122). All three antibodies can be used at a dilution of 1:1000 for fluorescence or 1:10,000 for DAB-based detection.

10. *Whole-brain imaging approaches.* We focus in this protocol on the analysis of labeled neurons in slices prepared from fixed brains. While sectioning offers a number of advantages—compatibility with confocal- and even super-resolution microscopy—it also has a number of drawbacks. These include sectioning artifacts and often difficulties in image alignment and consequently 3D reconstruction. Several "whole brain" imaging approaches are possible to overcome these shortfalls, such as serial two-photon tomography [16] and "ultramicroscopy" [17]. The latter requires chemical treatment to clear the tissue after the fixation. In our hands, RABV ΔG vectors expressing mCherry and eGFP are compatible with CUBIC-based clearing procedures [18]. The protocol may thus be adjusted to accommodate these eventual differences related to the ultimate imaging method.

11. *Use and storage of virus.* We dispose of the virus at the end of the day. While freeze-thawing or overnight storage in the refrigerator are both possible, they result in significant loss of titer. To reduce wastage, we find it easier to work with small aliquots (~5 μl).

12. *Glass capillaries.* As an alternative to injection needles, glass capillaries with a tip diameter of 20 μm can be used for the pressure injection. In this case the capillary is loaded after the injection coordinates have been selected, and the virus ejected by gentle constant pressure applied via a 20 ml syringe connected to the capillary via tubing. This approach may be advantageous for injections into the neocortex, for example, where the beveled tip of the injection needle makes it difficult to target a single layer.

13. *Sparse infection.* For sparse targeting it is possible to dilute the virus in HBSS to <10 infectious viral particles (as shown in ref. [2]). We recommend dilution of the vector rather than reducing the injection volume to below 50 nl as a means of controlling the number of injected particles.

14. *Diffusion characteristics of the RABV.* One of the big advantages of our approach is that we are using two viral vectors of the same type/origin. The size of a viral vector (or a fluorescent dye) plays a major role in determining its diffusion characteristics within the nervous tissue. We would like to point out that the radius of diffusion from the injection site is large for AAVs (diameter 20 nm), much smaller for lentiviruses (diameter 100 nm), and even more restricted for RABV (length 200 nm, diameter 80 nm). This spatially limited diffusion is advantageous for many applications, for example when trying to target neurons within a certain cortical layer or column. If neuronal infection of a wider area is desired, one can increase the ejection pressure.

15. *Fixation procedure.* For certain imaging approaches or subsequent immunofluorescent staining it may be necessary to modify the fixative or perfusion times. For example, for clearing using the CUBIC protocol [18], perfusion over a period of 15–20 min is necessary to ensure appropriate levels of fixation. For other clearing methods, perfusion with a mixture of acrylamide and formaldehyde may be required (e.g., [19]). We have also found that fixation in a 2 % PFA/0.2 % picric acid mix is useful for certain immunostaining procedures. Likewise, for certain antibodies it may be necessary to reduce the post-fixation perfusion time.

16. *Infection time window.* In our hands the optimal time-window for fluorescence expression is 6 days post infection. It is however possible to wait longer (up to 12 days post infection) to reach a more intense labeling or to shorten the time window (i.e., 4–5 days post infection). In between that range we find the morphology of infected cells healthy.

17. *Nuclear staining.* Counterstaining with nuclear stains such as DAPI or TO-PRO®-3 (Invitrogen) permit the visualization of gross-scale anatomical features in a slice. The choice of the counterstain will depend on the imaging method and the configuration of the imaging setup. For wide-field epifluorescence detection, we recommend DAPI because of the overlapping excitation spectra of TO-PRO®-3 and mCherry. While this combination may be possible with confocal microscopy, we routinely employ DAPI to permit imaging with both approaches.

18. *Mosaic imaging.* As an alternative to the use of a slide scanner, several other microscope systems are available that permit mosaic scanning acquisition (or "tiling"). The relative disadvantage is that more operator involvement is required. Other alternatives involving "whole brain" imaging are also possible (*see* **Note 10**).

19. *Optimal spatial resolution.* For a diffraction-limited confocal microscope, the maximum resolution is determined by Abbe's law ($d = \lambda/2NA$) and will be 179 nm for a wavelength of 500 nm and a numerical aperture (NA) of 1.4. While the resolution of the image can in fact not be further improved with conventional methods it may in particular cases be useful to increase the sampling density to achieve an 'idealized s to achieve an "idealized" sampling rate as defined by the Nyquist–Shannon sampling theory. In the above example the "idealized" sampling rate would be 43 nm in x/y (and 130 nm in z). The respective Nyquist limit can be calculated online (e.g., http://www.svi.nl/NyquistCalculator). The advantages and drawbacks should be considered carefully or tested in an initial experiment. As a rough estimate, a 4× increase in sampling rate is 4× more time-consuming (−), causes 4× more bleaching of the sample (−), improves the visual appearance of the final image (less grainy) (+), and several pixels are coding the same feature with graduation towards its edges (+). This can be advantageous in cases where signals are strong but the structures are fine (e.g., well-labeled small-caliber axons) and when the region-of-interest is small.
20. *Further application examples.* We have used the viral vectors described here for a range of experimental questions and approaches including the automated reconstruction of axon arborizations and dendritic trees [2, 20], analysis of fine-scale morphology using superresolution microscopy, for mapping the location of retrogradely labeled neurons and for volume analysis.
21. *Automated reconstruction approaches.* Most automated or semiautomated reconstruction approaches require a high signal-to-noise ratio to distinguish labeled morphological features. In this respect, RABV ΔG is highly suited to the purpose.
22. *Software solutions for reconstruction.* There are numerous software solutions available for digitally reconstructing the morphological features of neurons (reviewed in ref. [21]). However, most of them require substantial operator involvement to ensure that the anatomical features are faithfully represented in the subsequent reconstruction. Nonetheless, a number of good semiautomated approaches are available, permitting, for example, the long-range 3D reconstruction of axonal arbors (e.g., [22]) or reconstruction of an entire dendritic tree (e.g., as shown in Fig. 2). The latter approach, employing the APP2 neuron tracing algorithm [23] is rapid and can be performed in an almost fully automated manner within minutes with only small adjustments for the specific dataset. These, and several other algorithms are open-source and can be modified and extended for a particular purpose.

23. *Conversion of image files to an alternative format.* For image acquisition, we use the NanoZoomer slide scanner, a Hamamatsu slide scanner, which produces files in NDPI format. This format permits the compression of large data sets (in comparison to, for example, TIFF format), but unfortunately this format cannot be read by most analysis programs. For images obtained using other acquisition systems, which are already in TIFF format, this step may be omitted.
24. *Assigning brain coordinates to labeled neurons.* This type of analysis requires the user to create an overlay of images from the brain atlas and the respective brain splice. This task is typically performed manually, since matching individual slices to the corresponding image from the brain atlas is a complex task, based on the localization of numerous landmarks visualized on DAPI-labeled sections. Labeled neurons are then counted with respect to each functionally defined region of the brain. Although automated procedures for the alignment of atlas and brain slice images are possible, this task is often complicated by distortions arising from tissue handling as well as sectioning artifacts. To overcome these problems, the use of alternative imaging approaches are possible (e.g., [16, 17]; *see* **Note 10**), but such approaches have their own restrictions due to, for example, the need to handle large data files.

References

1. Wickersham IR, Finke S, Conzelmann K-K, Callaway EM (2007) Retrograde neuronal tracing with a deletion-mutant rabies virus. Nat Methods 4:47–49
2. Haberl MG, Viana da Silva S, Guest JM et al. (2015) An anterograde rabies virus vector for high-resolution large-scale reconstruction of 3D neuron morphology. Brain Struct Funct 220:1369. doi:10.1007/s00429-014-0730-z.
3. Mebatsion T, Schnell MJ, Cox JH, Finke S, Conzelmann KK (1996) Highly stable expression of a foreign gene from rabies virus vectors. Proc Natl Acad Sci U S A 93:7310–7314
4. Ginger M, Haberl M, Conzelmann K-K, Schwarz MK, Frick A (2013) Revealing the secrets of neuronal circuits with recombinant rabies virus technology. Front Neural Circuits 7:2
5. Ginger M, Bony G, Haberl MG, Frick A (2015). Use of rhabdoviruses to study neural circuitry. In: Pattnaik AK, Witt MA (eds), Biology and pathogenesis of rhabdo- and filoviruses, 1st edn. Chapter 11, pages: 263–287
6. Paxinos G, Franklin KBJ (2008) The mouse brain in stereotaxic coordinates. Academic, San Diego
7. Cetin A, Komai S, Eliava M, Seeburg PH, Osten P (2007) Stereotaxic gene delivery in the rodent brain. Nat Protoc 1:3166–3173
8. Gerfen CR (2003) Basic neuroanatomical methods. Curr Protoc Neurosci Chapter 1:Unit 1.1
9. Osakada F, Mori T, Cetin AH, Marshel JH, Virgen B, Callaway EM (2011) New rabies virus variants for monitoring and manipulating activity and gene expression in defined neural circuits. Neuron 71:617–631
10. Yonehara K, Farrow K, Ghanem A, Hillier D, Balint K, Teixeira M, Jüttner J, Noda M, Neve RL, Conzelmann K-K, Roska B (2013) The first stage of cardinal direction selectivity is localized to the dendrites of retinal ganglion cells. Neuron 79:1078–1085
11. Wickersham IR, Sullivan HA, Seung HS (2013) Axonal and subcellular labelling using modified rabies viral vectors. Nat Commun 4:2332
12. Wickersham IR, Sullivan HA, Seung HS (2010) Production of glycoprotein-deleted rabies viruses for monosynaptic tracing and

high-level gene expression in neurons. Nat Protoc 5:595–606

13. Osakada F, Callaway EM (2013) Design and generation of recombinant rabies virus vectors. Nat Protoc 8:1583–1601
14. Ghanem A, Kern A, Conzelmann K-K (2012) Significantly improved rescue of rabies virus from cDNA plasmids. Eur J Cell Biol 91:10–16
15. Kelly RM, Strick PL (2000) Rabies as a trans-neuronal tracer of circuits in the central nervous system. J Neurosci Methods 103:63–71
16. Ragan T, Kadiri LR, Venkataraju KU, Bahlmann K, Sutin J, Taranda J, Arganda-Carreras I, Kim Y, Seung HS, Osten P (2012) Serial two-photon tomography for automated ex vivo mouse brain imaging. Nat Methods 9:255–258
17. Dodt H-U, Leischner U, Schierloh A, Jährling N, Mauch CP, Deininger K, Deussing JM, Eder M, Zieglgänsberger W, Becker K (2007) Ultramicroscopy: three-dimensional visualization of neuronal networks in the whole mouse brain. Nat Methods 4:331–336
18. Susaki EA, Tainaka K, Perrin D, Kishino F, Tawara T, Watanabe TM, Yokoyama C, Onoe H, Eguchi M, Yamaguchi S, Abe T, Kiyonari H, Shimizu Y, Miyawaki A, Yokota H, Ueda HR (2014) Whole-brain imaging with single-cell resolution using chemical cocktails and computational analysis. Cell 157:726–739
19. Chung K, Wallace J, Kim S-Y, Kalyanasundaram S, Andalman AS, Davidson TJ, Mirzabekov JJ, Zalocusky KA, Mattis J, Denisin AK, Pak S, Bernstein H, Ramakrishnan C, Grosenick L, Gradinaru V, Deisseroth K (2013) Structural and molecular interrogation of intact biological systems. Nature 497:332–337
20. Ginger M, Broser P, Frick A (2013) Three-dimensional tracking and analysis of ion channel signals across dendritic arbors. Front Neural Circuits 7:61
21. Parekh R, Ascoli GA (2013) Neuronal morphology goes digital: a research hub for cellular and system neuroscience. Neuron 77:1017–1038
22. Dercksen VJ, Hege H-C, Oberlaender M (2014) The filament editor: an interactive software environment for visualization, proof-editing and analysis of 3D neuron morphology. Neuroinformatics 12:325–339
23. Xiao H, Peng H (2013) APP2: automatic tracing of 3D neuron morphology based on hierarchical pruning of a gray-weighted image distance-tree. Bioinformatics 29:1448–1454

Chapter 22

Mapping Synaptic Inputs of Developing Neurons Using Calcium Imaging

Johan Winnubst and Christian Lohmann

Abstract

Studying changing synaptic activity patterns during development provides a wealth of information on how activity-dependent processes shape synaptic connectivity. In this chapter we introduce a method that combines whole-cell electrophysiology with calcium imaging to map functional synaptic sites on the dendritic tree and follow their activity over time. The key strength of this method lies in its ability to distinguish between synaptic and non-synaptic calcium signaling by their coincidence with synaptic currents measured at the soma. Next to the required materials and protocols that are necessary to perform these experiments, we thoroughly discuss how the acquired data can be analyzed. Since this method can be employed in many neuronal systems we believe that it can be a valuable tool to study developmental changes in synaptic connectivity.

Key words Calcium imaging, Whole-cell electrophysiology, Spontaneous activity, Development, Synaptome mapping

1 Introduction

During development highly specific synaptic connectivity patterns emerge that are not only guided by molecular factors but also by activity dependent processes [1–3]. In many developing networks this activity is present in the form of spontaneously generated network events that occur before any sensory information is received [4–6]. These spontaneous network events are highly synchronized as large groups of neurons become activated at the same time. The resulting correlated activity is thought to be crucial in shaping synaptic connectivity through the engagement of Hebbian-like plasticity mechanisms [7–9].

Spontaneous activity in developing networks is often studied using calcium imaging whereby the action potential firing of up to hundreds of cells is monitored simultaneously over time. However, this network imaging technique is inadequate for studying activity-dependent refinement on the level of individual synapses. In this

Alexandros Poulopoulos (ed.), *Synapse Development: Methods and Protocols*, Methods in Molecular Biology, vol. 1538,
DOI 10.1007/978-1-4939-6688-2_22,

chapter we will describe an imaging method that combines whole-cell electrophysiology and high speed calcium imaging to study synaptic activity patterns during development in vitro. The strength of this approach lies in the fact that these two techniques complement each other in the types of data that they collect. Specifically, electrophysiological measurements can detect synaptic activity with high temporal precision but do not provide any information on where these inputs are located on the dendrite. Conversely, while local calcium signals on the dendrite can be observed with micrometer resolution it is not possible to identify the source of those signals (e.g., calcium influx through activated NMDA channels or calcium release from internal store). By correlating the occurrence of local calcium transients on the dendrite with coincident synaptic currents measured at the soma, we devised a measure to quantitatively separate synaptic from non-synaptic calcium activity. Furthermore, by voltage clamping a cell at its resting potential we can prevent cell-wide calcium increases caused by backpropagating action potentials that would normally mask synaptic activity on the dendrite. Using this approach we have previously mapped a large portion of functional synapses on the apical dendrites of CA3 pyramidal neurons and followed their activity over time [10]. Observing how specific activity patterns affect the fate of individual synapses promises to answer many outstanding questions about activity-dependent synaptic refinement during development. Here we discuss the necessary equipment and provide detailed protocols on how to successfully perform these experiments.

2 Materials

The required materials for imaging synaptic activity are listed below and include the reagents for patching and labeling neuronal cells with a calcium indicator dye. A majority of the described instrumentation will be present on most electrophysiology or live cell imaging setups. However, utilizing additional specialized equipment will help you get the most out of your experiments.

2.1 Reagents

1. Hank's balanced salt solution (HBSS) 10× (Life Technologies).
2. Recording medium: 4.17 mM NaH_2CO_3, 2.65 mM $CaCl_2$ in HBSS.
3. Trolox stock solution: 500 mM Trolox in DMSO.
4. Internal solution: 12 mM KCl, 130 mM Potassium D-gluconate, 10 mM HEPES, 4 mM Mg-ATP, 8 mM NaCl, pH 7.2 with KOH osmolarity 290 mOsm.
5. Calcium indicator dye: Oregon Green Bapta-1 hexapotassium salt (Life technologies).

2.2 Instrumentation

1. Upright microscope stage (BX51WI, Olympus).
2. 40× and 0.8 NA water dipping objective.
3. Piezo-stepper between the microscope and objective (P-721. LLQ, Physik Instrumente).
4. High speed and low noise CCD camera (e.g., Andor iXon+).
5. Excitation light source, in 488 nm range (PE-1, CoolLed).
6. Microscope temperature control system (TC05, Luigs and Neumann).
7. Fire polished borosilicate glass with filament, 0.86 × 1.50 × 80mm (Science products).
8. Micro-manipulator (Luigs-Neumann).
9. Microelectrode Amplifier, Axon MultiClamp 700B (Molecular Devices).
10. Electrophysiology digitizer, Digidata 1440A (Molecular Devices).
11. Culture inserts 30 mm, hydrophilic PTFE, 0.4 μm (Millicell).
12. Software for image acquisition, controlling the Piezo-stepper and electrophysiology acquisition.
13. Software for offline image processing and visualization (ImageJ, Matlab).

2.3 Optical Setup

Calcium imaging requires an optical system that is free of noise and allows for stable imaging. The reliability of your data analysis depends largely on the achieved imaging speeds and signal-to-noise ratio (SNR).

2.3.1 Microscope

The method described here requires an upright microscope since accessibility with a patch pipette is restricted to cells in the top layer of a slice. Our aim is to image as much of the dendritic tree as possible and we therefore make use of a standard fluorescence microscope. In our experience, the associated faster imaging speeds and SNRs are preferred to any increase in spatial resolution. In fact, since the calcium transients associated with synaptic activation follow a Gaussian distribution it is actually possible to infer the signals origin with sub-pixel resolution. Furthermore, the lower z-resolution allows us to image additional portions of the dendritic tree even if they are not perfectly in focus.

We used a 40 × 0.8 NA water dipping objective together with a LED excitation light system (505 nm wavelength). In principle any light source can be used that efficiently excites the calcium indicator fluorophore. Most calcium indicators have excitation/emission spectra comparable to GFP and this should therefore not pose a problem for most imaging systems. An important benefit of a LED light source is that one can precisely set the excitation intensity (0–100%), something that is not always possible with other systems. This is very useful considering that the excitation intensity should be carefully adjusted to achieve good SNRs while preventing bleaching and phototoxicity. It is therefore advisable to use a set of neutral density filters when exciting with light sources that do not have this capability. Another important step in limiting phototoxicity is to set the field aperture of your excitation light to tightly fit the field of view of your CCD camera. Otherwise, surrounding tissue may be unnecessarily excited and cause unneeded phototoxicity or stray light scattering.

2.3.2 High-Speed CCD Camera

The camera is one of the most important pieces of equipment for functional calcium imaging and will have a large effect on the quality of your recordings. When choosing a camera it is important to keep in mind the existing tradeoff between the SNR and imaging speed. For this reason we used an electron multiplier CCD (EMCCD) camera, which have a higher quantum efficiency then standard CMOS cameras. In our experiments we used a 4x4 binning during pixel read-out so that we could image full frame (250 × 250 pixels) at 32 ms with low read-out noise. This recording speed is fast enough to image calcium influx caused by even unitary synaptic activations (hundreds of ms). Most EMCCD cameras can also operate in "Crop Mode" where only a small subset of pixels are read out at high speed. This can be especially useful for single synapse imaging at frequencies higher than 100 Hz. Many cameras also have shutter signal output that indicates when the CCD sensor is being exposed. We record this signal on a separate channel on our electrophysiology system to later align the imaging data with the electrophysiology recordings.

2.3.3 Piezo Stepper

An inherent problem with imaging dendrites is that they can spread out over a large 3D volume often traversing large depth ranges. In order to maximize the dendritic area that can be imaged at the same time we attached a piezo-stepper between the microscope and the objective. This piezo-electric motor consists of a material that expands or shrinks based on the applied electric field. The z-position of the objective can hereby be regulated with nanometer precision and millisecond responsiveness. The piezo-stepper is controlled by the shutter signal of the CCD camera which moves the objective to the next set Z-position at the beginning of each frame. We typically record from three different focal planes within 100 ms. Images from these three planes are then collapsed into a maximum projection (Fig. 1) resulting in a total temporal resolution of 10 frames per second. In theory the number of focal planes could even be further increased since the calcium transients typically last a few hundreds of milliseconds. However, in our experience it is usually better not to image at speeds lower than 10 Hz since after this point the decay phase of the calcium transient can no longer be reliably recorded. This makes it much harder to later separate true calcium activity from imaging noise.

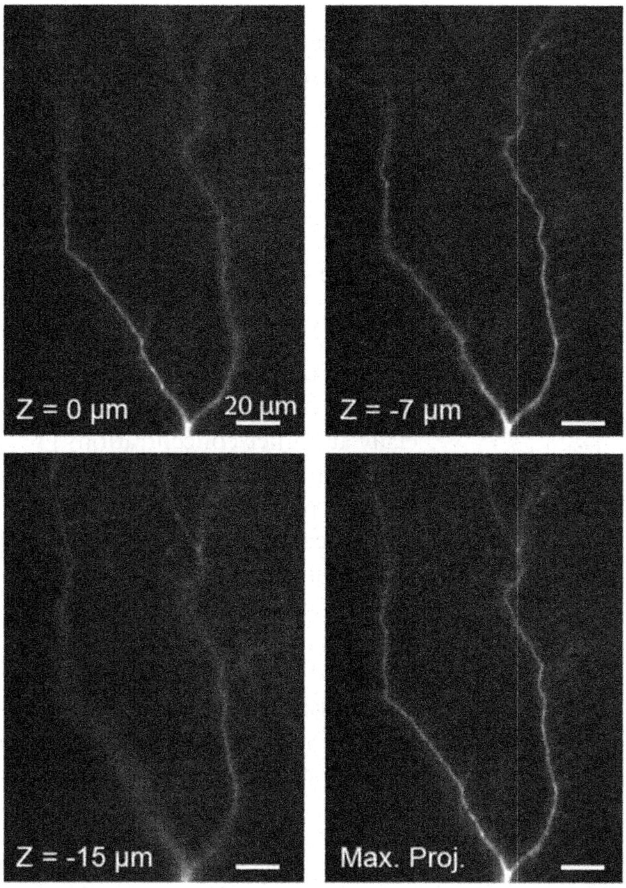

Fig. 1 High speed and large volume imaging of the dendritic tree. Apical dendrite of a CA3 pyramidal neuron imaged at different depths by moving the objective with a piezo-electric motor. *Last panel* shows the collapsed maximum projection of all three imaging planes

2.3.4 Recording Chamber

Long term imaging of synaptic activity requires that there are no large movements of the imaged dendrite over time. Especially temperature-induced expansion or shrinking of the slice will result in z-drift over time that cannot be corrected post hoc. Maintaining a stable temperature of the medium in the recording chamber is therefore critical for movement-free imaging. This can be achieved with a good chamber heating element with a precise temperature controller. Furthermore, the pulse-perfusion pump connected to the chamber should be properly calibrated and provide a constant perfusion of the slice without introducing sudden movements.

2.4 Choice of Calcium Indicator

The number of commercially available chemical Ca^{2+}-indicators has steadily increased over time and they are now available in a wide range of excitation and emission wavelengths. When choosing the most appropriate calcium indicator for an experiment it is important to consider the functional and spectral characteristics of the calcium indicator [11, 12] together with the limitations of your imaging system. First, calcium indicators can vary greatly in their affinity to bind free Ca^{2+} ions. This is often expressed as the inversely related disassociation constant (k_d), the molar concentration of Ca^{2+} at which half of the indicator molecules are bound with Ca^{2+}. In general, higher affinity indicators (e.g., OGB-1) are necessary to detect small changes in Ca^{2+} concentration associated with synaptic activity. However, these high affinity indicators also lengthen the decay times of the calcium signals since they are effectively acting as a calcium buffer. While this buffering activity makes identification of synaptic activity easier, it is important to remember that it can affect synaptic calcium signaling and plasticity, especially at higher concentrations [13, 14]. It is therefore advisable to take some time to determine the lowest concentration at which you can reliably detect synaptic activity.

Recently, great advances have been made with genetically encoded calcium indicators (GECIs) and they are now a viable alternative for imaging synaptic activity [15]. Especially GCaMP6s, with its high affinity and slower kinetics, stands out in this regard [16]. The improved response characteristics of GCaMP6s are in part due to a lower baseline fluorescence, thereby creating a larger relative fluorescence increase upon Ca^{2+} binding. However, due to this lower baseline fluorescence it may be necessary to co-express a separate, nonoverlapping, fluorescent protein for morphological identification.

3 Methods

3.1 Sample Preparation

The following protocol describes the procedure for preparing organotypical hippocampal slices from Postnatal day 0–5 (P0-5) rodent pups for patching and imaging. However, they could in theory be modified to work with acute slices or cultured slices from other brain regions.

3.1.1 Organotypical Slice Culture Preparation

1. Organotypical slices are prepared from P0-5 rats or mice following the Stoppini method described previously [17].
2. Up to four slices are placed on a porous membrane insert (30 mm, 0.4 μm) and incubated in serum containing medium at 37 °C in 5 % CO_2.
3. Cultured slices will attach to the membrane over time and can be used after 2 days in vitro (DIV).

3.1.2 Preparing for Imaging

1. To protect the cells against phototoxicity (*see* **Note 1**) add Trolox stock solution at 1:5000 (final 100 μM Trolox) to the Recording medium and finally set the osmolarity to 290 mOsm.
2. Individual slices are cut out of the membrane using small scissors and placed in the prepared recording medium.

3. Individual slices are placed in the recording chamber of the microscope and weighed down with a platinum ring. All slices are first visually inspected and the ones with clear identifiable cell bodies in the top layer of the slice are selected for later use.
4. The orientation of the slice in the chamber is adjusted so that the pyramidal dendrites of the targeted neurons can later be imaged without the patch pipette blocking the field-of-view.

3.2 Acquisition

3.2.1 Patching Procedure

1. A patch pipette with a resistance between 3 and 4 MΩ is pulled using Borosilicate glass.
2. One pipette is filled with 10 μl internal solution containing Calcium indicator (33–100 μM Oregon Green Bapta-1, *see* **Note 1**) and is attached to the preamplifier.
3. A cell is chosen and the electrode is moved towards it under visual control while applying positive pressure to prevent clogging of the pipette tip. A test current is applied at regular intervals to monitor any changes in the resistance of the pipette.
4. When the electrode is in close proximity to the soma the positive pressure is removed and replaced by a small amount of negative pressure. The electrode should now form a very tight connection to the cell membrane with a measured resistance above 1 GΩ ("Giga-seal"). If a Giga-seal is not formed, then the pipette is immediately replaced and the procedure is restarted.
5. Before an attempt is made to break open the cell membrane the pipette potential is first clamped to −65 mV to prevent any sudden depolarization of the cell upon break-in.
6. The membrane under the pipette tip is then opened by applying a strong but brief amount of negative pressure. A successful break-in can be identified by the current response to the test pulse as a relatively large capacitive transient should appear.
7. A strong and stable connection to the interior of the cell is required in order to reliably measure all synaptic events and for sufficient dye diffusion to take place. For this reason, only cells with an access resistance below 20 MΩ (*see* **Note 2**) and a leak current that does not exceed 100 pA are used.

3.2.2 Imaging Procedure

1. The patched cell is left undisturbed for the first 10 min in order for the calcium indicator in the pipette to diffuse into the cell.
2. The excitation light is turned down to its lowest level to make a visual inspection of the cell and to select a region where a large portion of the dendrite can be imaged. Extra care is taken to keep exposure of the soma and the pipette to a minimum in order to reduce phototoxicity (*see* **Note 1**). If necessary, the excitation light intensity is adjusted to accommodate for the intensity of the calcium indicator signal in the dendrites.

3. A total of three focal planes are chosen, using the piezo-stepper software, in such a way that large continuous stretches of dendrite can be captured in focus. The piezo-stepper is then set to loop between these three focal planes on the start of each imaging frame.
4. Before the start of the recording, the access resistance and leak current are measured and written down. If these parameters are within acceptable limits then the electrophysiology software is set to start recording upon the trigger signal from the camera (*see* **Note 3**). The recording is then initiated by starting the image acquisition.
5. Between each recording the focus of the objective is adjusted to correct for any movement of the slice (*see* **Note 4**).

3.3 Analysis

Detection of local calcium transients from raw recordings can be a computationally intensive process. Furthermore, these signals must be aggregated and grouped according to their spatial characteristics in order to identify putative synaptic sites. Similarly, synaptic inputs measured in the electrophysiological recordings must be detected and aligned to the imaging data in order to determine whether detected calcium signals are synaptic or non-synaptic. Here we discuss in detail the automated analysis we first described in Kleindienst et al. [10].

3.3.1 Detection of Local Calcium Signals and Putative Synaptic Sites

First, the images from the three focal planes are collapsed into a single image using a maximum projection. In this image each pixel value represents the maximum value of that pixel in all planes (i.e., the pixel that was most in focus). In order to detect any changes in fluorescence a F_0 image is created in which each pixel represents the median value of that pixel over time. A top-hat filter is then applied to the F_0 image (radius 3 pixels) in order to remove any unevenness in the background illumination (Fig. 2a). Pixels in this F_0 image are determined to belong to a dendrite if their intensity is higher than two times the standard deviation of the entire image and have a shared area larger than 200 connected pixels (Fig. 2b).

To show the percentage increase in fluorescence a $\Delta F/F_0$ image stack is calculated by subtracting and dividing each pixel with its counterpart in the F_0 image (Fig. 2c). Since synaptic calcium signals are characterized by a fast increase in fluorescence, we calculate the derivative of the $\Delta F/F_0$ image stack using a convolution filter. This derivative stack is first average filtered (radius, 3 pixels) then eroded with a disk-shaped structure (radius, 2 pixels) and thresholded at 15% $\Delta F/F_0 s^{-1}$ (Fig. 2d). Any signal with more than 10 connected pixels on the dendrite (in both space and time) is considered to be a real calcium signal.

In order to determine the origin of a calcium signal on the dendrite we first identify the center of a signal as the pixel with the

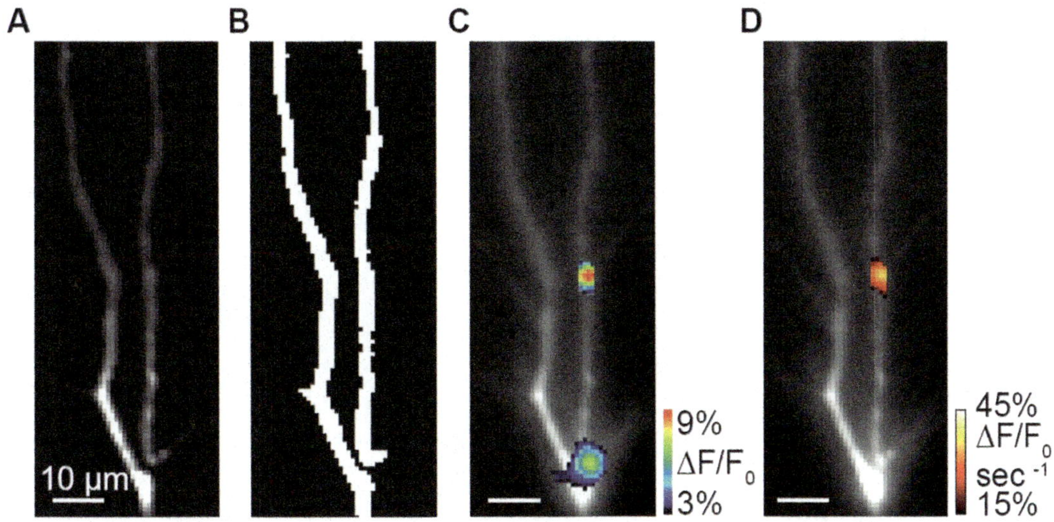

Fig. 2 Detection of synaptic calcium signaling. (**a**) Top-hat filtered image of an apical dendrite. (**b**) Detection of the dendritic morphology by thresholding the image shown in **a**. (**c**) Example of a spontaneous local calcium transient superimposed onto the dendrite. (**d**) Derivative over time calculated from the same signal

largest relative fluorescence increase. If necessary these coordinates are shift corrected between recordings in cases of large movement or drift. Center coordinates of all detected signals are then aggregated and clustered into putative synaptic sites. For this we use a furthest distance clustering algorithm where the weighted center of all signals determines the center coordinate.

3.3.2 Discriminating Between Synaptic and Non-synaptic Sites

Next, we investigate when synaptic currents occur in the electrophysiology recording to see if detected calcium signals coincide with synaptic inputs arriving at the soma. The recorded current trace is first average filtered (5 ms window) and convoluted to calculate the derivative. The onset of a synaptic current is then determined as the time point where the derivative is higher than six times the standard deviation (Fig. 3a, b). A shutter signal from the camera, recorded in parallel on a separate channel, is used to relate each signal to a specific imaging frame (Fig. 3c, d). Using this information, we calculate for each detected site what proportion of their calcium activity coincides with synaptic currents measured at the soma. Since a calcium signal can occur simultaneously with a synaptic current purely by chance we compared the measured proportion of correlated activity to the proportion expected by chance. If the coincidence of calcium transients with synaptic currents exceeded the chance level 1.5 times then a site was considered to be synaptic. Previously, we confirmed that this method can accurately identify synaptic sites as bath application of APV almost completely blocked activity at these sites [10].

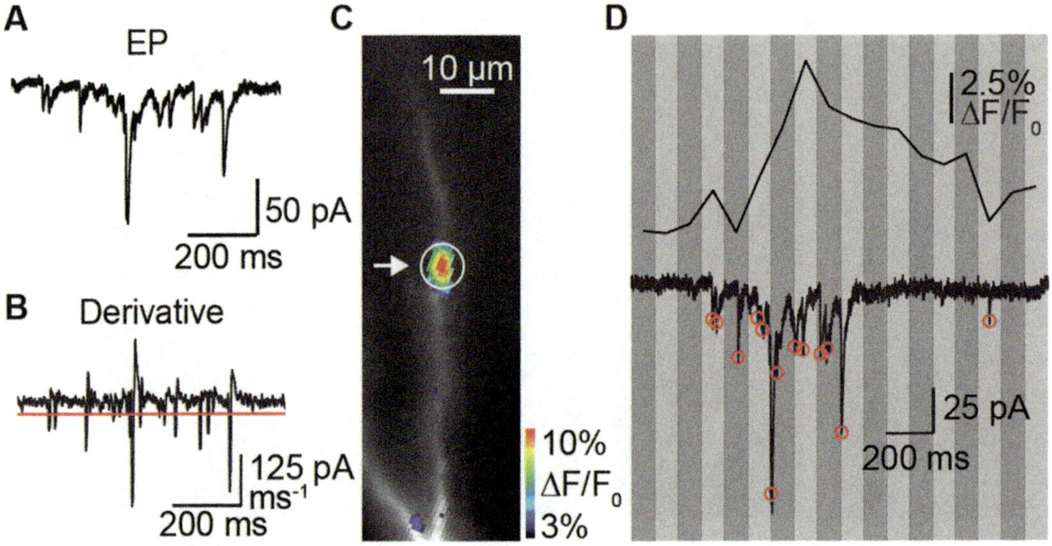

Fig. 3 Coincidence of a dendritic calcium transient with a synaptic burst. (**a**) Whole-cell voltage clamp recording reveals a spontaneous burst of synaptic inputs arriving at the soma. (**b**) Derivative of the electrophysiology trace shown in **a**. *Red line* indicates the threshold used for detecting the onset of synaptic currents. (**c**) Spontaneous calcium transient superimposed on the apical dendrite. The white circle indicates the area used for the $\Delta F/F_0$ traces shown in **d**. (**d**) *Traces* show changes in calcium concentration over time as $\Delta F/F_0$ (*top*) together with the aligned voltage-clamp recording (*bottom*). *Light* and *dark grey bars* represent the duration of an imaging frame. Detected synaptic currents are indicated by a *red circle*. The onset of the dendritic calcium transient coincides with the start of the synaptic burst suggesting that the observed calcium increase is due to synaptic activation

4 Notes

1. Most problems with calcium imaging experiments are related to reduced cell viability caused by induced phototoxicity. It is therefore highly advised to invest some time identifying what calcium indicator concentrations and excitation intensities are sufficient to reliably detect most calcium signals. This is crucial as higher intensity levels of excitation can negatively affect the health of your cell. Measuring how much bleaching of the baseline fluorescence takes place is a good guideline for this process as it is directly related to the amount of created free radicals. Most importantly, exposure of the soma should be kept to a minimum as it is especially sensitive to oxidative stress.

2. Sometimes the calcium indicator will not properly diffuse into the cell and the dendrites are not clearly visible. Instead of increasing the concentration, consider lowering the resistance of your pipette (should be between 2.5 and 3.5 MΩ). Remember that the series resistance, as a readout of your access to the cell, largely determines how freely calcium indicator will diffuse into the cell.

3. Great care should be taken to assure that the imaging and electrophysiological data are properly aligned. Simply starting the recordings at the same time is not sufficient as variability in frame timing of the camera can increase over time and cause the recordings to be out of sync. The camera used here has a TTL trigger signal that signals the opening of the shutter with microsecond precision. With some software packages it is also possible to send a synchronization signal each time the computer sends a command to the camera to take the next frame.
4. Under certain conditions the imaged dendrite can slowly drift out of focus during a recording. This is mostly due to temperature changes that shrink or expand the slice in the recording chamber. In cases where this becomes a problem, for instance when longer continuous recordings are required, it can help to leave the slice in the chamber for a period of time before starting your experiments.

Acknowledgments

We thank Thomas Kleindienst for his significant contribution to the development of the methods described here. The work in laboratory was supported by the Netherlands Organization for Scientific Research (NWO, TOP ZonMw, No 912.10.009; ALW Open Program, No 822.02.006; Vici, No 865.12.001).

References

1. Sanes JR, Yamagata M (2009) Many paths to synaptic specificity. Annu Rev Cell DevBiol 25:161–195
2. Cline H (2003) Sperry and Hebb: oil and vinegar? Trends Neurosci 26:655–661
3. Goodman CS, Shatz CJ (1993) Developmental mechanisms that generate precise patterns of neuronal connectivity. Cell 72:77–98
4. Ben Ari Y (2001) Developing networks play a similar melody. Trends Neurosci 24:353–360
5. Galli L, Maffei L (1988) Spontaneous impulse activity of rat retinal ganglion cells in prenatal life. Science 242:90–91
6. Ben-Ari Y, Cherubini E, Corradetti R, Galarsa J-L (1989) Giant synaptic potentials in immature rat CA3 hippocampal neurones. J Physiol 416:303–325
7. Stellwagen D, Shatz CJ (2002) An instructive role for retinal waves in the development of retinogeniculate connectivity. Neuron 33:357–367
8. Shatz CJ, Stryker MP (1988) Prenatal tetrodotoxin infusion blocks segregation of retinogeniculate afferents. Science 242:87–89
9. Xu HP, Furman M, Mineur YS, Chen H, King SL et al (2011) An instructive role for patterned spontaneous retinal activity in mouse visual map development. Neuron 70:1115–1127
10. Kleindienst T, Winnubst J, Roth-Alpermann C, Bonhoeffer T, Lohmann C (2011) Activity-dependent clustering of functional synaptic inputs on developing hippocampal dendrites. Neuron 72:1012–1024
11. Grienberger C, Konnerth A (2012) Imaging calcium in neurons. Neuron 73:862–885
12. Paredes RM, Etzler JC, Watts LT, Zheng W, Lechleiter JD (2008) Chemical calcium indicators. Methods 46:143–151
13. Cummings JA, Mulkey RM, Nicoll RA, Malenka RC (1996) Ca2+ signaling requirements for long-term depression in the hippocampus. Neuron 16:825–833
14. Nabavi S, Kessels HW, Alfonso S, Aow J, Fox R et al (2013) Metabotropic NMDA receptor function is required for NMDA receptor-dependent long-term depression. Proc Natl Acad Sci 110:4027–4032

15. Looger LL, Griesbeck O (2012) Genetically encoded neural activity indicators. Curr Opin Neurobiol 22:18–23
16. Chen T-W, Wardill TJ, Sun Y, Pulver SR, Renninger SL et al (2013) Ultrasensitive fluorescent proteins for imaging neuronal activity. Nature 499:295–300
17. Stoppini L, Buchs P-A, Muller D (1991) A simple method for organotypic cultures of nervous tissue. J Neurosci Methods 37:173–182

Chapter 23

Monosynaptic Tracing in Developing Circuits Using Modified Rabies Virus

Laura Cocas and Gloria Fernandez

Abstract

An attenuated rabies virus that expresses fluorescent protein has made it possible to analyze retrograde (presynaptic) monosynaptic connections in vivo. By combining attenuated rabies virus with a Cre-loxP based system to target cells in a subtype-specific fashion, it is possible to examine neuronal input in vivo onto any class of neuron, in development and in the mature brain. We describe here the methods to amplify deletion mutant, pseudotyped rabies virus, selectively target cells of interest using genetic and viral approaches, as well as the stereotaxic procedures required to target neuronal subtypes of interest in vivo.

Key words Monosynaptic circuit tracing, Rabies virus, Neuronal circuit development, Viral amplification, Stereotaxic surgery

1 1. Introduction

In the last 5 years work from Callaway, Wickersham, and colleagues has revealed that monosynaptic circuit tracing using deletion-mutant rabies virus and clever targeting of the viral proteins necessary for infection and spread is a powerful tool to analyze precise circuits in vivo [1–4]. The approach relies on the ability to target neuronal subtypes in live animals, typically through the use of two helper viruses to deliver the viral proteins necessary for infection and spread, along with a third deletion mutant rabies virus that when complemented with the helper viruses can move retrograde to presynaptic partner cells [1, 2, 4–6]. Due to the long timelines required for the helper viruses to function prior to transsynaptic labeling, these approaches are not useful for tracing early circuits in developmental studies. Further, low levels of fluorescent protein expression from helper virus can make it difficult to identify initially infected versus connected cells (unpublished observations and [7]).

We sought a method that would allow analysis throughout life, and that would be fast enough to allow short time windows for analyses of developing circuits. After using one of the approaches

Alexandros Poulopoulos (ed.), *Synapse Development: Methods and Protocols*, Methods in Molecular Biology, vol. 1538, DOI 10.1007/978-1-4939-6688-2_23, © Springer Science+Business Media LLC 2017

below to target specific cells with G and TVA, as well as a fluorescent label, these mice are then injected with deletion mutant avian envelope protein pseudotyped rabies virus. Using any of these approaches, only the cells of interest are infected, and their presynaptic partner cells are retrogradely labeled, resulting in monosynaptic labeling of presynaptic cells and their postsynaptic partners. We describe here our current process to monosynaptically label neuronal circuits with rabies virus, including the amplification, pseudotyping, and ultracentrifugation of the virus, and the stereotaxic injection of both AAV and RabV virus. Amplification and pseudotyping of deletion mutant virus has been discussed in an excellent methods paper in detail; our methods here outline our modifications on the original approach [4]. General stereotaxic surgery has been extensively described previously. Here we outline our approach, modifying an existing protocol for stereotaxic injection of lentivirus [8]. We also discuss the main strategies for targeting viral proteins to cells of interest, using one of three approaches to express G and TVA in a cell subtype specific fashion: (1) Cre/LoxP dependent cell targeting with transgenics, (2) Cre/tTA-dependent targeting with transgenics, or (3) helper virus-dependent Cre/LoxP viral targeting (Fig. 1).

2 Materials

2.1 Viral Amplification

1. Supplemented DMEM: Dulbecco's modified Eagle's medium high glucose, 10% heat-inactivated FBS, 1% GlutaMAX, 1% penicillin–streptomycin.
2. 10 cm tissue culture plate.
3. Syringe filters, 0.2 μm pore size.
4. 250 ml Stericup, 0.45 μm pore size.
5. 15 cm tissue culture plate.
6. Sucrose solution: 20% sucrose in PBS. Filter-sterilize with 0.2 μm filter before use or storage. After filter sterilization, the solution can be stored at 4 °C indefinitely.

Fig. 1 (continued) sites, allowing expression of GRab and TVA proteins postnatally. Infection postnatally with SAD-ΔG-mCherry (EnvA) pseudotyped virus allows selective targeting in cells expressing both G and TVA proteins. MCherry-labeled virus spreads upstream to cells directly synaptically connected to virus-infected GRab- and TVA- expressing neurons of interest. *2.* Genetically regulated expression of tTA (tissue-specific Cre; ROSA-tTA; ROSA-GFP) allows doxycycline-dependent control of GRab and TVA using the pTRE-Bi-G-TVA mouse line as well as a fluorescent reporter to label initially infected, presynaptic neurons. Infection and spread of SAD-ΔG-mCherry (EnvA) labels initially infected cells and their inputs as above. *3.* Floxed, Cre-dependent GRab, TVA, and GFP are delivered with AAV-Flex helper virus. At least 2 weeks later, infection and spread of SAD-ΔG-mCherry (EnvA) labels initially infected cells and their inputs as above

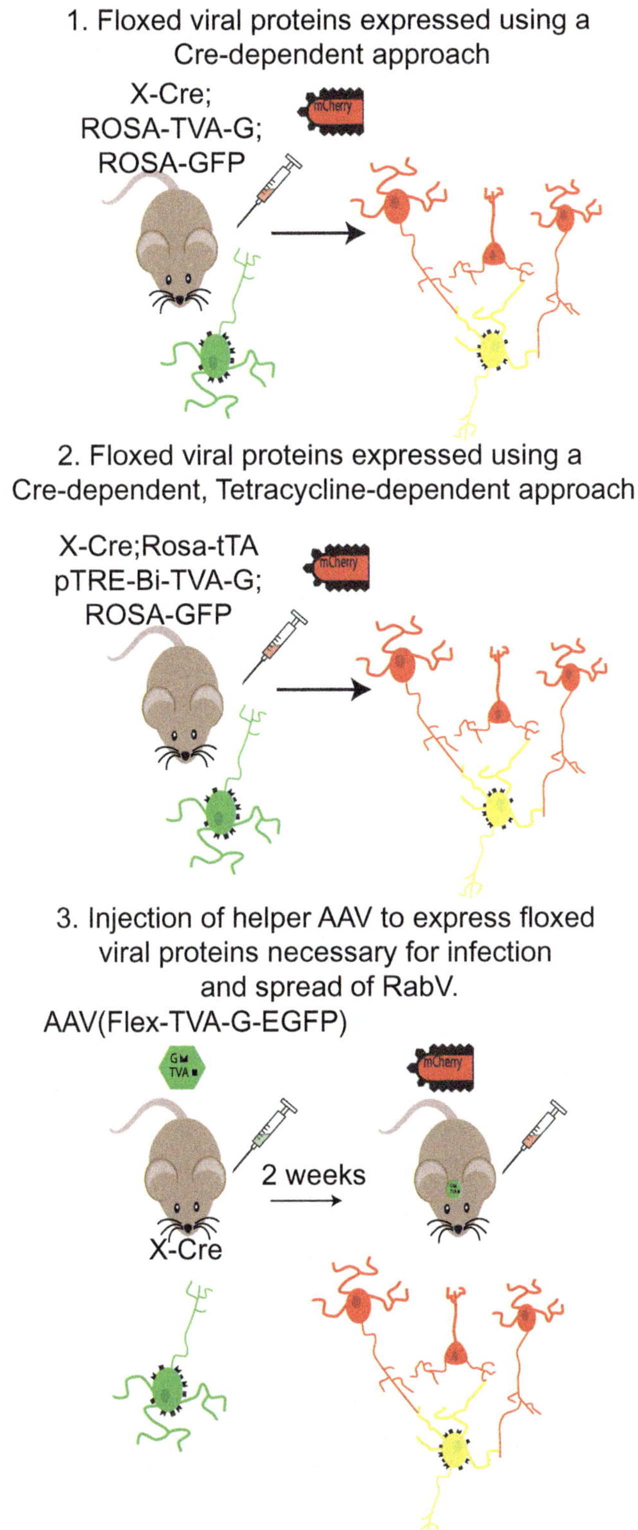

Fig. 1 Three strategies to target specific presynaptic connections in neurons in vivo use a modified deletion-mutation rabies virus. *1*. Genetically regulated expression of TVA and G using a tissue-specific promoter-Cre mouse crossed to floxed ROSA-TVA-G and ROSA-GFP mice. In these mice, cells of interest express the rabies glycoprotein GRab, and avian sarcoma virus leukosis receptor (ASLV) TVA proteins necessary for targeted infection and monosynaptic spread in a Cre-targeted subgroup of neurons. Cre recombines and excises LoxP

7. PBS (sterile).
8. Starting stock of RabV virus (available through Salk Viral Vector Core).
9. 3T3 cells.
10. BHK-19G cells [4].
11. BHK-EnvA cells [4].
12. 293T-TVA-800 cells (can be requested for Dr. John Young, jyoung@salk.edu).
13. Low temp/low CO_2 incubator: Standard cell culture incubator set to 35 °C and 3% CO_2.
14. Ultracentrifuge.
15. Swing rotor and buckets (e.g., Beckman SW28).
16. Polyallomer centrifuge tubes (e.g., Beckman, cat. no. 326823).

2.2 Delivery of G and TVA in a Cell-Type Specific Manner

1. pAAV-synP-FLEX-splitTVA-EGFP-B19G (Addgene, plasmid #52473).
2. pTRE-Bi-G-TVA mice [9].
3. ROSA-tTA mice (Jackson, strain name; B6.Cg-*Gt(ROSA)26 Sortm1(rtTA,EGFP)Nagy*/J.
4. ROSA-Ai14 mice (Jackson, strain name: B6;129S6-*Gt(ROSA) 26Sortm14(CAG-tdTomato)Hze*/J.
5. ROSA-TVA-G mice (Jackson B6;129P2-*Gt(ROSA) 26Sortm1(CAG RABVgp4,-TVA)Arenk*/J).

2.3 Stereotaxic Surgery Reagents

1. Stainless steel, 33 G hypodermic tubing (type 304), pre-cut to 34 mm.
2. Stainless steel, 26 G hypodermic tubing (type 304), pre-cut to 27 mm.
3. Polyethylene tubing, PE-20, 0.015″ (inner diameter) °ø 0.043″ (outer diameter).
4. Quick-setting epoxy (Super Glue).
5. Stereotaxic alignment system base with mouse adaptor.
6. Alignment indicator.
7. Micromanipulator with three axis, 1 μm resolution and digital display readout.
8. Centering microscope with 20× magnification.
9. Stereotaxic drill unit.
10. Dual cannula insertion tool (e.g., David Kopf, Model 1973).
11. Infusion pump (e.g., PHD 22/2000, Harvard Apparatus, Holliston, MA).
12. 70% ethanol.

13. Sterile water.
14. Hair trimmer.
15. Electric heating pad.
16. Drill bit for stereotaxic drill, #85, 0.011″ diameter.
17. Tuberculin syringes with 26 G needle.
18. Betadine solution: 10% povidone–iodine topical solution.
19. Bupivacaine: 0.25% Bupivacaine HCl (Sensocardine-MPF).
20. Isoflurane inhalation system: Isoflurane, isoflurane vaporizer, supply gas oxygen, supply gas regulator, flowmeter, induction chamber, tubing and valves, facemask, and charcoal filter.
21. Veterinary ophthalmic ointment.
22. Cyanoacrylate tissue glue (e.g., Vetbond tissue adhesive, Henry Schein).
23. Cotton tip applicators, 6″.
24. Hamilton gastight syringe, 10 μL; N; Gauge: 26 s; 2 in. (51 mm); Point style: 3 (e.g., Hamilton Model 80075).
25. Stainless steel utility scissors.
26. Adson forceps, straight, 12 cm.
27. Nondissolvable sutures.

3 Methods

3.1 Production of Unpseudotyped Glycoprotein-Deficient Rabies Virus

1. Virus can be amplified by recovering from DNA, or amplifying from existing stocks, using either supernatant or ultracentrifuged stock. We discuss here only the modifications of the amplification approach. For full details on the original methods, *see* ref. 4.

3.1.1 Rabies Virus Amplification

1. Thaw BHK-19G cells and plate into Supplemented DMEM in 1 10 cm dish.
2. After cells have reached confluency (1–3 days later), split BHK-19G cells into 2 10 cm dishes so that cells will reach a confluency of 60% the following day.
3. In a Biosafety Level 2 (BSL-2) hood, infect cells with desired starting stock of RabV virus (RabV-GFP, RabV-mCherry, etc., initial aliquot stocks are usually 1–2 ml) (*see* **Note 1**). Pipette virus onto one 10 cm dish, and add Supplemented DMEM to a total volume of 8.5 ml. Once virus has been applied, return plate to low temp/low CO_2 incubator (set to 35 °C/3% CO_2) (Use the second 10 cm dish to amplify additional BHK-19G cells for subsequent steps).
4. 5–6 h later, remove the virus and replace with 10 ml of Supplemented DMEM, and return to low temp/low CO_2 incubator.

5. Monitor infected neurons daily for confluence and fluorescence. When 90 % of cells are fluorescent, the virus is ready to be passaged to new cells. If cells have not reached 90 % fluorescence by day 5–6, but are confluent and becoming pyknotic or floating in the dish in large numbers, passage the virus rather than waiting for 90 % infection.
6. Filter the viral supernatant and add the entire supplement to one new 15 cm dish with BHK-19G cells at 60 % confluency.
7. Remove the viral supernatant 5–6 h later, replace with 20 ml fresh Supplemented DMEM.
8. Wait 5–6 days, until cells have reached 90 % fluorescence or have become too confluent to continue to grow, as above.
9. When virus is ready, filter, and place 5 ml of viral supernatant and 13 ml of media onto new dishes, for a total of four 15 cm dishes producing virus.
10. Monitor infected neurons daily for confluence and fluorescence; when 90 % of cells are fluorescent (or cells are over-confluent but not yet at 90 % infection), remove supernatant, filter, and place 5 ml of viral supernatant onto eight new dishes.
11. Extra filtered supernatant can be aliquoted and frozen in 5 ml cryotubes and stored at −80C.
12. 5–6 days later, when virus is ready, pool and filter supernatant from all eight dishes for ultracentrifugation.

3.1.2 Virus Particle Concentration

13. To ultracentrifuge virus, insert plastic tubes into buckets for swinging rotor in a BSL-2 safety hood.
14. Place 25 ml of filtered supernatant into each of the six tubes. Gently add 5 ml of Sucrose Solution to the bottom of each tube containing supernatant.
15. Close and weigh buckets, weighing each pair of buckets and adjusting the volumes until each pair are balanced.
16. Decontaminate buckets before removing from hood.
17. Spin in an ultracentrifuge at 50,000 × *g* for 2 h at 4 °C. If using a Beckman SW28 rotor, 20,000 RPM is the equivalent speed.
18. After centrifugation, in a BSL-2 hood, aspirate supernatant and pipette 100 μl sterile PBS onto pellet inside the plastic tube insert. Recap buckets, decontaminate, and place buckets on a shaker overnight at 1 HZ at 4 °C.
19. The following day, pool the virus from all six buckets, briefly vortex, briefly spin down in a tabletop centrifuge, and aliquot into 50 μl aliquots. Decontaminate tubes and store at −80 °C. Save 10 μl for titering virus.
20. Titer virus on 3T3 cells. Titer can be measured using flow cytometry or manual quantification, *see* ref. 4 for discussion of differing titer techniques.

3.2 Production of EnvA-Pseudotyped Glycoprotein-Deficient Rabies Virus

1. Thaw BHK-EnvA cells and plate in a 10 cm dish.
2. When cells are 80% confluent, split cells and replate into three 15 cm dishes.
3. When 2 BHK-EnvA dishes have reached 80% confluency, combine cells from two dishes into 15 ml of medium, and replate 1 ml of cells into each of 15 new 15 cm dishes with 20 ml of medium. (The goal is to have EnvA cells with 60% confluency within 24 h; this can generally be done by splitting 1 15 cm plate 1:6).
4. Using the titer calculated on 3T3 cells above, infect 15 plates containing BHK-EnvA cells with the ultracentrifuged glycoprotein-deleted virus at a Multiplicity of Infection ratio (MOI) of 2.
5. Add 20 ml medium and store in the low temp/low CO_2incubator.
6. The following day, aspirate the supernatant and wash cells 3× with 12 ml PBS. Carefully wash both bottom and sides of each dish, being careful not to disturb cells and adding the PBS and aspirating from the same area of the dish to minimize disturbing the cells (*see* **Note 2**).
7. Add 12 ml of fresh Supplemented DMEM. Monitor cells for fluorescence.
8. 2 days after infection, aspirate the supernatant (the first supernatant is contaminated with parent unpseudotyped virus) and replace with 12 ml of fresh Supplemented DMEM. Cells should begin expressing fluorescence.
9. 3 days after infection, the majority of EnvA cells should be expressing the viral fluorophore of interest. Pool supernatant from all 15 plates, filter, and pipette into rotor buckets as above.
10. Add 5 ml Sucrose Solution as above, and spin in an ultracentrifuge at 50,000 ×*g* for 2 h at 4 °C.
11. After centrifugation, in a BSL-2 hood, aspirate supernatant and pipette 100 μl sterile PBS onto pellet inside plastic tube insert. Recap buckets, decontaminate, and place buckets on a shaker overnight at 1 HZ at 4 °C.
12. The following day, pool the virus from all six buckets, briefly vortex, briefly spin down in a tabletop centrifuge, and aliquot into 20 μl aliquots (*see* **Note 3**).
13. Decontaminate tubes and store at −80 °C. Save 10 μl for titering virus.
14. Titer virus on TVA-800 cells and control 3T3 cells; EnvA pseudotyped virus should not infect 3T3 cells. Titer can be measured using flow cytometry or manual quantification, *see* ref. 4 for discussion of differing titer techniques. Titers achieved

using this approach range drastically between unpseudotyed and pseudotyped virus, from 2.9×10^5 to 1.6×10^{10}, *see* ref. 10. If viral titer is low, or virus is contaminated with parent stock (3T3 cells are infected during titering) (*see* **Note 4**).

3.3 Approaches to Target Cells of Interest with GRab and TVA Proteins Necessary for Monosynaptic Infection and Spread of Rabies Virus

We utilize one of three different approaches to selectively target cells of interest, depending upon the timing and the experimental design; each approach is outlined here.

Approach 1: We combine a cell-type-specific Cre driver mouse with a ROSA-G-TVA reporter mouse containing a floxed bicistronic construct coding for the avian envelope protein receptor (TVA) and the rabies glycoprotein (GRab), as well as a fluorescent reporter (e.g., ROSA-YFP) [11, 12] This is especially useful for studying large populations of neurons, and is used for developmental studies, as helper virus requires a 2 week delay in analysis.

Approach 2: We combine a cell-type specific Cre line with a tetracycline-dependent mouse that expresses TVA and GRab (pTRE-Bi-G-TVA), requiring a tetracycline transactivator (ROSA-tTA) mouse and a fluorescent (e.g.) ROSA- YFP reporter mouse [9, 11, 13]. The tetracycline allows doxycycline-regulated control of GRab and TVA, and is useful for developmental studies, as AAV expression is too slow for short developmental time points.

Approach 3: We inject an AAV helper virus expressing GRab, TVA, and GFP. This virus contains the Addgene construct made by Ian Wickersham, pAAV-synP-FLEX-EGFP-B19G (Plasmid #59333). We have the AAV produced by the UNC Vector Core; we use AAV serotype 8. The AAV must be injected at least 2 weeks prior to rabies virus injection [7] (*see* Fig. 2). This approach is very

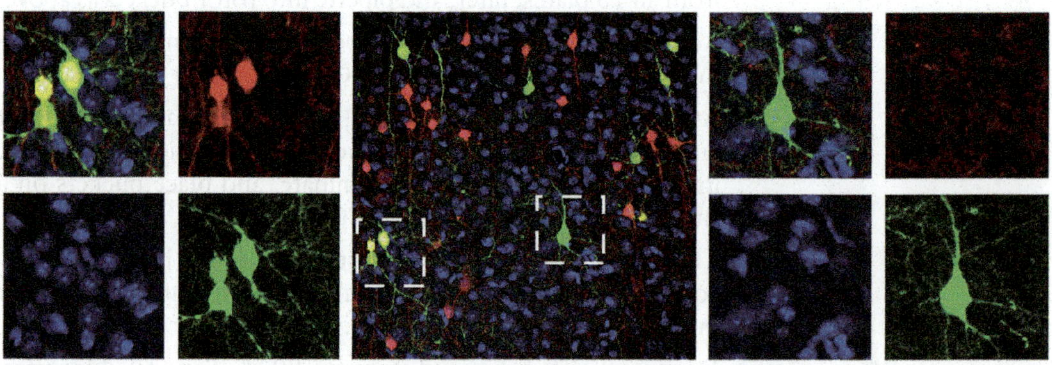

Fig. 2 Viral circuit tracing using genetic labeling of neurons for infection and spread. Example images of viral circuit tracing. Injection with Dlx5/6-Cre; ROSA-tTA; ROSA-Ai14; pTRE-Bi-G-TVA and pseudotyped (EnvA) SADΔGGFP, using Dlx5/6-Cre to target interneuron progenitors. Presynaptic Dlx5/6-lineage cells targeted with virus are *red* and *green* (*yellow*, expressing Tomato from ROSA locus and GFP from (EnvA) SADΔGGFP virus), and connected neurons are labeled by (EnvA) SADΔGGFP only in *green*. Higher magnification optical sections illustrate synaptically connected neurons labeled with virus

useful when labeling small numbers of neurons in adult brain, but requires careful targeting by reinjecting the same region with AAV and RabV at least 2 weeks apart.

3.4 Viral Injection Procedures

For viral tracing experiments, P12 and older mice can be used on the stereotaxis frame. Younger animals require a custom modified frame insert, available from Kopf. Viral proteins necessary for infection and spread (glycoprotein and TVA, the avian envelope protein receptor) can be delivered with any of the methods outlined above. Cre helper virus systems are discussed elsewhere [2, 14]. If using AAV virus, the AAV-Flex-G-TVA-GFP virus must be delivered at least 2 weeks prior to injection with EnvA pseudotyped glycoprotein-deleted virus ultracentrifuged as above. If using transgenics, the expression of G and TVA will be driven by the cell or tissue-specific Cre promoter and will therefore be dependent upon the promoter of interest. If using AAV, volumes of 0.2–1 μl are appropriate, depending upon region-of-interest and the desired target size. Post injection, label bregma with black marker and Vetbond to allow ease of targeting for surgery 2 weeks later. As the procedure is otherwise identical, stereotaxic surgery will be discussed for EnvA pseudotyped deletion mutant rabies virus that has been ultracentrifuged and titered above.

3.4.1 Injection Needle Preparation

1. Prepare injection needles at least one day prior to surgery. Briefly, lightly coat one 34 mm 33 G stainless steel tube with epoxy, and then quickly insert the tube into one 27 mm 26 G stainless steel tube, leaving approximately 5 mm of the 33 G tube sticking out. The 33 G contains a small thread inside; move back and forth briefly to keep the epoxy from clogging the needle but do not remove. Allow them to dry overnight.
2. The following day, use a ruler to bend 10 mm of the side opposite the 5 ml 33 G overhang at a 90° angle, creating an l-shaped tube with a thin 5 ml overhang; this small overhang will be inserted directly into the brain.
3. Remove the thread and fit with a small piece of polyurethane tubing. Using a standard injection needle filled with water, inject water through the newly made needle. If water is unable to flow, or if the needle leaks in between the two tubes, discard. If water flows through freely, it is ready for use.
4. Flush briefly with 70 % ethanol and store in a 50 ml conical tube containing 70 % ethanol; needles can be reused indefinitely.

3.4.2 Stereotactic Injection

Virus should be handled according to institutional BSL-2 guidelines, and animal protocols followed for each institution's animal care and safety guidelines. The stereotaxis frame and surgery preparation and suturing area should be prepared for surgery. Tools should be sterilized prior to each injection, and the needle/injector prepared for injection.

5. Fit a gastight syringe with approximately 30 cm of polyethylene tubing. Attach the homemade needle from above to the opposite end of the polyethylene tubing.
6. Secure the needle into the dual cannula insertion tool by unscrewing the plastic cover and fitting the needle into the groove in the insertion tool. Replace the plastic cover and secure with the screw, making sure that the long side of the "L" of the needle containing the 33 G 5 ml section is facing down; this is the side of the needle that will enter the brain.
7. Place the gastight syringe securely into the infusion pump, and remove the plunger.
8. Back fill the syringe with water using a 10 ml disposable tuberculin syringe. Ensure that the water flows continuously through the gastight syringe, polyethylene tubing, and needle and continue until there are no bubbles in the entire line; bubbles will prevent proper volume measurements during viral injection (*see* **Note 5**).
9. Replace the plunger, and gently pull the plunger until a 1 cm gap of air is present in the tubing next to the needle; this will ensure that when virus is added, it will not mix with the water in the tube.
10. The mouse should be anesthetized with isoflurane inhalation and the head shaved prior to surgery; the flow rate to initially anesthetize the mouse in the chamber should be between 3 and 5 %; once the mouse is on the stereotaxic frame the flow rate is reduced to 1.5 % (*see* **Note 6**).
11. Once unconscious, the mouse's tongue is gently placed to the side with blunt forceps and the bite bar inserted into the mouth; this will prevent gagging. The bite bar is secured into its holder on the stereotaxic frame, and the ear bars are placed inside each ear, and tightened until the head is secure. It is important to balance the ear bar distances between the left and right ears, and to ensure that the ear bars are placed anterior to the ear canals, just touching the zygomatic arch.
12. Once the mouse is secured onto the stereotaxic frame, the head can be prepared for surgery. Clean the skull with Betadine and 100 % ethanol, and then inject 100 μl Bupivacaine beneath the skin. Cover each eye with vet eye ointment.
13. Make an incision from the top of the skull near the nose to the base, just past lambda, and gently clean underneath the skin with a sterile cotton applicator.
14. Attach the microscope to the stereotaxic frame and determine the location of bregma and lambda.
15. Attach the alignment indicator to the frame. Gently lower the alignment indicator until it touches the skull and at least one of

the two needles on the alignment indicator read zero. If both alignment needles do not reach zero, the X alignment is off, and will need to be adjusted. Loosen the screw on the M/L knob on the stereotaxic frame and adjust the tilt of the skull until both needles read 0. Then tighten the M/L knob, and raise the tool above the skull.

16. Loosen the knob on the alignment indicator, turning it 90° to determine the Z alignment. Lower the alignment indicator again until one of the alignment needles reach zero. If both needles are not at 0, loosen the screw on the A/P knob until the needles both read 0. Tighten the screw on the A/P knob and raise the alignment indicator above the skull.
17. Repeat the alignment of the X-axis as above; sometimes Z alignment will affect the original X alignment.
18. Once the X alignment is confirmed, remove the alignment indicator and replace with the microscope tool. Find bregma and set the digital manipulator to 0 in X and Y.
19. Remove the microscope, and replace with the drill tool, fitted with a 0.011″ diameter drill bit. Move the drill to the desired X and Y coordinates, which are visible on the digital manipulator.
20. Gently lower the drill onto the skull, and slowly drill a hole through the skull, being careful to stop as soon as the skull is pierced.
21. Clean the area with a cotton applicator. Remove the drill bit and replace with the cannula insertion tool. Raise the insertion tool so that there is at least a 1 cm gap between the tool and the skull.
22. Using the cap of a PCR tube, pipette 3 μl of deletion mutant G-deleted ultracentrifuged virus into the corner of the cap. Place the cap under the needle and lower the needle until it is touching the meniscus of the viral droplet. Gently pull the plunger of the syringe in the infusion pump to back fill the needle and polyethylene tube with virus; an air gap should remain between the virus and the water. Fill the tube with approximately 1 μl of virus, taking care not to create any bubble in the virus in the tube.
23. Set the plunger against the infusion pump, and set the infusion pump to inject at 5 μl/min to test the flow of the system. As soon as a droplet appears at the tip of the needle, stop the infusion pump, and reset the flow rate to 1 μl/min (*see* **Note 7**).
24. Choose a target volume of 100 nl to 500 nl, depending upon the area of interest (*see* **Note 8**). Mark the polyethylene tube at the edge of the virus with a permanent marker; this will help determine whether the virus is flowing properly during the injection.

25. Gently wipe off the needle with ethanol, and move the cannula insertion tool to the 0 point of the digital reader. This should be at bregma; lower the needle until it touches the skull at bregma, then press 0 and set to zero the Z-axis on the digital manipulator. Return to the original desired X and Y coordinates; this should be located directly over the drill hole.
26. Lower the needle through the hole into the brain until the desired depth is reached, then lower the needle an additional 0.1 μm, and raise the needle back to the desired depth. This will create a small space for the virus to flow and will prevent clogging.
27. Press start on the infusion pump, and monitor the virus movement, comparing the distance of the virus to the marker label to determine that the virus is flowing properly.
28. Once the pump has finished the injection (generally 2–5 min depending upon volume) wait an additional 10 min before beginning to raise the needle out of the brain; this will allow the virus to dissipate and will prevent it from being pulled back out with the needle retraction.
29. After 10 min, slowly raise the needle until it is outside of the brain; removing the needle should occur at an approximate rate of 0.3 μm/min.
30. Use forceps to suture the skull skin closed with nondissolvable sutures (or as directed by institutionally approved protocol). 3–4 drops of Vetbond placed over each suture will ensure that the sutures will stay closed.
31. Give mouse first buprenorphine dose (0.05–0.1 mgs/kg). Return the mouse to a clean cage placed on a 37 °C heating pad, and allow to recover fully, giving mouse second buprenorphine dose (0.05–0.1 mg/kg) 4–6 h later or as directed by institutionally approved protocol).
32. 5–14 days later, the mouse can be anesthetized and transcardially perfused, and the brain removed and prepared for histology.

4 Notes

1. The Salk Vector Core provides aliquots of ultracentrifuged unpseudotyped virus that can be used for amplification. If starting with a 10 μl ultracentrifuged aliquot from the Salk Vector Core, also begin with one 10 cm dish and amplify virus as outlined above. Supernatant stocks from this first amplification can be stored for future preps.
2. It is incredibly important to rinse the EnvA dishes for making EnvA pseudotyped virus carefully and thoroughly with PBS. Overwashing results in cell loss; underwashing results in contaminated stocks of unpseudotyped viral stock.

3. Viral stocks can be stored indefinitely at −80 °C. Avoid freeze–thaw of virus, which can result in lower titers.
4. If viral titers are lower than desired, and the viral MOI used to infect cells is already maximized at 2, it is possible to increase titers by doubling the number of producer plates of virus and centrifuging serially, layering new supernatant over the first viral spin; remove supernatant after first spin and instead of adding PBS, replace with more filtered supernatant, and layer with sucrose solution.
5. It is important to avoid clogging the injection needle. If the needle clogs prior to injection, remove the virus and flush with water prior to continuing. Avoid clogging during the surgery by checking that viral flow is fluid prior to lowering the needle. Some bleeding may occur after the initial drilling through the skull; blot the drill site and ensure that bleeding has stopped prior to beginning the injection to avoid clogging the needle with blood during the lowering of the needle into the injection site.
6. In our hands, we find that isoflurane anesthesia is critical for consistent survival: we have also used Nembutal or ketamine–xylazine cocktails and find that the survival rates are lower, compared to isoflurane anesthesia alone.
7. The volume injected can range from 100 nl to 1 μl; it is important to inject slowly (1 μl/min) and allow recovery of at least 10 min before removing the needle to avoid tissue damage and to prevent the virus from moving to the surface during needle removal.
8. The number of cells that are targeted in vivo can be affected by several factors: A. The efficiency of the Cre promoter and number of cells it targets; B. The titer and volume of the AAV used; C. The titer and volume of RabV. If large numbers of targeted cells are desired, genetic approaches to target cells with GRab and TVA yield larger numbers of initially infected cells. If fewer numbers of targeted cells are desired, AAV helper virus produces fewer initially infected cells. Using tamoxifen-dependent Cre lines can further decrease the number of cells targeted.

References

1. Wickersham IR, Lyon DC, Barnard RJ, Mori T, Finke S, Conzelmann KK, Young JA, Callaway EM (2007) Monosynaptic restriction of transsynaptic tracing from single, genetically targeted neurons. Neuron 53:639–647. doi:10.1016/j.neuron.2007.01.033, S0896-6273(07)00078-5 [pii]
2. Wall NR, Wickersham IR, Cetin A, De La Parra M, Callaway EM (2010) Monosynaptic circuit tracing in vivo through Cre-dependent targeting and complementation of modified rabies virus. Proc Natl Acad Sci U S A 107:21848–21853. doi:10.1073/pnas.1011756107
3. Larsen DD, Wickersham IR, Callaway EM (2007) Retrograde tracing with recombinant rabies virus reveals correlations between projection targets and dendritic architecture in layer 5 of mouse barrel cortex. Front Neural Circuits 1:5. doi:10.3389/neuro.04.005.2007

4. Wickersham IR, Finke S, Conzelmann KK, Callaway EM (2007) Retrograde neuronal tracing with a deletion-mutant rabies virus. Nat Methods 4:47–49. doi:10.1038/nmeth999, nmeth999 [pii]
5. Sun Y, Nguyen AQ, Nguyen JP, Le L, Saur D, Choi J, Callaway EM, Xu X (2014) Cell-type-specific circuit connectivity of hippocampal CA1 revealed through cre-dependent rabies tracing. Cell Rep 7:269–280. doi:10.1016/j.celrep.2014.02.030
6. Marshel JH, Mori T, Nielsen KJ, Callaway EM (2010) Targeting single neuronal networks for gene expression and cell labeling in vivo. Neuron 67:562–574. doi:10.1016/j.neuron.2010.08.001
7. Kohara K, Pignatelli M, Rivest AJ, Jung H-Y, Kitamura T, Suh J, Frank D, Kajikawa K, Mise N, Obata Y, Wickersham IR, Tonegawa S (2014) Cell type-specific genetic and optogenetic tools reveal hippocampal CA2 circuits. Nat Neurosci 17:269–279. doi:10.1038/nn.3614
8. Lasek AW, Azouaou N (2010) Virus-delivered RNA interference in mouse brain to study addiction-related behaviors. Methods Mol Biol 602:283–298. doi:10.1007/978-1-60761-058-8_17
9. Weible AP, Schwarcz L, Wickersham IR, Deblander L, Wu H, Callaway EM, Seung HS, Kentros CG (2010) Transgenic targeting of recombinant rabies virus reveals monosynaptic connectivity of specific neurons. J Neurosci 30:16509–16513. doi:10.1523/JNEUROSCI.2442-10.2010 [pii]
10. Osakada F, Mori T, Cetin AH, Marshel JH, Virgen B, Callaway EM (2011) New rabies virus variants for monitoring and manipulating activity and gene expression in defined neural circuits. Neuron 71:617–631. doi:10.1016/j.neuron.2012.03.019
11. Srinivas S, Watanabe T, Lin CS, William CM, Tanabe Y, Jessell TM, Costantini F (2001) Cre reporter strains produced by targeted insertion of EYFP and ECFP into the ROSA26 locus. BMC Dev Biol 1:4. doi:10.1186/1471-213X-1-4
12. Li Y, Stam FJ, Aimone JB, Goulding M, Callaway EM, Gage FH (2013) Molecular layer perforant path-associated cells contribute to feed-forward inhibition in the adult dentate gyrus. Proc Natl Acad Sci U S A 110:9106–9111. doi:10.1073/pnas.1306912110
13. Wang L, Sharma K, Deng H-X, Siddique T, Grisotti G, Liu E, Roos RP (2008) Restricted expression of mutant SOD1 in spinal motor neurons and interneurons induces motor neuron pathology. Neurobiol Dis 29:400–408. doi:10.1016/j.nbd.2007.10.004
14. Miyamichi K, Amat F, Moussavi F, Wang C, Wickersham I, Wall NR, Taniguchi H, Tasic B, Huang ZJ, He Z, Callaway EM, Horowitz MA, Luo L (2010) Cortical representations of olfactory input by trans-synaptic tracing. Nature 472:191–196. doi:10.1038/nature09714, nature09714 [pii]

INDEX

Alexandros Poulopoulos (ed.), *Synapse Development: Methods and Protocols*, Methods in Molecular Biology, vol. 1538, DOI 10.1007/978-1-4939-6688-2, © Springer Science+Business Media LLC 2017

D

E

R

S

CPSIA information can be obtained
at www.ICGtesting.com
Printed in the USA
LVOW02*1741230217
525239LV00004B/78/P